Mathematics of Big Data

MIT Lincoln Laboratory Series

Mathematics of Big Data: Spreadsheets, Databases, Matrices, and Graphs, Jeremy Kepner and Hayden Jananthan

Perspectives in Space Surveillance, edited by Ramaswamy Sridharan and Antonio F. Pensa

Perspectives on Defense Systems Analysis: The What, the Why, and the Who, but Mostly the How of Broad Defense Systems Analysis, William P. Delaney

Ultrawideband Phased Array Antenna Technology for Sensing and Communications Systems, Alan J. Fenn and Peter T. Hurst

Decision Making Under Uncertainty: Theory and Practice, Mykel J. Kochenderfer

Applied State Estimation and Association, Chaw-Bing Chang and Keh-Ping Dunn

MIT Lincoln Laboratory is a federally funded research and development center that applies advanced technology to problems of national security. The books in the *MIT Lincoln Laboratory Series* cover a broad range of technology areas in which Lincoln Laboratory has made leading contributions. The books listed above and future volumes in this series renew the knowledge-sharing tradition established by the seminal *MIT Radiation Laboratory Series* published between 1947 and 1953.

Mathematics of Big Data
Spreadsheets, Databases, Matrices, and Graphs

Jeremy Kepner and Hayden Jananthan

The MIT Press
Cambridge, Massachusetts
London, England

This book was set in LaTeX by the authors.

Printed and bound in the United States of America.

This material is based upon work supported by the National Science Foundation under Grant No. DMS-1312831. Any opinions, findings, and conclusions or recommendations expressed in this material are those of the authors and do not necessarily reflect the views of the National Science Foundation.

This work is sponsored by the Assistant Secretary of Defense for Research and Engineering under Air Force Contract FA8721-05-C-0002 and/or FA8702-15-D-0001. Opinions, interpretations, recommendations and conclusions are those of the authors and are not necessarily endorsed by the United States Government.

Library of Congress Cataloging-in-Publication Data is available.

ISBN: 978-0-262-03839-3
10 9 8 7 6 5 4 3 2

for

Alix

Jemma

Lekha

Contents

Foreword

I remember the day that Jeremy Kepner walked into my office and challenged me. I was on sabbatical at MIT Lincoln Laboratory, and Jeremy was my host.

"How do you multiply two arbitrary matrices?" he quizzed me.

"What do you mean two arbitrary matrices?" I responded. "In order to multiply two matrices, the number of columns of the first matrix must equal the number of rows of the second matrix."

Everyone who has taken a linear algebra course knows that basic mathematical fact. I, myself, have known it since high school. Why, my own textbook on algorithms states that matrices must be "compatible" for them to be multiplied.

"No, they don't," Jeremy countered.

"Okay," I said. "Explain."

And that was my first introduction to associative arrays.

Associative arrays are marvelous! Why should we assume that matrices rows and columns are indexed from 1 (or 0) by consecutive integers? Why shouldn't we index them with arbitrary character strings? Then, we can not only do everything we want with consecutive integers, we can do amazing things, like relational-database joins! Wait! Not just database operations. Associative arrays work on graphs and spreadsheets, too! Wow!

Once you know about associative arrays, it's tempting to dismiss them as obvious. Yes, they are obvious in hindsight, but what are the implications? What do associative arrays do to linear algebra? What do they do to relational-database theory? What do they do to graphs and spreadsheets? Why are they an essential tool in dealing with Big Data?

In this book, Jeremy Kepner and Hayden Jananthan will lead you on a voyage to relearn everything you know about matrices, graphs, databases, and spreadsheets, viewing their tight interrelationships through the lens of associative arrays. Or, if you're new to these topics, they will lead you on a journey of discovery, teaching them to you without the arbitrary constraints of traditional educational narratives. Each of the four data representations offers its own richness to the exploration of a data set. The notion of associative arrays unifies them conceptually, providing a simple model in which all four points of view can be represented mathematically.

And their study is not just an abstract odyssey of mathematics. Associative arrays make computational sense. Data are entered in a spreadsheet, related though a database, analyzed with a matrix computation, and visualized as a graph. Through the D4M software, all four perspectives can be represented within the same framework, simplifying data processing in this era of Big Data.

With numerous examples drawn principally from art and music, this book illustrates the theory and practice of associative arrays. It demonstrates how the D4M software can be leveraged to yield practical insights into large data sets. It provides the mathematical foundations for the structure and transformation of associative arrays. It shows how linear algebra can provide analytical tools for understanding data sets.

It is hard to make complicated things simple. But that is what this book does. As big data sets present themselves as targets for analysis and machine learning, we need a unifying framework to deal with them. Associative arrays can provide such a unifying framework, and with the mathematical and computational tools described in this volume, they can support an unprecedented understanding of our world. With Jeremy Kepner and Hayden Jananthan as your guides, the voyage starts here.

Charles E. Leiserson
Edwin Sibley Webster Professor
Professor of Computer Science and Engineering
Massachusetts Institute of Technology

Preface

Big is not absolute; it is relative. A person can be big relative to other people. A person is tiny compared to a mountain and gigantic compared to an ant. The same is true of data. A few rows of data in a spreadsheet can be big if existing approaches rely on human visual inspection. Likewise, all data ever typed by humans can be small for a system designed to handle all data flowing over all communication networks.

Big Data describes a new era in the digital age in which the volume, velocity, and variety of data are rapidly increasing across a wide range of fields, such as internet search, healthcare, finance, social media, wireless devices, and cybersecurity. These data are growing at a rate well beyond our ability to analyze them. Tools such as spreadsheets, databases, matrices, and graphs have been developed to address these challenges. The common theme amongst these tools is the need to store and operate on data as whole sets instead of as individual data elements. This book describes the common mathematical foundations of these data sets (associative arrays) that apply across many applications and technologies. Associative arrays unify and simplify data, leading to rapid solutions to volume, velocity, and variety problems. Understanding the mathematical structure of data will allow the reader to see past the differences that lie on the surface of these tools and to use their mathematical similarities to solve the hardest big data challenges. Specifically, understanding associative arrays reduces the effort required to pass data between steps in a data processing system, allows steps to be interchanged with full confidence that the results will be unchanged, and makes it possible to recognize when steps can be simplified or eliminated.

A modern professional career spans decades. It is normal to work in many fields, with an ever-changing set of tools applied to a variety of data. The goal of this book is to provide you, the reader, with the concepts and techniques that will allow you to adapt to increasing data volume, velocity, and variety. The ideas discussed are applicable across the full spectrum of data sizes. Specific software tools and online course material are referred to in this book and are freely available for download [1, 2]. However, the mathematical concepts presented are independent of the tools and can be implemented with a variety of technologies. This book covers several of the primary technological viewpoints on data (spreadsheets, databases, matrices, and graphs) that encompass a large part of human activity. Spreadsheets are used by more than 100 million people every day. Databases are used in nearly every digital transaction on Earth. Matrices and graphs are employed in most data analysis.

The purpose of collecting data is not to fill archives, but to generate insight that leads to new solutions of practical problems. Nothing handles big like mathematics. Mathematics is at ease with both the infinite and the infinitesimal. For this reason, a mathematical approach to data lies at the very heart of the scientific method

$$\text{theory} + \text{experiment} = \text{discovery}$$

Mathematics makes theory computable. Likewise, data are the principal products of experiments. A mathematical approach to data is the quickest path to bringing theory and experiment together. Computers are the primary tools for this merger and are the "+" in the above formula that transforms mathematics into operations and data into computer bits.

This book will discuss mathematics, data, and computations that have been proven on real-world applications in science, engineering, bioinformatics, healthcare, banking, finance, computer networks, text analysis, social media, electrical networks, transportation, and building controls. The most interesting data sets that provide the most enthralling examples are extremely valuable and extremely private. Companies are interested in these data sets so they can sell you the products you want. Using these data, companies, stores, banks, hospitals, utilities, and schools aim to provide goods and services that are tailored specifically to you. Such data are not readily available to be distributed by anyone who wishes to write a book on the topic. Thus, while it is possible to talk about the results of analyzing such data in general terms, it will not be possible to use the data that are most compelling to you and to the global economy. In addition, such examples would quickly become outdated in this rapidly moving field. The examples in the book will be principally drawn from art and music. These topics are both compelling, readily shared, and have a long history of being interesting. Finally, it is worth mentioning that big data is big. It is not possible to use realistically sized examples given the limitations of the number of characters on a page. Fortunately, this is where mathematics comes to the rescue. In mathematics one can say that

$$\mathbf{c}(i) = \mathbf{a}(i) + \mathbf{b}(i)$$

for all $i = 1,...,n$ and know this to be true. The ability to exactly predict the large-scale emergent behavior of a system from its small-scale properties is one of the most powerful properties of mathematics. Although the examples in this book are tiny compared to real applications, by learning the key mathematical concepts, the reader can be confident that they apply to data at all scales. The observation that a few mathematical concepts can span a diverse set of applications over many sizes is the most fundamental idea in this book.

This book is divided into three parts: I – Applications and Practice, II – Mathematical Foundations, and III – Linear Systems. The book will unfold so that a variety of readers can find it useful. Wherever possible, the relevant mathematical concepts are introduced in the context of big data to make them easily accessible. In fact, this book is a practical introduction to many of the more useful concepts found in matrix mathematics, graph theory, and abstract algebra. Extensive references are provided at the end of each chapter. Wherever possible, references are provided to the original classic works on these topics, which provide added historical context for many of these mathematical ideas. Obtaining some of these older texts may require a trip to your local university library.

Part I – Applications and Practice introduces the concept of the associative array in practical terms that are accessible to a wide audience. Part I includes examples showing how associative arrays encompass spreadsheets, databases, matrices, and graphs. Next, the associative array manipulation system D4M (Dynamic Distributed Dimensional Data Model) is described along with some of its successful results. Finally, several chapters describe applications of associative arrays to graph analysis and machine learning systems. The goal of Part I is to make it apparent that associative arrays are a powerful tool for creating interfaces to data processing systems. Associative array-based interfaces work because of their strong mathematical foundations that provide rigorous properties for predicting the behavior of a data processing system.

Part II – Mathematical Foundations provides a mathematically rigorous definition of associative arrays and describes the properties of associative arrays that emerge from this definition. Part II begins with definitions of associative arrays in terms of sets. The structural properties of associative arrays are then enumerated and compared with the properties of matrices and graphs. The ability to predict the structural properties of associative arrays is critical to their use in real applications because these properties determine how much data storage is required in a data processing system.

Part III – Linear Systems shows how concepts of linearity can be extended to encompass associative arrays. Linearity provides powerful tools for analyzing the behavior of associative array transformations. Part III starts with a survey of the diverse behavior of associative arrays under a variety of transformations, such as contraction and rotation, that are the building blocks of more complex algorithms. Next, the mathematical definitions of maps and bases are given for associative arrays to provide the foundations for understanding associative array transformations. Eigenvalues and eigenvectors are then introduced and discussed. Part III ends with a discussion of the extension of associative arrays to higher dimensions.

In recognition of the severe time constraints of professional readers, each chapter is mostly self-contained. Forward and backward references to other chapters are limited, and key terms are redefined as needed. The reader is encouraged to consult the table of contents and the index to find more detailed information on concepts that might be covered in less detail in a particular chapter. Each chapter begins with a short summary of its content. Specific examples are given to illustrate concepts throughout each chapter. References are also contained in each chapter. This arrangement allows professionals to read the book at a pace that works with their busy schedules.

While most algorithms are presented mathematically, when working code examples are required, these are expressed in D4M. The D4M software package is an open-source toolbox that runs in the MATLAB, GNU Octave, and Julia programming languages. D4M is the first practical implementation of associative array mathematics and has been used in diverse applications. D4M has a complete set of documentation, example programs, tutorial slides, and many hours of instructional videos that are all available online (see d4m.mit.edu). The D4M examples in the book are written in MATLAB, and some familiarity with MATLAB is helpful, see [3–5] for an introduction. Notationally, associative arrays and their corresponding operations that are specifically referring to the D4M use of associative arrays will be written using sans serif font, such as

$$\mathsf{C} = \mathsf{A} + \mathsf{B}$$

Likewise, associative arrays and their corresponding operations that are specifically referring to the mathematical use of associative arrays will be written using serif font, such as

$$\mathbf{C} = \mathbf{A} \oplus \mathbf{B}$$

A complete summary of the notation in the book is given in the Appendix.

This book is suitable as either the primary or supplemental book for a class on big data, algorithms, data structures, data analysis, linear algebra, or abstract algebra. The material is useful for engineers, scientists, mathematicians, computer scientists, and software engineers.

References

[1] J. Kepner, "D4M: Dynamic Distributed Dimensional Data Model." http://d4m.mit.edu.

[2] J. Kepner, "D4M: Signal Processing on Databases – MIT OpenCourseWare online course." https://ocw.mit.edu/resources/res-ll-005-d4m-signal-processing-on-databases-fall-2012, 2011.

[3] D. J. Higham and N. J. Higham, *MATLAB Guide*. SIAM, 2005.

[4] C. B. Moler, *Numerical Computing with MATLAB*. SIAM, 2004.

[5] J. Kepner, *Parallel MATLAB for Multicore and Multinode Computers*. SIAM, 2009.

About the Authors

Jeremy Kepner is a MIT Lincoln Laboratory Fellow. He founded the Lincoln Laboratory Supercomputing Center and pioneered the establishment of the Massachusetts Green High Performance Computing Center. He has developed novel big data and parallel computing software used by thousands of scientists and engineers worldwide. He has led several embedded computing efforts, which earned him a 2011 R&D 100 Award. Dr. Kepner has chaired SIAM Data Mining, the IEEE Big Data conference, and the IEEE High Performance Extreme Computing conference. Dr. Kepner is the author of two best-selling books, *Parallel MATLAB for Multicore and Multinode Computers* and *Graph Algorithms in the Language of Linear Algebra*. His peer-reviewed publications include works on abstract algebra, astronomy, astrophysics, cloud computing, cybersecurity, data mining, databases, graph algorithms, health sciences, plasma physics, signal processing, and 3D visualization. In 2014, he received Lincoln Laboratory's Technical Excellence Award. Dr. Kepner holds a BA degree in astrophysics from Pomona College and a PhD degree in astrophysics from Princeton University.

Hayden Jananthan is a mathematics educator. He is a certified mathematics teacher and has taught mathematics in Boston area public schools. He has also taught pure mathematics in a variety of programs for gifted high school students at MIT and at other institutions of higher learning. Hayden has been a researcher at MIT Lincoln Laboratory, supervising undergraduate researchers from MIT and CalTech, and authored a number of peer-reviewed papers on the application of mathematics to big data problems. His work has been instrumental in defining the mathematical foundations of associative array algebra and its relationship to other branches of pure mathematics. Hayden holds a BS degree in mathematics from MIT and is pursuing a PhD in pure mathematics at Vanderbilt University.

About the Cover

This book presents a detailed description of how associative arrays can be a rigorous mathematical model for a wide range of data represented in spreadsheets, databases, matrices, and graphs. The goal of associative arrays is to provide specific benefits for building data processing systems. Some of these benefits are

Common representation — reduces data translation between steps

Swapping operations — allows improved ordering of steps

Eliminating steps — shortens the data processing system

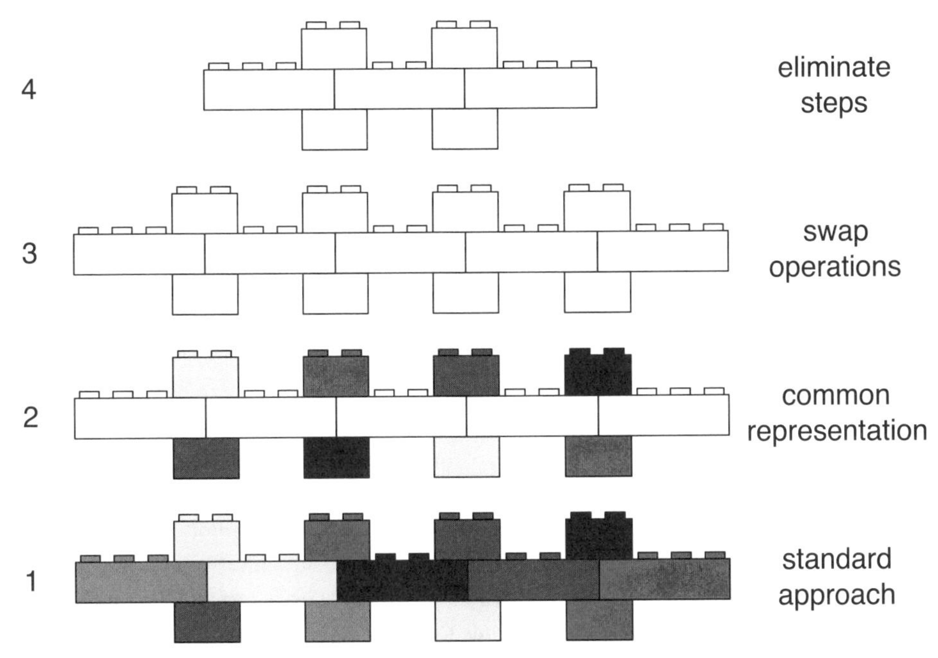

Figure 1
Structures of various complexity with various simplifications. Level 1 is the most complex and is analogous to the standard approach to building a data processing system. Levels 2, 3, and 4 apply additional simplifications to the structure that are analogous to the benefits of applying associative arrays to data processing systems.

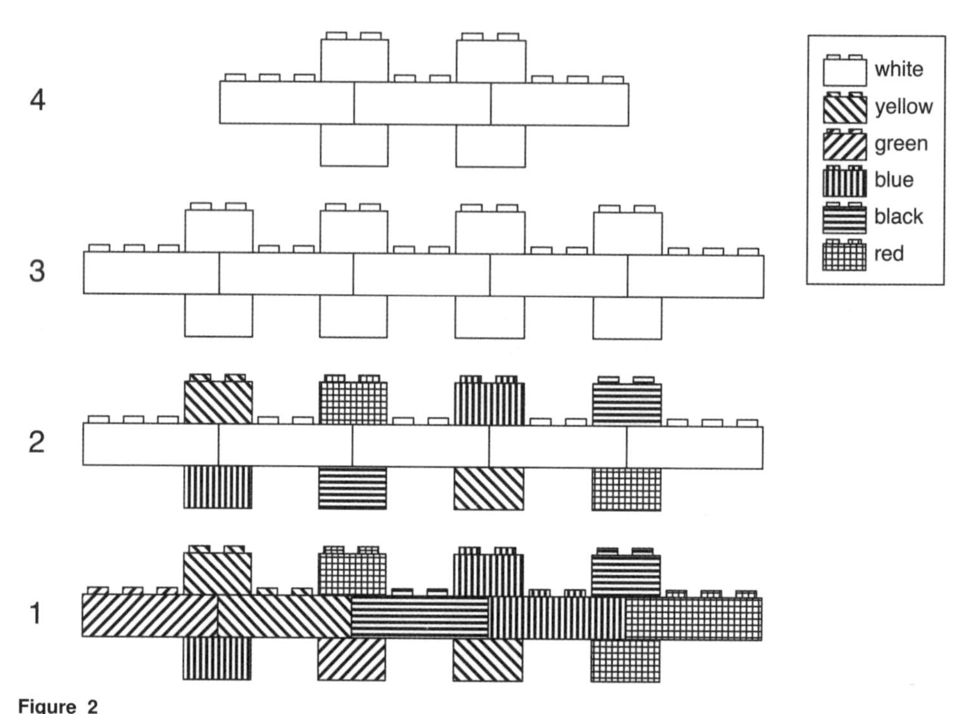

Figure 2
Plans for building different structures. The plans specify the required color used for each building block. A builder uses a collection of building blocks and the corresponding plan to build the structure.

Each of the above benefits can significantly reduce the effort required to build a data processing system. The benefits are cumulative. A common representation allows the swapping of operations that further enables the elimination of steps.

Although these benefits are realized by exploiting the rigorous mathematics of associative arrays, they can be understood through a simple everyday analogy. Consider the task of assembling a structure with a child's toy, such as LEGO® bricks. Figure 1 shows four examples of such a structure arranged from most complicated at the bottom (level 1) and to least complicated at the top (level 4). Between each of the structures a simplification has been made. At the bottom (level 1), the structure is made with a complex set of different colored pieces, and this structure is analogous to the standard approach to building a data processing system in which every piece must be precisely specified. A simplification of the structure can be achieved by making all the middle pieces common, see level 2. Likewise, the structure is further simplified if the edge pieces holding the middle pieces can be swapped, see level 3. Finally, eliminating pieces simplifies the structure, see level 4.

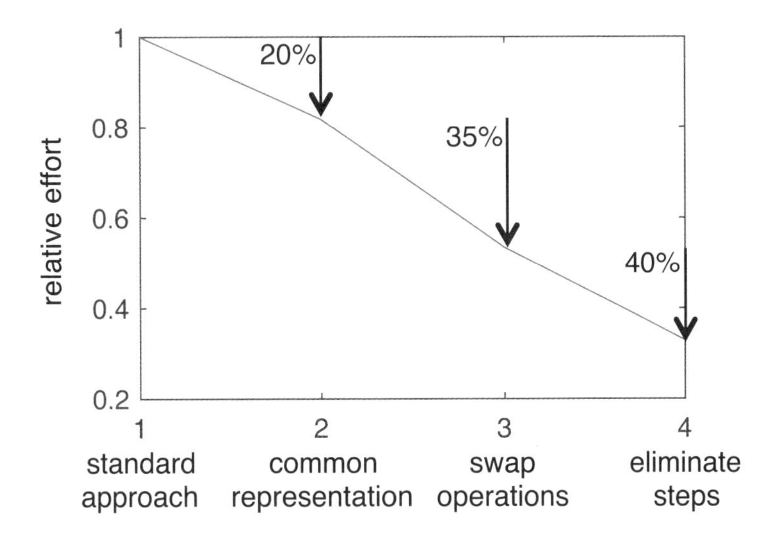

Figure 3
The relative effort required to build the different structures shown in Figure 1 from the plans shown in Figure 2. As expected, simpler structures require less effort to build.

Intuitively, it is apparent that a simpler structure is easier to build. This intuition can be confirmed by a simple experiment. Figure 2 shows the plans for each of the structures in Figure 1. Given a set of pieces and the corresponding plan from Figure 2, it is possible to time the effort required for a person to assemble the pieces from the plan. Figure 3 shows representative relative efforts for these tasks and confirms our intuition that simpler structures require less effort to build. Of course, it is a huge conceptual leap to go from building a structure out of a child's toy to the construction of a data processing system. However, it is hoped that the mathematics presented in this text allows the reader to experience similar benefits.

Acknowledgments

There are many individuals to whom we are indebted for making this book a reality. It is not possible to mention them all, and we would like to apologize in advance to those we may not have mentioned here due to accidental oversight on our part. The development of the *Mathematics of Big Data* has been a journey that has involved many colleagues who have made important contributions along the way. This book marks an important milestone in that journey with the broad availability and acceptance of a tabular approach to data. Our own part in this journey has been aided by numerous individuals who have directly influenced the content of this book.

This work would not have been possible without extensive support from many leaders and mentors at MIT and other institutions. We are particularly indebted to S. Anderson, R. Bond, J. Brukardt, P. Burkhardt, M. Bernstein, A. Edelman, E. Evans, S. Foster, J. Heath, C. Hill, B. Johnson, C. Leiserson, S. Madden, D. Martinez, T. Mattson, S. Pritchard, S. Rejto, V. Roytburd, R. Shin, M. Stonebraker, T. Tran, J. Ward, M. Wright, and M. Zissman.

The content of this book draws heavily upon prior work, and we are deeply grateful to our coauthors S. Ahalt, C. Anderson, W. Arcand, N. Arcolano, D. Bader, M. Balazinska, M. Beard, W. Bergeron, J. Berry, D. Bestor, N. Bliss, A. Buluç, C. Byun, J. Chaidez, N. Chiu, A. Conard, G. Condon, R. Cunningham, K. Dibert, S. Dodson, J. Dongarra, C. Faloutsos, J. Feo, F. Franchetti, A. Fuchs, V. Gadepally, J. Gilbert, V. Gleyzer, B. Hendrickson, B. Howe, M. Hubbell, D. Hutchinson, A. Krishnamurthy, M. Kumar, J. Kurz, B. Landon, A. Lumsdaine, K. Madduri, S. McMillan, H. Meyerhenke, P. Michaleas, L. Milechin, B. Miller, S. Mohindra, P. Monticciolo, J. Moreira, J. Mullen, H. Nguyen, A Prout, S. Reinhardt, A. Reuther, D. Ricke, A. Rosa, M. Schmidt, V. Shah, A. Shcherbina, D. Sherrill, W. Song, S. Sutherland, P. Wolfe, C. Yee, and A. Yerukhimovich.

The production of this book would not have been possible without the efforts of D. Granchelli, M. Lee, D. Ryan, and C. Savage. The authors would also like to thank the many students, colleagues, and anonymous reviewers whose invaluable comments significantly improved this book.

Finally, we would like to thank our families for their support and patience throughout this journey.

I APPLICATIONS AND PRACTICE

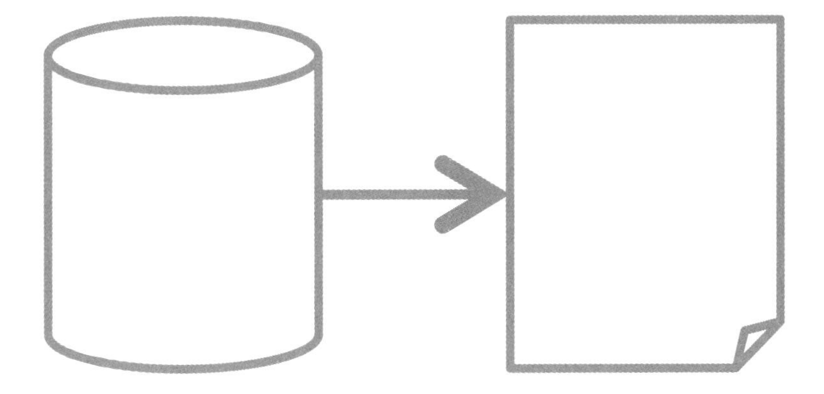

1 Introduction and Overview

Summary

Data are stored in a computer as sets of bits (0's and 1's) and transformed by data processing systems. Different steps of a data processing system impose different views on these sets of bits: spreadsheets, databases, matrices, and graphs. These views have many similar mathematical features. Making data rigorous mathematically means coming up with a rigorous definition of sets of bits (associative arrays) with corresponding operations (addition and multiplication) and showing that the combination is a reasonable model for data processing in the real world. If the model is accurate, then the same mathematical representations can be used across many steps in a data processing system, thus simplifying the system. Likewise, the mathematical properties of associative arrays can be used to swap, reorder, and eliminate data processing steps. This chapter presents an overview of these ideas that will be addressed in greater detail in the rest of the book.

1.1 Mathematics of Data

While some would suggest that data are neither inherently good nor bad, data are an essential tool for displacing ignorance with knowledge. The world has become "data driven" because many decisions are obvious when the correct data are available. The goal of data—to make better decisions—has not changed, but how data are collected has changed. In the past, data were collected by humans. Now, data are mostly collected by machines. Data collected by the human senses are often quickly processed into decisions. Data collected by machines are dormant until the data are processed by machines and acted upon by humans (or machines). Data collected by machines are stored as bits (0's and 1's) and processed by mathematical operations. Today, humans determine the correct mathematical operations for processing data by reasoning about the data as sets of organized bits.

Data in the world today are sets of bits stored and operated on by diverse systems. Each data processing system has its own method for organizing and operating on its sets of bits to deliver better decisions. In most systems, both the sets of bits and the operations can be described precisely with mathematics. Learning the mathematics that describes the sets of bits in a specific processing system can be time-consuming. It is more valuable to learn the mathematics describing the sets of bits that are in many systems.

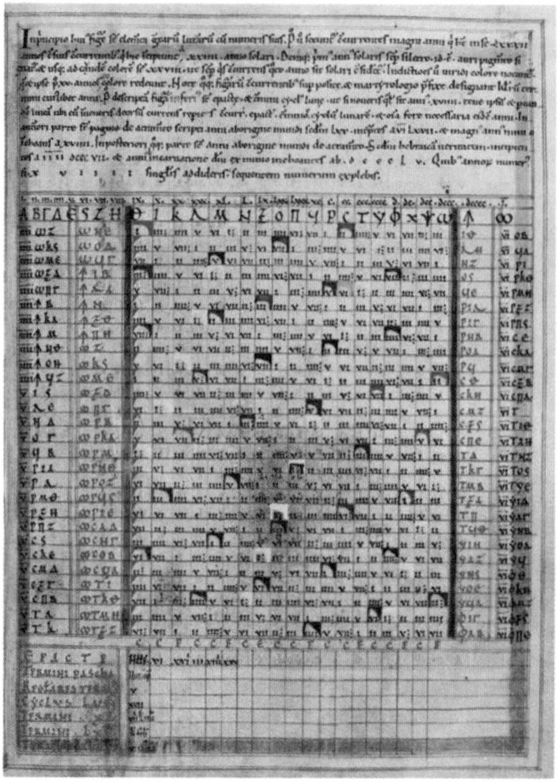

Figure 1.1
Tables have been used since antiquity as demonstrated by the Table of Dionysius Exiguus (MS. 17, fol. 30r, St. John's College, Oxford University) from the Thorney Computus, a manuscript produced in the first decade of the 12th century at Thorney Abbey in Cambridgeshire, England. ©2011 IEEE. Reprinted, with permission, from [1]

The mathematical structure of data stored as sets of bits has many common features. Thus, if individual sets of bits can be described mathematically, then many different sets of bits can be described using similar mathematics. Perhaps the most intuitive way to organize a set of bits is as a table or an associative array. Associative arrays consisting of rows, columns, and values are used across many data processing systems. Such arrays (see Figure 1.1) have been used by humans for millennia [2] and provide a strong foundation for reasoning about whole classes of sets of bits.

Making data rigorous mathematically means combining specific definitions of sets of bits, called associative arrays, with the specific operations of addition and multiplication, and showing that the combination makes sense. Informally, "makes sense" means that the combination of associative arrays and operations behave in ways that are useful. Formally, "makes sense" means that the combination of associative arrays has certain mathematical properties that are useful. In other words, utility is the most important aspect of making

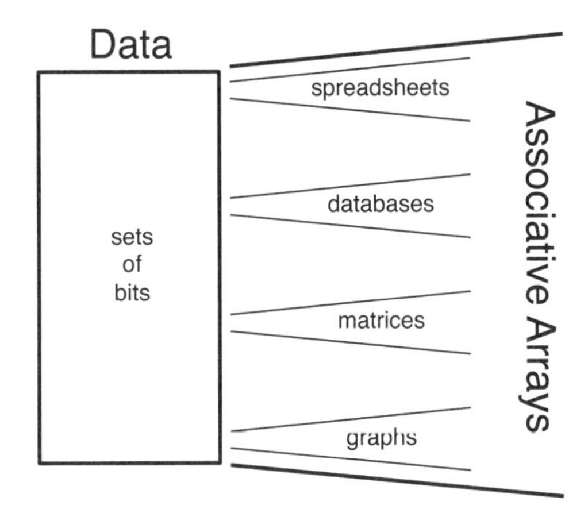

Figure 1.2
Data are sets of bits in a computer. Spreadsheets, databases, matrices, and graphs provide different views for organizing sets of bits. Associative arrays encompass the mathematical properties of these different views.

data rigorous. This fact should be kept in mind as these ideas are developed throughout the book. All the mathematics of data presented here have been demonstrated to be useful in real applications.

1.2 Data in the World

Data in the world today can be viewed from several perspectives. Spreadsheets, database tables, matrices, and graphs are commonly used ways to view data. Most of the benefits of these different perspectives on data can be encompassed by associative array mathematics (or algebra). The first practical implementation of associative array mathematics that bridges these perspectives on data can be found in the Dynamic Distributed Dimensional Data Model (D4M). Associative arrays can be understood textually as data in tables, such as a list of songs and the various features of those songs. Likewise, associative arrays can be understood visually as connections between data elements, such as lines connecting different elements in a painting.

Perspectives on Data

Spreadsheets provide a simple tabular view of data [3]. It is estimated that more than 100 million people use a spreadsheet every day. Almost everyone has used a spreadsheet at one time or another to keep track of their money, analyze data for school, or plan a schedule for an activity. These diverse uses in everyday life give an indication of the power and flexibility of the simple tabular view of data offered by spreadsheets.

As the number of rows and columns in a table grows beyond what can be easily viewed, then a database [4] can be used to store and retrieve the same tabular information. Databases

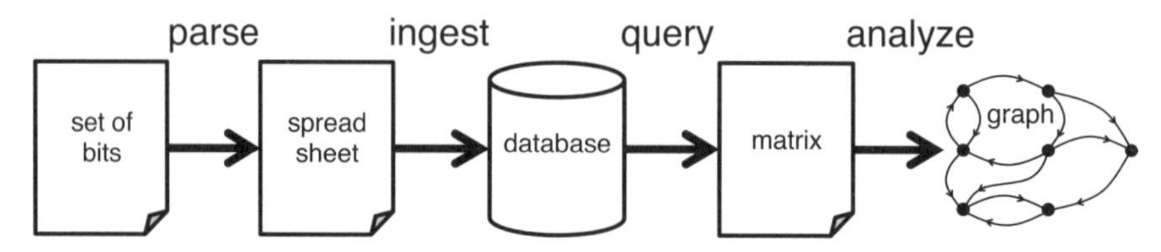

Figure 1.3
The standard data processing steps often require different perspectives on the data. Associative arrays enable a common mathematical perspective to be used across all the steps.

that organize data into large related tables are the most commonly used tool for storing and retrieving data in the world. Databases allow separate people to view the data from different locations and conduct transactions based on entries in the tables. Databases play a role in most purchases. In addition to transactions, another important application of databases is the analysis of many rows and columns in a table to find useful patterns. For example, such analysis is used to determine if a purchase is real or fake.

Mathematics also uses a tabular view to represent numbers. This view is referred to as a matrix [5], a term first coined by English mathematician James Joseph Sylvester in 1848 while working as an actuary with fellow English mathematician and lawyer Arthur Cayley [6]. The term matrix was taken from the Latin word for "womb." In a matrix (or womb of numbers), each row and column is specified by integers starting at 1. The values stored at a particular row and column can be any number. Matrices are particularly useful for comparing whole rows and columns and determining which ones are most alike. Such a comparison is called a correlation and is useful in a wide range of applications. For example, matrix correlations can determine which documents are most like other documents so that a person looking for one document can be provided a list of similar documents. Or, if a person is looking for one kind of document, a correlation can be used to estimate what products they are most likely to buy.

Mathematical matrices also have a concept of sparsity whereby numbers equal to zero are treated differently from other numbers. A matrix is said to be sparse if lots of its values are zero. Sparsity can simplify the analysis of matrices with large numbers of rows and columns. It is often useful to rearrange (or permute) the rows and columns so that the groups of nonzero entries are close together, clearly showing the rows and columns that are most closely related.

Humans have an intuitive ability to understand visual relationships among data. A common way to draw these relationships is a through a picture (graph) consisting of points (vertices) connected by lines (edges). These pictures can readily highlight data that are connected to lots of other data. In addition, it is also possible to determine how closely connected two data elements are by following the edges connecting two vertices. For example, given a person's set of friends, it is possible to suggest likely new friends from their friends' friends [7].

A	Artist	Date	Duration	Genre
053013ktnA1	Bandayde	2013-05-30	5:14	Electronic
053013ktnA2	Kastle	2013-05-30	3:07	Electronic
063012ktnA1	Kitten	2010-06-30	4:38	Rock
082812ktnA1	Kitten	2012-08-28	3:25	Pop

Figure 1.4
Tabular arrangement of a collection of songs and the features of those songs arranged into an associative array **A**. That each row label (or row key) and each column label (or column key) in **A** is unique is what makes it an associative array.

Interestingly, graphs can also be represented in a tabular view using sparse matrices. Furthermore, the same correlation operation that is used to compare rows and columns in a matrix can also be used to follow edges in a graph. The duality between graphs and matrices is one of the many interesting mathematical properties that can be found among spreadsheets, databases, matrices, and graphs. Associative arrays are a tool that encompasses the mathematical properties of these different views of data (see Figure 1.2). Understanding associative arrays is a valuable way to learn the mathematics that describes data in many systems.

Dynamic Distributed Dimensional Data Model

The D4M software (d4m.mit.edu) [8, 9] is the first practical implementation of the full mathematics of associative arrays that successfully bridges spreadsheets, databases, matrices, and graphs. Using associative arrays, D4M users are able to implement high performance complex algorithms with significantly less effort. In D4M, a user can read data from a spreadsheet, load the data into a variety of databases, correlate rows and columns with matrix operations, and visualize connections using graph operations. These operations correspond to the steps necessary to build an end-to-end data processing system (see Figure 1.3). Often, the majority of time spent in building a data processing system is in the defining of the interfaces between the various steps, which normally requires a conversion from one mathematical perspective of the data to another. By using the same mathematical abstraction across all steps, the construction time of a data processing system is significantly reduced. The success of D4M in building real data processing systems has been a prime motivation for formalizing the mathematics of associative arrays. By making associative arrays mathematically rigorous, it becomes possible to apply associative arrays in a wide range of programming environments (not just D4M).

Associative Array Intuition: Text

Associative arrays derive much of their power from their ability to represent data intuitively in easily understandable tables. Consider the list of songs and the various features of those songs shown in Figure 1.4. The tabular arrangement of the data shown in Figure 1.4 is an associative array (denoted **A**). This arrangement is similar to those widely used in

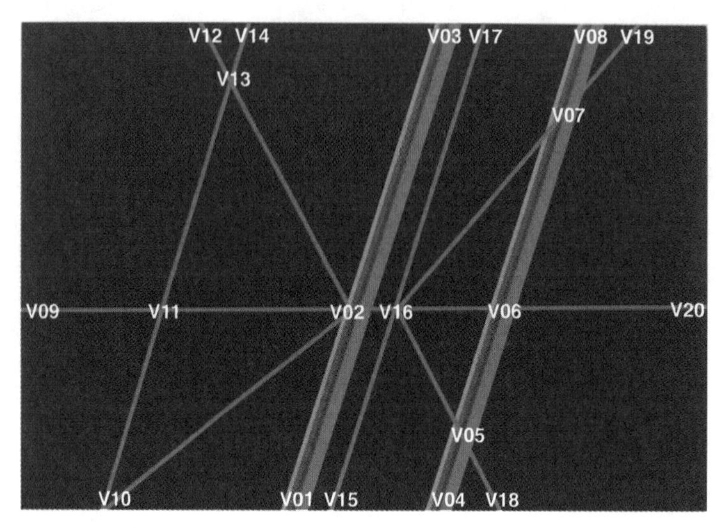

Figure 1.5
Abstract line painting (*XCRS* by Ann Pibal) showing various colored lines. The intersections and terminations of the lines are labeled vertices (V01,...,V20) and have been superimposed onto the painting in white letters.

spreadsheets and databases. Figure 1.4 does contain one property that distinguishes it from being an arbitrary arrangement of data in a two-dimensional grid. Specifically, each row key and each column key in **A** is unique. This property is what makes **A** an associative array and allows **A** to be manipulated as a spreadsheet, database, matrix, or graph.

An important aspect of Figure 1.4 that makes **A** an associative array is that each row and column is identified with a string called a key. An entry in **A** consists of a triple with a row key, a column key, and a value. For example, the upper-left entry in **A** is

$$\mathbf{A}(\text{'053013ktnA1 '},\text{'Artist '}) = \text{'Bandayde '}$$

In many ways, associative arrays have similarities to matrices where each entry has a row index, column index and a value. However, in an associative array the row keys and the column keys can be strings of characters and are not limited to positive integers as they are in a matrix. Likewise, the values in an associative array are not just real or complex numbers and can be numbers or strings or even sets. Typically, the rows and columns of an associative array are represented in sorted order such as alphabetical ordering. This ordering is a practical necessity to make retrieval of information efficient. Thus, in practice, associative array row keys and column keys are orderable sets.

Associative Array Intuition: Graphics

Associative arrays can be visualized as relations between data elements, which are depicted as lines connecting points in a painting (see Figure 1.5). Such a visual depiction of

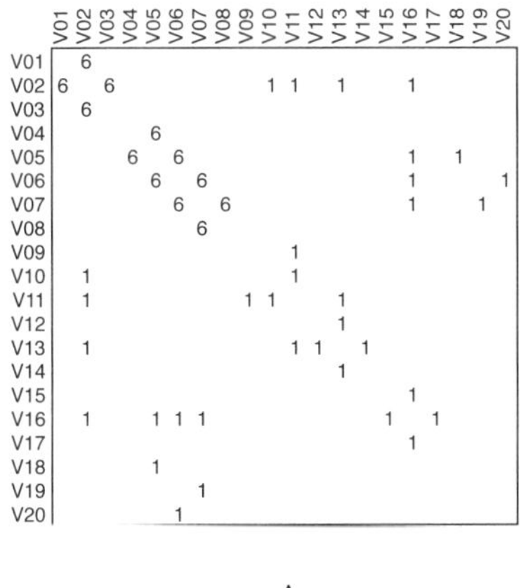

Figure 1.6
Square symmetric associative array representation of the edges depicted in Figure 1.5; each value represents the number of edges connecting each pair of vertices.

relationships is referred to mathematically as a graph. The points on the graph are called the vertices. In the painting, the intersections and terminations of the lines (or edges) are called vertices and are labeled V01,...,V20. An associative array can readily capture this information (see Figure 1.6) and allow it to be manipulated as a spreadsheet, database, matrix, or graph. Such analysis can be used to identify the artist who made the painting [10] or used by an artist to suggest new artistic directions to explore.

In Figure 1.6, each value of the associative array stores the count of edges going between each pair of vertices. In this case, there are six pairs of vertices that all have six edges between them

$$(\mathsf{V01},\mathsf{V02}),(\mathsf{V02},\mathsf{V03}),(\mathsf{V04},\mathsf{V05}),(\mathsf{V05},\mathsf{V06}),(\mathsf{V06},\mathsf{V07}),(\mathsf{V07},\mathsf{V08})$$

This value is referred to as the edge weight, and the corresponding graph is described as a weighted-undirected graph. If the edge weight is the number of edges between two vertices, then the graph is a multi-graph.

1.3 Mathematical Foundations

Data stored as sets of bits have many similar mathematical features. It makes sense that if individual types of sets can be described mathematically, then many different sets can be described using similar mathematics. Perhaps the most common way to arrange a set of bits is as a table or an associative array. Associative arrays consisting of rows, columns, and values are used across many data processing systems. To understand the mathematical structure of associative arrays requires defining the operations of addition and multiplication and then creating a formal definition of associative arrays that is consistent with those operations. In addition, the internal structure of the associative array is important for a range of applications. In particular, the distribution of nonzero entries in an array is often used to represent relationships. Finally, while the focus of this book is on two-dimensional associative arrays, it is worth exploring those properties of two-dimensional associative arrays that extend into higher dimensions.

Mathematical Operations

Addition and multiplication are the most common operations for transforming data and also the most well studied. The first step in understanding associative arrays is to define what adding or multiplying two associative arrays means. Naturally, addition and multiplication of associative arrays will have some properties that are different from standard arithmetic addition

$$2 + 3 = 5$$

and standard arithmetic multiplication

$$2 \times 3 = 6$$

In the context of diverse data, there are many different functions that can usefully serve the role of addition and multiplication. Some common examples include max and min

$$\max(2,3) = 3$$
$$\min(2,3) = 2$$

and union, denoted $\cup$, and intersection, denoted $\cap$

$$\{2\} \cup \{3\} = \{2,3\}$$
$$\{2\} \cap \{3\} = \emptyset$$

To prevent confusion with standard addition and multiplication, $\oplus$ will be used to denote associative array element-wise addition and $\otimes$ will be use to denote associative array element-wise multiplication. In other words, given associative arrays **A**, **B**, and **C**, that represent spreadsheets, database tables, matrices, or graphs, this book will precisely define corresponding associative array element-wise addition

$$\mathbf{C} = \mathbf{A} \oplus \mathbf{B}$$

associative array element-wise multiplication

$$\mathbf{C} = \mathbf{A} \otimes \mathbf{B}$$

and associative array multiplication that combines addition and multiplication

$$\mathbf{C} = \mathbf{A}\mathbf{B}$$

The above array multiplication can also be denoted $\oplus.\otimes$ to highlight its special use of both addition and multiplication

$$\mathbf{C} = \mathbf{A} \oplus.\otimes \mathbf{B}$$

Finally, array transpose is used to swap rows and columns and is denoted

$$\mathbf{A}^{\mathsf{T}}$$

That these operations can be defined so that they make sense for spreadsheets, databases, matrices, and graphs is what allows associative arrays to be an effective tool for manipulating data in many applications. The foundations of these operations are basic concepts from abstract algebra that allow the ideas of addition and multiplication to be applied to both numbers and words. It is a classic example of the unforeseen benefits of pure mathematics that ideas in abstract algebra from the 1800s [11] are beneficial to understanding data generated over a century later.

Formal Properties

It is one thing to state what associative arrays should be able to represent and what operations on them are useful. It is another altogether to prove that for associative arrays of all shapes and sizes that the operations hold and maintain their desirable properties. Perhaps the most important of these properties is coincidentally called the *associativity* property, which allows operations to be grouped arbitrarily. In other words,

$$(\mathbf{A} \oplus \mathbf{B}) \oplus \mathbf{C} = \mathbf{A} \oplus (\mathbf{B} \oplus \mathbf{C})$$
$$(\mathbf{A} \otimes \mathbf{B}) \otimes \mathbf{C} = \mathbf{A} \otimes (\mathbf{B} \otimes \mathbf{C})$$
$$(\mathbf{A}\mathbf{B})\mathbf{C} = \mathbf{A}(\mathbf{B}\mathbf{C})$$

The associativity property allows operations to be executed in any order and is extremely useful for data processing systems. The ability to swap steps or to change the order of processing in a system can significantly simplify its construction. For example, if arrays of data are entering a system one row at a time and the first step in processing the data is to perform an operation across all columns and the second requires processing across all rows, this switching can make the system difficult to build. However, if the processing steps possess the property of associativity, then the first and second steps can be performed in a different order, making it much easier to build the system. [Note: the property of associativity should not be confused with the adjective *associative* in associative array; the similarity is simply a historical coincidence.]

Another powerful property is *commutativity*, which allows arrays in an operation to be swapped

$$\mathbf{A} \oplus \mathbf{B} = \mathbf{B} \oplus \mathbf{A}$$
$$\mathbf{A} \otimes \mathbf{B} = \mathbf{B} \otimes \mathbf{A}$$
$$(\mathbf{A}\mathbf{B})^{\mathsf{T}} = \mathbf{B}^{\mathsf{T}}\mathbf{A}^{\mathsf{T}}$$

If operations in data processing systems are commutative, then this property can be directly translated into systems that will have fewer deadlocks and better performance when many operations are run simultaneously [12].

To prove that associative arrays have the desired properties requires carefully studying each aspect of associative arrays and verifying that it conforms to well-established mathematical principles. This process pulls in basic ideas from abstract algebra, which at first glance may feel complex, but when presented in the context of everyday concepts, such as tables, can be made simple and intuitive.

Special Arrays and Graphs

The organization of data in an associative array is important for many applications. In particular, the placement of nonzero entries in an array can depict relationships that can also be shown as points (vertices) connected by lines (edges). These diagrams are called graphs. For example, one such set of relationships is those genres of music that are performed by particular musical artists. Figure 1.7 extracts these relationships from the data in Figure 1.4 and displays it as both an array and a graph.

Certain special patterns of relationships appear frequently and are of sufficient interest to be given names. Modifying Figure 1.7 by adding and removing some of the relationships (see Figure 1.8) produces a special array in which each row corresponds to exactly one column. Likewise, the graph of these relationships shows the same pattern, and each genre vertex is connected to exactly one artist vertex. This pattern of connections is referred to as the *identity*.

Modifying Figure 1.7 by adding relationships (see Figure 1.9) creates a new array in which each row has a relationship with every column. Likewise, the graph of these relationships shows the same pattern, and each genre vertex is connected to all artist vertices. This arrangement is called a biclique.

In addition, to the identity and the biclique patterns, there are a variety of other patterns that are important because of their special properties. For example, the square-symmetric pattern (see Figure 1.6), where the row labels and the column labels are the same and the pattern of values is symmetric around the diagonal, indicates the presence of an undirected graph. Understanding how these patterns manifest themselves in associative arrays makes it possible to recognize these special patterns in spreadsheets, databases, matrices, and graphs. In a data processing system, recognizing that the data have one of these special

Figure 1.7
Relationships between genres of music and musical artists taken from the data in Figure 1.4. The array on the left shows how many songs are performed for each genre and each artist. The graph on the right shows the same information in visual form.

Figure 1.8
Modifying Figure 1.7 by removing some of the relationships results in a special array where each row corresponds to exactly one column. The graph of these relationships has the same pattern, and each genre vertex connects to exactly one artist vertex. This pattern is referred to as the identity.

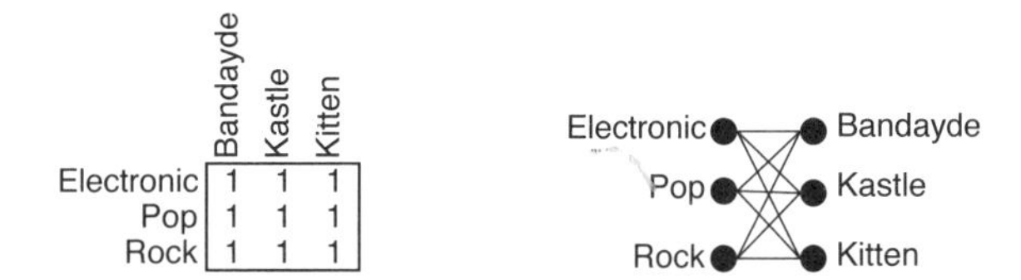

Figure 1.9
Modifying Figure 1.7 by adding relationships produces a special array in which each row has a relationship with every column. The graph of these relationships shows the same pattern, and each genre vertex is connected to all artist vertices. This collection is referred to as a biclique.

patterns can often be used to eliminate or simplify a data processing step. For example, data with the identity pattern shown in Figure 1.7 simplifies the task of looking up an artist, given a specific genre, or a genre, given a specific artist, because there is a 1-to-1 relationship between genre and artist.

Higher Dimensions

The focus of this book is on two-dimensional associative arrays because of their natural connection to spreadsheets, databases, matrices, and graphs, which are most commonly two-dimensional. It is worth examining the properties of two-dimensional associative arrays that also work in higher dimensions. Figure 1.10 shows the data from Figures 1.7,

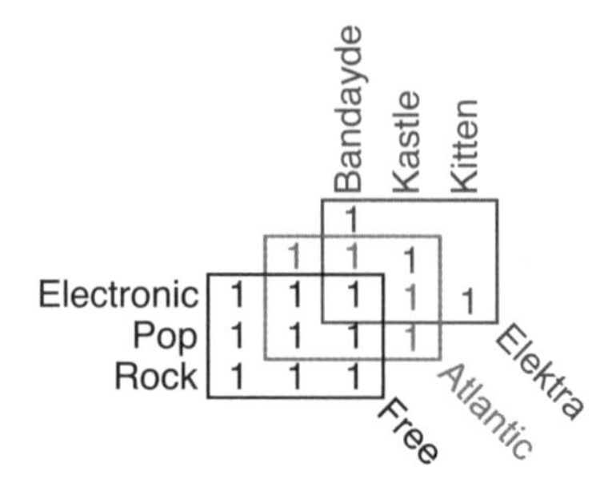

Figure 1.10
Data from Figures 1.7, 1.8, and 1.9 arranged in a three-dimensional array, or tensor, using an additional dimension corresponding to how the music was distributed.

1.8, and 1.9 arranged in a three-dimensional array, or tensor, using an additional dimension corresponding to how the music was distributed. Many of the structural properties of arrays in two dimensions also apply to higher-dimensional arrays.

1.4 Making Data Rigorous

Describing data in terms of rigorous mathematics begins with combining descriptions of sets of bits in the form of associative arrays with mathematical operations, such as addition and multiplication, and proving that the combination makes sense. When a combination of sets and operations is found to be useful, it is given a special name so that the combination can be referred to without having to recall all the necessary definitions and proofs. The various named combinations of sets and operations are interrelated through a process of specialization and generalization. For example, the properties of the real numbers

$$\mathbb{R} = (-\infty, \infty)$$

are in many respects a specialization of the properties of the integers

$$\mathbb{Z} = \{\ldots, -1, 0, 1, \ldots\}$$

Likewise, associative arrays $\mathbb{A}$ are a generalization that encompasses spreadsheets, databases, matrices, and graphs. To prove this generalization requires building up associative arrays from more fundamental combinations of sets and operations with well-established mathematical properties. These combinations include well-defined sets and operations, such as matrices, addition and multiplication of matrices, and the generalization of matrix entries to words and numbers.

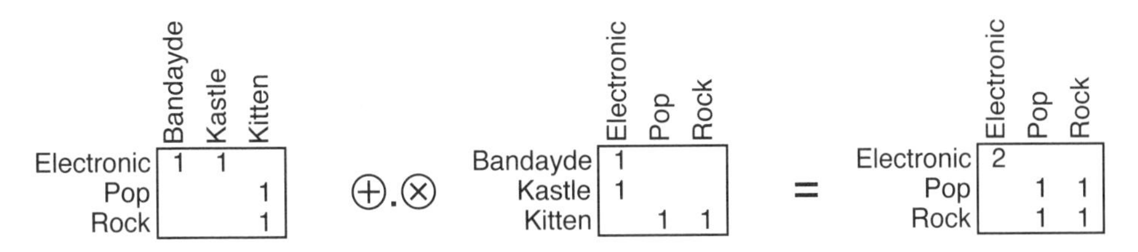

Figure 1.11
Correlation of different musical genres using associative array multiplication ⊕.⊗.

Matrices, Combining Matrices, and Beyond

If associative arrays encompass the matrices, then many of the useful behaviors that are found in matrices may also be found in associative arrays. A matrix is a two-dimensional arrangement of numbers with specific rules on how matrices can be combined using addition and multiplication. The property of associativity allows either addition or multiplication operations on matrices to be performed in various orders and to produce the same results. The property of distributivity provides a similar benefit to certain combinations of multiplications and additions. For example, given matrices (or associative arrays) **A**, **B**, and **C**, these matrices are distributive over addition ⊕ and multiplication ⊗ if

$$\mathbf{A} \otimes (\mathbf{B} \oplus \mathbf{C}) = (\mathbf{A} \otimes \mathbf{B}) \oplus (\mathbf{A} \otimes \mathbf{C})$$

An even stronger form of the property of distributivity occurs when the above formula also holds for the matrix multiplication that combines addition and multiplication

$$\mathbf{A}(\mathbf{B} \oplus \mathbf{C}) = (\mathbf{AB}) \oplus (\mathbf{AC})$$

As with the associativity property, the distributivity property enables altering the order of steps in a data processing system and can be used to simplify its construction.

The property of distributivity has been proven for matrices in which the values are numbers and the rows and columns are labeled with positive integers. Associative arrays generalize matrices to allow the values, rows, and columns to be numbers or words. To show that a beneficial property like distributivity works for associative arrays requires rebuilding matrices from the ground up with a more general concept for the rows, columns, and values.

Multiplication

Multiplication of associative arrays is one of the most useful data processing operations. Associative array multiplication can be used to correlate one set of data with another set of data, transform the row or column labels from one naming scheme to another, and aggregate data into groups. Figure 1.11 shows how the different musical genres can be correlated by artist using associative array multiplication.

For associative array multiplication to provide these benefits requires understanding how associative array multiplication will behave in a variety of situations. One important situation occurs when associative array multiplication will produce a result that contains only zeros. It would be expected that multiplying one associative array by another associative array containing only zeros would produce only zeros. Are there other conditions under which this is true? If so, recognizing these conditions can be used to eliminate operations.

Another important situation is determining the conditions under which associative array multiplication will produce a result that is unique. If correlating musical genre by artist produces a particular result, will that result come about only with those specific associative arrays or will different associative arrays produce the same result? If multiplying by certain classes of associative arrays always produces the same result, this property can also be used to reduce the number steps.

Eigenvectors

Knowing when associative array multiplication produces a zero or unchanging result is very useful for simplifying a data processing system, but these situations don't always occur. If they did, associative array multiplication would be of little use. A situation that occurs more often is when associative array multiplication produces a result that projects one of the associative arrays by a fixed amount along a particular direction (or eigenvector). If a more complex processing step can be broken up into a series of simple eigenvector projection operations on the data, then it may be possible to simplify a data processing system.

1.5 Conclusions, Exercises, and References

This chapter has provided a brief overview of the remaining chapters in the book with the goal of making clear how the whole book ties together. Readers are encouraged to refer back to this chapter while reading the book to maintain a clear understanding of where they are and where they are going.

This book will proceed in three parts. Part I: Applications and Practice introduces associative arrays with real examples that are accessible to a variety of readers. Part II: Mathematical Foundations provides a mathematically rigorous definition of associative arrays and describes the properties of associative arrays that emerge from this definition. Part III: Linear Systems shows how concepts of linearity can be extended to encompass associative arrays.

Exercises

Exercise 1.1 — Refer to the array in Figure 1.4.

(a) Compute the number of rows m and number of columns n in the array.

(b) Compute the total number of entries mn.

(c) How many empty entries are there?

(d) How many filled entries are there?

Remember the row labels and column labels are not counted as part of the array.

Exercise 1.2 — Refer to the painting in Figure 1.5.

(a) Count the total number of vertices.

(b) One line passes through six vertices, list the vertices on this line.

(c) There are six triangles in the pictures, list the vertices in each triangle.

Exercise 1.3 — Refer to the associative array in Figure 1.6.

(a) Why is the array called square?

(b) Why is the array called symmetric?

(c) Sum the rows and the columns and find the row and column with the largest sum?

Exercise 1.4 — Refer to the array and graph in Figure 1.7.

(a) Compute the number of rows m and number of columns n in the array.

(b) How many genre vertices are in the graph? How many artist vertices are in the graph?.

(c) Compute the total number of entries mn in the array.

(d) How many empty entries are there in the array?

(e) How many filled entries are there in the array?

(f) How many edges are in the graph?

Exercise 1.5 — Consider arrays **A**, **B**, and **C** and element-wise addition denoted by $\oplus$ and element-wise multiplication denoted by $\otimes$.

(a) Write an expression that illustrates associativity among **A**, **B**, and **C**.

(b) Write an expression that illustrates commutativity among **A**, **B**, and **C**.

(c) Write an expression that illustrates distributivity among **A**, **B**, and **C**.

Exercise 1.6 — List some of the earliest examples of humans using tables to store information. Discuss how those instances are similar and different from how humans use tables today.

Exercise 1.7 — List some of the different perspectives that can be used to view data. How are these different perspectives similar and different?

Exercise 1.8 — What is the main goal of a data processing system? What are the advantages to a common mathematical view of data as it flows through a data processing system?

Exercise 1.9 — What are the two main mathematical operations that are performed on data? Speculate as to why these are the most important operations.

Exercise 1.10 — What is the main goal of rigorously defining mathematical operations on associative arrays?

References

[1] J. Kepner, W. Arcand, D. Bestor, B. Bergeron, C. Byun, V. Gadepally, M. Hubbell, P. Michaleas, J. Mullen, A. Prout, A. Reuther, A. Rosa, and C. Yee, "Achieving 100,000,000 database inserts per second using Accumulo and D4M," in *High Performance Extreme Computing Conference (HPEC)*, IEEE, 2014.

[2] F. T. Marchese, "Exploring the origins of tables for information visualization," in *15th International Conference on Information Visualisation (IV), 2011*, pp. 395–402, IEEE, 2011.

[3] D. J. Power, "A history of microcomputer spreadsheets," *Communications of the Association for Information Systems*, vol. 4, no. 1, p. 9, 2000.

[4] K. L. Berg, T. Seymour, and R. Goel, "History of databases," *International Journal of Management & Information Systems (Online)*, vol. 17, no. 1, p. 29, 2013.

[5] "Khan academy - intro to the matrices." http://www.khanacademy.org/math/algebra2/alg2-matrices/basic-matrix-operations-alg2/v/introduction-to-the-matrix. Accessed: 2017-04-08.

[6] A. Cayley, "A memoir on the theory of matrices," *Philosophical Transactions of the Royal Society of London*, vol. 148, pp. 17–37, 1858.

[7] M. Zuckerburg, "Facebook and computer science." Harvard University CS50 guest lecture, Dec. 7 2005.

[8] J. V. Kepner, "Multidimensional associative array database," Jan. 14 2014. US Patent 8,631,031.

[9] J. Kepner, W. Arcand, W. Bergeron, N. Bliss, R. Bond, C. Byun, G. Condon, K. Gregson, M. Hubbell, J. Kurz, A. McCabe, P. Michaleas, A. Prout, A. Reuther, A. Rosa, and C. Yee, "Dynamic Distributed Dimensional Data Model (D4M) database and computation system," in *2012 IEEE International Conference on Acoustics, Speech and Signal Processing (ICASSP)*, pp. 5349–5352, IEEE, 2012.

[10] C. R. Johnson, E. Hendriks, I. J. Berezhnoy, E. Brevdo, S. M. Hughes, I. Daubechies, J. Li, E. Postma, and J. Z. Wang, "Image processing for artist identification," *IEEE Signal Processing Magazine*, vol. 25, no. 4, 2008.

[11] R. Dedekind, "Über die komposition der binären quadratischen formen," in *Über die Theorie der ganzen algebraischen Zahlen*, pp. 223–261, Springer, 1964.

[12] A. T. Clements, M. F. Kaashoek, N. Zeldovich, R. T. Morris, and E. Kohler, "The scalable commutativity rule: Designing scalable software for multicore processors," *ACM Transactions on Computer Systems*, vol. 32, no. 4, p. 10, 2015.

2 Perspectives on Data

Summary

Spreadsheets, databases, matrices, and graphs all use two-dimensional data structures in which each data element can be specified with a triple denoted by a row, column, and value. Spreadsheets provide a simple tabular view of data and are the most commonly used tool for analyzing data in the world. Databases that organize data into large related tables are the world's most commonly used tool for storing and retrieving data. Databases allow separate people to view the data from different locations and conduct transactions based on entries in the tables. Relational databases are designed for highly structured transactions and often have hundreds of distinct tables that are connected by many relation tables. In a matrix, each row and column is specified by integers starting at 1. The values stored at a particular row and column can be any number. Matrices are particularly useful for comparing whole rows and columns and determining which ones are most alike. A common way to show visual relationships among data is through a picture consisting of vertices connected by edges. These graphs are often represented in spreadsheets and databases as sparse adjacency matrices or as sparse incidence matrices. This chapter reviews the historical development of spreadsheets, databases, matrices, and graphs, and lays the foundation for viewing all of these data structures as associative arrays.

2.1 Interrelations

As sets of bits move through a data processing system, they are viewed from different perspectives by different parts of the system. Data often are first parsed into a spreadsheet form, then ingested into a database, operated on with matrix mathematics, and presented as a graph of relationships. A large fraction of the effort of developing and maintaining a data processing system goes into sustaining these different perspectives. Thus, it is desirable to minimize the differences between these perspectives. Fortunately, spreadsheets, databases, matrices, and graphs are all two-dimensional representations whereby each data element is specified by a row, column, and value. Using this common reference point, many technologies have been developed to bridge the gaps between these different perspectives (see Figure 2.1). Array programming languages, beginning with A Programming Language (APL) [1] and branching out to MATLAB, Octave, R, Python, and Julia have become a standard approach for manipulating matrices (both dense [2, 3] and sparse [4])

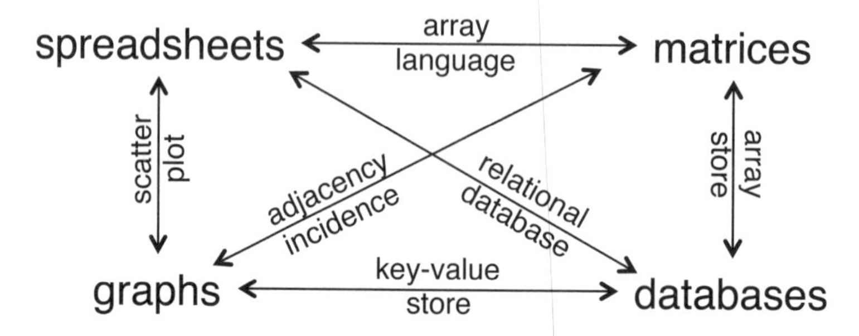

Figure 2.1
Spreadsheets, databases, matrices, and graphs have spawned many concepts that exploit the deeper underlying connections that exist between them.

since the 1990s. Many of these languages have had direct support for spreadsheet manipulation for nearly as long. An even stronger connection exists between spreadsheets and relational databases. A prime example is the SAP enterprise resource planning package (www.sap.com), which is the dominant software used for accounting and payroll management throughout the world. SAP relies on seamless integration between SQL databases and spreadsheets. More recently, spreadsheets have incorporated adjacency matrices to manipulate and visualize graphs by using their built-in plotting capabilities [5]. Perhaps the largest recent development has been the introduction of key-value store databases [6] that are specifically designed to store massive sparse tables and are ideal for storing graphs. Array store databases [7] have taken sparse tables a step further by also including first-class support of matrix operations on those data. The deep connection between graphs and sparse matrices [8] has been recognized to such an extent that it has led to the development of new standards for bringing these fields together [9–17].

That so much has been achieved by exploiting the underlying connections between spreadsheets, databases, matrices, and graphs is strong motivation for understanding their formal mathematical unification. The remainder of this chapter will present the core idea that each perspective uses to organize data. An emphasis will be placed on their common two-dimensional representations that will be the basis of their unification by associative arrays, which will be covered in subsequent chapters. Finally, an overview of some of the perspectives on data is presented.

2.2 Spreadsheets

Spreadsheets are the most widely used tool for analyzing data in the world because they provide an easy to understand tabular view of data. The history of electronic spreadsheets [18] begins at the dawn of the computer age in the early 1960s [19]. LANPAR, the LANguage for Programming Arrays at Random, was invented in 1969 and was patented in 1983

Figure 2.2
American computer magnates Steve Jobs (left), co-founder of Apple, and Bill Gates, co-founder of Microsoft, as they take questions at a press conference to announce Microsoft's Excel software program at Tavern on the Green, New York, New York, May 2, 1985. (Photo by Andy Freeberg/Getty Images – reproduced with permission).

[20]. Broadly available electronic spreadsheets began with the release of VisiCalc in the late 1970s [21, 22]. Lotus 1-2-3 developed by Mitch Kapor for the IBM PC became the most widely used spreadsheet in the early 1980s [23]. Microsoft Excel was first announced by Apple Computer founder Steve Jobs and Microsoft founder Bill Gates at New York's Tavern on the Green on May 2, 1985 (see Figure 2.2). "Excel has 16,000 rows and 256 columns," Gates said during the announcement [24] that may mark the dawn of the big data era.

It is estimated that more than 100 million people use a spreadsheet every day. Almost everyone has used a spreadsheet at one time or another to keep track of their money, analyze data for school, or plan a schedule for an activity. These diverse uses in everyday life give an indication of the power and flexibility of the tabular view of data offered by spreadsheets. Figure 2.3 shows some of the various applications of spreadsheets for analyzing two-dimensional data, such as personnel records, finances, and coordinates. What is even more impressive is that these diverse data can be stored in the same spreadsheet and still be easily understood and manipulated by most humans.

The flexibility of spreadsheets is derived from their representation of each data element as a triple consisting of an orderable row key, an orderable column key, and value that can be a number or word. This triple allows the spreadsheet to consistently and efficiently organize an enormous variety of two-dimensional data. Individual spreadsheet technologies often have their own row and column labeling system, such as 1A or R1C1. Common formats for exchanging spreadsheet data, such as comma-separated values (.csv files) and

	A	B	C	D	E	F	G	H	I	J	K
1	1	2	2	1		Code	Name	Job		Date	$ in bank
2	2	3	3	2		A0001	Alice	scientist		2000 Jan 01	$11,700
3	2	3	3	2		B0002	Bob	engineer		2001 Jan 01	$10,600
4	1	2	2	1		C0003	Carl	mathematician		2002 Jan 01	$10,200
5										2003 Jan 01	$8,600
6										2004 Jan 01	$10,400
7										2005 Jan 01	$10,600
8					y=10	6	12	18		2006 Jan 01	$10,900
9					y=8	5	10	15		2007 Jan 01	$12,300
10					y=6	4	8	12		2008 Jan 01	$12,600
11					y=4	3	6	9		2009 Jan 01	$9,000
12					y=2	2	4	6		2010 Jan 01	$10,600
13					y=0	1	2	3		2011 Jan 01	$11,700
14						x=0	x=5	x=10			

Figure 2.3
Diverse uses of spreadsheets for organizing arrays of two-dimensional data: a matrix (upper left), personnel records (upper middle), finances (right), and coordinates (lower middle).

tab-separated values (.tsv files), allow the row keys and column keys to be arbitrarily specified by the user. These formats have been around since the advent of digital computing [25].

The ability of spreadsheets to store diverse data is why they are often the perspective of choice for organizing raw sets of bits when they first enter a data processing system. Spreadsheets impose minimal requirements on the data, only that the data be organizable in small tabular chunks. The benefit of using spreadsheets in this first step is that the raw sets of bits fall within a well-defined perspective that is readily viewable and understandable.

2.3 Databases

As the number of rows and columns in a table grows beyond what can be easily viewed, a database can be used to store and retrieve the same tabular information. Databases that organize data into large related tables have been the most frequently used tool for storing and retrieving data in the world since the 1960s [26, 27]. Databases enable data to be viewed by different people at different locations and allow people to transact business by updating entries in shared tables. Databases play a role in most purchases. In addition to transactions, another important application of databases is the analysis of many rows and columns in a table to find useful patterns. Such analyses are routinely used to determine if a purchase is real or potentially fraudulent.

The modern development of databases effectively begins with relational or SQL (Structured Query Language) databases [28, 29], which have been the standard interface to databases since the 1980s and are the bedrock of electronic transactions around the world.

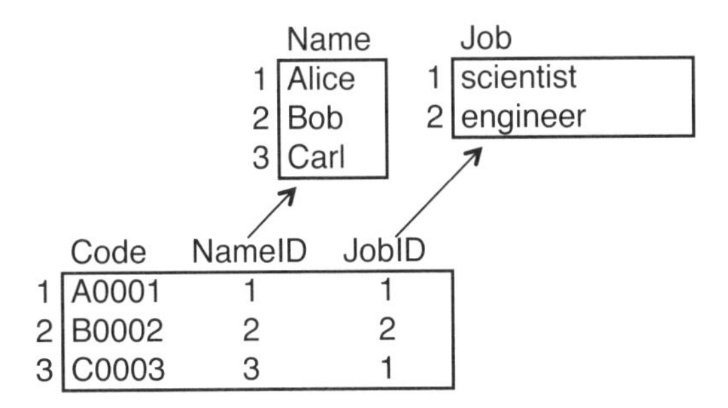

Figure 2.4
Relational databases use multiple tables to create relations between data. Here a list of names and a list of job titles are related by a third table.

More recently, key-value stores (NoSQL databases) [30] have been developed for representing large sparse tables to aid in the analysis of data for Internet search. As a result, the majority of the data on the Internet is now analyzed using key-value stores [31–33]. In response to the same data analysis challenges, the relational database community has developed a new class of array store (NewSQL) databases [34–37] to provide the features of relational databases while also scaling to very large data sets.

Relational Databases (SQL)

Relational databases store data in multiple tables in which each table contains rows or records. Each record consists of a set of values corresponding to the different columns in the table. Relational databases are designed to support multi-party transactions. They have rigorous mechanisms for ensuring that transactions are carried out in the proper order and that data are never lost. Relational databases also structure data by defining exactly the type of data that can go into each entry in the database.

Relational databases are well suited for financial data (right side of Figure 2.3). For personnel records (upper middle of Figure 2.3), a relational database could store the data as they are shown, but often the data would be broken up into three tables (see Figure 2.4). One table would hold a list of names and another table would hold a list of job titles. This arrangement allows both the names and the job titles to be tightly controlled, and only valid names and job titles can be inserted into these tables. A third table would then hold the *relations* between the names and the job titles. Each row of this third table would consist of the code, an index to a name, and an index to a job title. A common schema for relational databases is to have several distinct primary tables that are connected to each other by many relation tables.

Relational databases are most often accessed by using standard SQL syntax, which is inspired by relational algebra that defines mathematical operations for selection and for combining rows and columns from different tables. Interestingly, and not surprisingly,

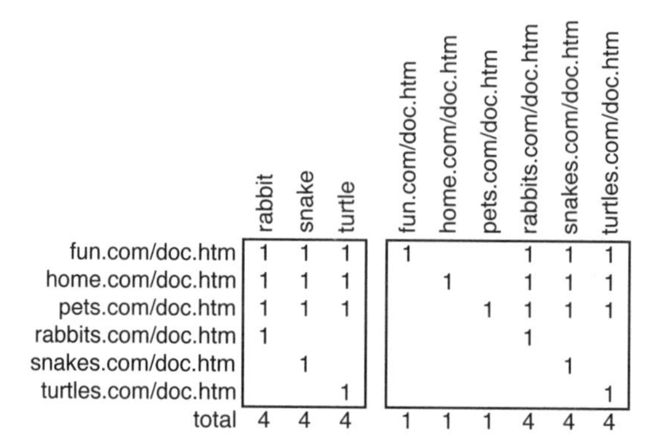

	rabbit	snake	turtle	fun.com/doc.htm	home.com/doc.htm	pets.com/doc.htm	rabbits.com/doc.htm	snakes.com/doc.htm	turtles.com/doc.htm
fun.com/doc.htm	1	1	1	1			1	1	1
home.com/doc.htm	1	1	1		1		1	1	1
pets.com/doc.htm	1	1	1			1	1	1	1
rabbits.com/doc.htm	1						1		
snakes.com/doc.htm		1						1	
turtles.com/doc.htm			1						1
total	4	4	4	1	1	1	4	4	4

Figure 2.5
Example of the kind of data used for searching large numbers of linked documents. Each row represents a document and each column represents either a word in the document or a link to another document.

advanced relational algebra [38] uses a number of mathematical concepts that are used in associative arrays.

Key-Value Stores (NoSQL)

Searching for information on the Internet created a need for a whole new kind of data storage system less focused on transactions and more focused on analysis of data. In particular, storage systems were needed that could ingest many documents and index the words in these documents and their links to other documents.

These key-value stores are often referred to as NoSQL databases because they differ in many ways from relational databases (SQL). Figure 2.5 shows one example of the kind of data employed in searching a large number of linked documents. Each row is a document and each column is either a word in the document or can be a link to another document. If a person wants to look for documents relating to the word rabbit, the search can be done by selecting the corresponding column in Figure 2.5 to find the rows (or documents) containing the word rabbit. However, to rank these documents requires looking at the links in the documents. In this simple example, the document with the largest number of links pointing toward it (rabbits.com/doc.html) could be ranked highest.

Key-value stores need to ingest and index very large quantities of document data. They are designed to be simple so they can be very fast and very large. Typically, a key-value store will keep a sorted list of keys and then have a set of values associated with those keys. In Figure 2.5, the keys correspond to the rows and columns, and the value is the value. For

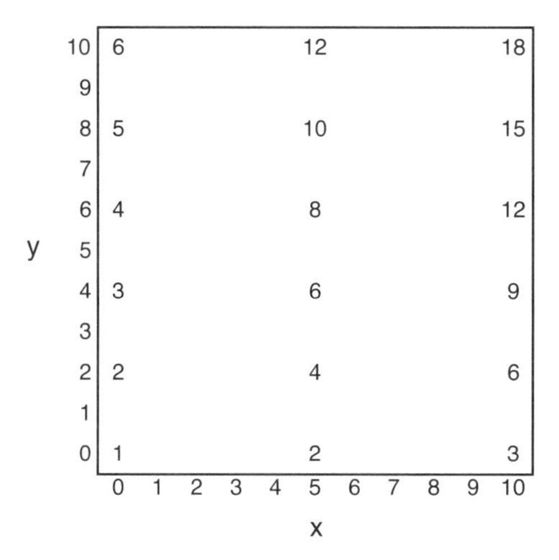

Figure 2.6
Coordinate data from Figure 2.3 (lower middle) organized in an array store. The array store table has two dimensions x and y ranging from 0 to 10. To conserve space, only the non-empty entries are stored.

example, the first few key-value pairs in Figure 2.5 could be

$$(\text{fun.com/doc.htm rabbit}, 1)$$
$$(\text{fun.com/doc.htm snake}, 1)$$
$$(\text{fun.com/doc.htm turtle}, 1)$$

In a key-value store, only the rows and columns with entries are stored, so the tables can be very sparse so that most rows and columns are mostly empty. Little structure is imposed, and keys and values can be any number or word. Furthermore, the type of analyses done on the data are often more complex than simply looking up rows or columns. Matrix algorithms are often used to compute the most relevant documents for a search.

Array Stores (NewSQL)
Sensor data from cameras, medical devices, and sound systems often consist of large quantities of numbers that are best organized in multidimensional arrays. Databases are well suited for organizing and accessing these data, and a new class of array store databases (NewSQL) has been developed to address these new patterns of data. Figure 2.6 shows how the coordinate data (lower middle) from Figure 2.3 (lower middle) could be organized in an array store. The array store table consists of two dimensions x and y that range from 0 to 10. Only the non-empty entries are stored.

2.4 Matrices

Mathematics also uses a tabular view to represent numbers. This view is referred to as a matrix. In a matrix, each row and column index is given by pairs of integers starting at $(1,1)$. The value stored at a given row and column index is typically a real number. Matrices are particularly valuable for comparing entire rows and columns and determining which rows or columns are most alike. Such a comparison is called a correlation and is useful in a wide range of applications. For example, matrix correlations can determine which documents are most like other documents so that a person looking for one document can also be given a list of similar documents.

The correlation of two matrices $\mathbf{A}$ and $\mathbf{B}$ is computed using matrix multiplication $\mathbf{AB}$ or more formally $\mathbf{A} \oplus.\otimes \mathbf{B}$. Because of its usefulness across many applications, much of matrix mathematics is devoted to the study of matrix multiplication. Consider the 2×2 matrix

$$\mathbf{A} = \begin{array}{c} \\ 1 \\ 2 \end{array} \begin{array}{c} \begin{array}{cc} 1 & 2 \end{array} \\ \begin{bmatrix} 0.5 & 1.1 \\ 1.1 & 0.5 \end{bmatrix} \end{array}$$

What happens to $\mathbf{A}$ when it is multiplied with other matrices is shown in Figure 2.7. $\mathbf{A}$ is depicted as the blue parallelogram on which the uppermost point (1.6,1.6) corresponds to the sum of the rows (or columns) of $\mathbf{A}$. The circle of radius 1 is depicted in green and is designated the unit circle.

Let $\mathbf{u}$ be any point lying on the green line. $\mathbf{Au}$ is the product of $\mathbf{A}$ with $\mathbf{u}$ and is shown as the green dashed ellipse. The green dashed ellipse represents how $\mathbf{A}$ distorts the unit circle. Note that the green dashed ellipse touches the corners of the blue parallelogram and is tangent to the sides of the blue parallelogram at that point. The long axis of the green dashed ellipse points directly at the upper rightmost point of the parallelogram.

The points on the unit circle that align with the axis of the green dashed ellipse are show by the black square. These points are given a special name called the *eigenvectors* of the matrix $\mathbf{A}$. Multiplication of $\mathbf{A}$ by any point along these eigenvectors will only stretch the eigenvectors. All other points, when multiplied by $\mathbf{A}$, will be stretched and rotated. The red box shows what happens when the eigenvectors of $\mathbf{A}$ are multiplied by $\mathbf{A}$. The red box corresponds exactly to the axis of the green dashed ellipse. The lengths of the sides of the red box determine how much stretching occurs along each direction of the eigenvector. The long axis implies that points along this eigenvector will become longer. The short axis indicates that points along this eigenvector will become shorter.

The ratio of the area of the red box to the black box is equal to the ratio of the area of the green dashed ellipse to the green solid circle. This ratio is equal to the area of the blue parallelogram and is called the *determinant*.

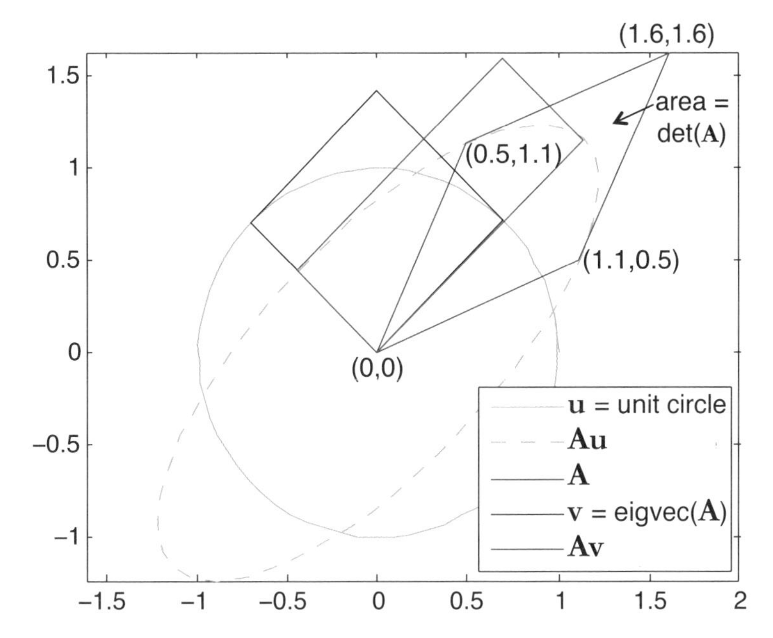

Figure 2.7

The matrix multiplication properties of the 2×2 matrix $\mathbf{A} = \begin{bmatrix} 0.5 & 1.1 \\ 1.1 & 0.5 \end{bmatrix}$.

While not all 2×2 matrices behave as shown in Figure 2.7, many do. In particular, many of the matrices that are relevant to the kinds of data represented with spreadsheets, databases, and graphs have the properties shown Figure 2.7. Furthermore, these properties hold as the matrix grows (they just get harder to draw).

2.5 Graphs

Humans have an intuitive ability to understand visual relationships. A common way to draw these relationships is through a picture (graph) consisting of points (vertices) connected by lines (edges). These pictures can readily highlight data that are connected to lots of other data. In addition, it is also possible to determine how closely connected two data elements are by following the edges connecting two vertices. For example, given a person's set of friends, it is possible to suggest likely new friends from their friends' friends.

Graphs are one of the most pervasive representations of data and algorithms [39]. Graphs are often stored in spreadsheets and databases as various sparse matrices. The two most common types are the adjacency matrix and the incidence matrix. In an adjacency matrix $\mathbf{A}$, each row and column represents vertices in the graph, and setting $\mathbf{A}(i, j) = 1$ denotes an edge from vertex i to vertex j. The same correlation operation that is used to compare rows

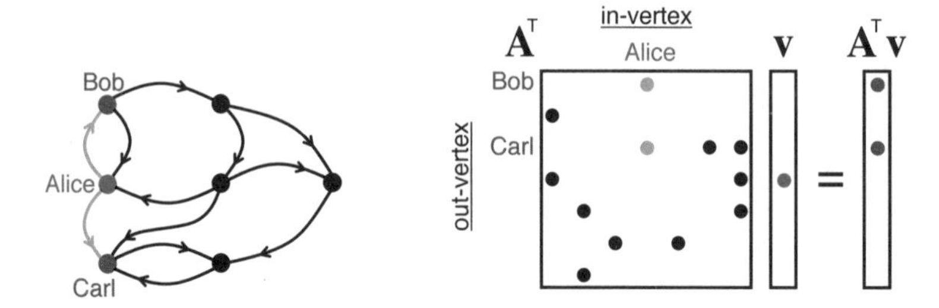

Figure 2.8
The relationships among 7 vertices and 12 edges depicted as a graph (left) and its corresponding adjacency matrix (right). The neighbors of the vertex Alice can be found by following the edges (left) or by multiplying the adjacency matrix with the appropriate vector (right).

and columns in a matrix can also be used to follow edges in a graph. Figure 2.8 shows this graph/matrix duality. On the left of Figure 2.8 is a graph with 7 vertices and 12 edges. The neighbors of the vertex Alice can be found by following all the edges that lead out of Alice to the vertices Bob and Carl. This graph operation is called breadth-first search (BFS). On the right of Figure 2.8 is an adjacency matrix representation of the same graph. Multiplying the transpose of adjacency matrix $\mathbf{A}^\mathsf{T}$ by a vector $\mathbf{v}$ containing a nonzero entry only at Alice results in another vector whose nonzero entries correspond to the neighbors of Alice: Bob and Carl. Many graph algorithms leverage the close link between graphs and matrices to exploit matrix mathematics in their algorithms [8]. The most well-known example is the Google PageRank algorithm [40], which rates web pages on the basis of the first eigenvector of the Internet weblinks adjacency matrix.

Graph Construction

Graph construction, a fundamental operation in the data processing pipeline, is typically done by multiplying the incidence associative array representations of a graph, $\mathbf{E}_{\text{out}}$ and $\mathbf{E}_{\text{in}}$, to produce an adjacency associative array of the graph that can be processed with a variety of machine learning clustering techniques. Suppose G is a (possibly directed) weighted graph with a set of edges K and vertex set

$$K_{\text{out}} \cup K_{\text{in}}$$

where K_{out} is the set of vertices with out edges and K_{in} is the set of vertices with in edges. In addition, let the set of possible values associated with the edge be drawn from a set V.

The two incidence associative arrays of G are the *out incidence array*

$$\mathbf{E}_{\text{out}} : K \times K_{\text{out}} \to V$$

and the *in incidence array*

$$\mathbf{E}_{\text{in}} : K \times K_{\text{in}} \to V$$

To be an incidence array requires

$$\mathbf{E}_{\text{out}}(k, k_{\text{out}}) \neq 0$$

if edge k goes out of vertex k_{out} and

$$\mathbf{E}_{\text{in}}(k, k_{\text{in}}) \neq 0$$

if edge k goes into vertex k_{in}.

Given a graph G with vertex set $K_{\text{out}} \cup K_{\text{in}}$, an associative array

$$\mathbf{A} : K_{\text{out}} \times K_{\text{in}} \to V$$

is an *adjacency array* of the graph G if and only if

$$\mathbf{A}(k_{\text{out}}, k_{\text{in}}) \neq 0$$

implies k_{out} is adjacent to k_{in}. In a directed graph, this means that

$$\mathbf{A}(k_{\text{out}}, k_{\text{in}}) \neq 0$$

if there is an edge starting at vertex k_{out} and ending at vertex k_{in}.

Let G be a graph with out and in incidence arrays $\mathbf{E}_{\text{out}}$ and $\mathbf{E}_{\text{in}}$, respectively. An entry of

$$\mathbf{A} = \mathbf{E}_{\text{out}}^{\mathsf{T}} \mathbf{E}_{\text{in}}$$

called

$$\mathbf{A}(k_{\text{out}}, k_{\text{in}})$$

for $k_{\text{out}} \in K_{\text{out}}$ and $k_{\text{in}} \in K_{\text{in}}$ may be expressed as

$$\mathbf{A}(k_{\text{out}}, k_{\text{in}}) = \bigoplus_{k \in K} \mathbf{E}_{\text{out}}^{\mathsf{T}}(k_{\text{out}}, k) \otimes \mathbf{E}_{\text{in}}(k, k_{\text{in}})$$

where ^{T} denotes the transpose operation. For $\mathbf{E}_{\text{out}}^{\mathsf{T}} \mathbf{E}_{\text{in}}$ to be the adjacency array of G, the entry $\mathbf{A}(k_{\text{out}}, k_{\text{in}})$ must be nonzero exactly when vertex k_{out} has an edge to vertex k_{in}.

2.6 Map Reduce

One of the simplest and most popular approaches to analyzing data is to *map* the same function to different inputs, such as files, and then to take the different outputs and *reduce* them together into a single output. This *map reduce* paradigm [41, 42] has been widely used in many systems and can naturally be expressed in terms of array multiplication

$$\mathbf{C} = \mathbf{A}\mathbf{B} = \mathbf{A} \oplus.\otimes \mathbf{B}$$

In this notation, $\otimes$ approximately corresponds to the map applied to each pair of inputs, and $\oplus$ approximately corresponds to the reduce function applied to the outputs. This approach can be extended to performing computations on database analysis systems that have table reader functions and table writer functions [43]. The table reader function corresponds to reading data and performing the map function, while the table writer function corresponds to the reduce function.

2.7 Other Perspectives

The four perspectives described here are by no means the only ways to view data. Spreadsheets, databases, matrices, and graphs have been combined in a variety of ways in a range of technologies. Some of the most richly implemented examples include Linda tuplespaces, SPARQL resource description frameworks (RDFs), R labeled matrices, MATLAB table objects, and the Spark resilient distributed data set (RDD).

The Linda [44] tuplespace data model and parallel programming paradigm is one of the most innovative systems in data modeling and computation. In Linda, data values are tuples that have many similarities to associative array triples. For example, two edges in Figure 2.8 could be represented as the tuples (Alice,Bob,1) and (Alice,Carl,1). Tuples with more than three attributes appear similar to a dense associative array with each row representing a distinct tuple and each column a distinct dimension. Linda defines four primary operations on its tuplespaces: rd, in, out, and eval. rd selects a specified tuple from a tuplespace. rd is similar to looking up an entry in an associative array. in selects a tuple from a tuplespace and returns it and removes it from the tuplespace. in looks up an entry in an associative array and then substracts that entry from the associative array after it is has been found. out inserts a specified tuple into a tuplespace. out is similar to adding a new entry to an associative array via addition. eval selects a specified tuple and then performs operations on the values from that tuple. eval is similar to a user-defined $\oplus$ or $\otimes$ operation on an associative array.

RDF was developed by the W3 (World Wide Web) Consortium [45] to provide a general mechanism for modeling relationships between objects typically found on the Internet. In RDF, a relationship is specified with a triple denoted (subject, predicate, object). In graph terminology, the subject is the start vertex, the object is the end vertex, and the predicate labels the edge. In database terminology, a table row is the subject, the table column name is the object, and the predicate is the table value. For example, if the two edges in Figure 2.8 represented author citations, they could be represented using RDF triples as (Alice,cited,Bob) and (Alice,cited,Carl). The Simple Protocol And RDF Query Language (SPARQL) [46] was likewise developed by the W3 Consortium to provide a mechanism for querying RDF data in a manner similar to how SQL can be used to query a database table.

The concept of assigning the string row and column keys to a matrix has been used in a number of environments. In the R programming language [47] the colnames and rownames functions allow users to assign these values to a matrix. The MATLAB language has a similar concept in its table objects. Likewise, in the Spark RDD [48], the filename and row number form an implicit row key, and the columns can be explicitly labeled.

Furthermore, nearly all modern programming languages have a mechanism for associating one set of strings with another. Examples from more recent programming environments include Perl associative arrays [49], C++ maps, Python dictionaries [50], and Java hashmaps [51]. Other earlier instances include SNOBOL (StriNg Oriented and symBOlic Language) tables [52], MUMPS (Massachusetts General Hospital Utility Multi-Programming System) arrays [53], REXX (REstructured eXtended eXecutor) associative arrays [54], and Lua tables [55]. The mathematically rigorous associative arrays described in this book encompass and extend all of these representations.

2.8 Conclusions, Exercises, and References

This chapter described the primary perspectives used in data processing systems for organizing data. Spreadsheets, databases, matrices, and graphs all heavily use two-dimensional perspectives on data that can be captured by row, column, and value triples. A variety of technologies merge aspects of these perspectives. For example, array languages merge the matrix and spreadsheet perspectives. Associative arrays provide the mathematical basis for merging all four of these perspectives. The Dynamic Distributed Dimensional Data Model (D4M) is the first practical implementation of associative arrays and will be described in the next chapter.

Exercises

Exercise 2.1 — Figure 2.3 shows four arrays of two-dimensional data: a matrix (upper left), personnel records (upper middle), finances (right), and coordinates (lower middle).

(a) Compute the number of rows m and number of columns n in the array.

(b) Compute the total number of entries mn.

(c) How many empty entries are there?

(d) What are the types of entries in each cell. Are they real numbers, integers, words, dollars, ... ?

Remember the row labels and column labels are not counted as part of the array.

Exercise 2.2 — Refer to Figure 2.4.

(a) What is the Name associated with NameID = 2?

(b) What is the Job associated with JobID = 1?

Exercise 2.3 — Refer to Figure 2.5.

(a) Compute the number of rows m and number of columns n in the array.

(b) Compute the total number of entries mn.

(c) How many empty entries are there?

(d) How many non-empty entries are there?

Exercise 2.4 — Refer to Figure 2.6.

(a) Compute the number of rows m and number of columns n in the array.

(b) Compute the number of non-empty rows $\underline{m}$ and number of non-empty columns $\underline{n}$ in the array.

(c) List the non-empty row keys and the non-empty column keys.

(d) Compute the total number of entries mn.

(e) Compute the total number of entries in the non-empty rows and columns $\underline{mn}$.

(f) How many empty entries are there?

(g) How many non-empty entries are there?

(h) List the triple corresponding to the largest value in the array.

Exercise 2.5 — Refer to Figure 2.7.

(a) Describe a relation of the blue parallelogram to the red rectangle.

(b) Describe a relation of the red rectangle to the black square.

(c) Describe a relation of the red rectangle to the green dashed ellipse.

(d) Describe a relation of the green dashed ellipse to the blue parallelogram.

Exercise 2.6 — Refer to Figure 2.8.

(a) How many edges go into the vertex Alice and how many edges leave the vertex Alice?

(b) How many edges go into the vertex Bob and how many edges leave the vertex Bob?

(c) How many edges go into the vertex Carl and how many edges leave the vertex Carl?

Exercise 2.7 — From your own experience, describe data you have put into a spreadsheet and how you used those data. Why was a spreadsheet the right tool for that task?

Exercise 2.8 — Why were key-value stores developed? How do key-value stores differ from relational databases?

Exercise 2.9 — Describe one method for understanding how a specific matrix behaves during matrix multiplication.

Exercise 2.10 — Draw the graph represented by the adjacency matrix in Figure 2.5.

Exercise 2.11 — Compare and contrast spreadsheets, databases, matrices, and graphs.

References

[1] K. E. Iverson, "A programming language," in *Proceedings of the May 1-3, 1962, Spring Joint Computer Conference*, pp. 345–351, ACM, 1962.

[2] C. Moler, *MATLAB Users' Guide*. University of New Mexico, 1980.

[3] C. B. Moler, *Numerical Computing with MATLAB*. SIAM, 2004.

[4] J. R. Gilbert, C. Moler, and R. Schreiber, "Sparse matrices in MATLAB: Design and implementation," *SIAM Journal on Matrix Analysis and Applications*, vol. 13, no. 1, pp. 333–356, 1992.

[5] M. A. Smith, B. Shneiderman, N. Milic-Frayling, E. Mendes Rodrigues, V. Barash, C. Dunne, T. Capone, A. Perer, and E. Gleave, "Analyzing (social media) networks with nodexl," in *Proceedings of the Fourth International Conference on Communities and Technologies*, pp. 255–264, ACM, 2009.

[6] A. Cordova, B. Rinaldi, and M. Wall, *Accumulo: Application Development, Table Design, and Best Practices*. O'Reilly Media, Inc., 2015.

[7] P. Cudré-Mauroux, H. Kimura, K.-T. Lim, J. Rogers, R. Simakov, E. Soroush, P. Velikhov, D. L. Wang, M. Balazinska, J. Becla, J. Becla, D. DeWitt, B. Heath, D. Maier, S. Madden, J. Patel, M. Stonebraker, and S. Zdonik, "A demonstration of SciDB: A science-oriented DBMS," *Proceedings of the VLDB Endowment*, vol. 2, no. 2, pp. 1534–1537, 2009.

[8] J. Kepner and J. Gilbert, *Graph Algorithms in the Language of Linear Algebra*. SIAM, 2011.

[9] T. Mattson, D. Bader, J. Berry, A. Buluc, J. Dongarra, C. Faloutsos, J. Feo, J. Gilbert, J. Gonzalez, B. Hendrickson, J. Kepner, C. Leiseron, A. Lumsdaine, D. Padua, S. Poole, S. Reinhardt, M. Stonebraker, S. Wallach, and A. Yoo, "Standards for graph algorithm primitives," in *High Performance Extreme Computing Conference (HPEC)*, IEEE, 2013.

[10] T. Mattson, "Motivation and mathematical foundations of the GraphBLAS," in *International Parallel and Distributed Processing Symposium Workshops (IPDPSW)*, IEEE, 2014.

[11] J. Gilbert, "Examples and applications of graph algorithms in the language of linear algebra," in *International Parallel and Distributed Processing Symposium Workshops (IPDPSW)*, IEEE, 2014.

[12] R. S. Xin, D. Crankshaw, A. Dave, J. E. Gonzalez, M. J. Franklin, and I. Stoica, "Graphx: Unifying data-parallel and graph-parallel analytics," in *International Parallel and Distributed Processing Symposium Workshops (IPDPSW)*, IEEE, 2014.

[13] D. Mizell and S. Reinhardt, "Effective graph-algorithmic building blocks for graph databases," in *International Parallel and Distributed Processing Symposium Workshops (IPDPSW)*, IEEE, 2014.

[14] J. Kepner and V. Gadepally, "Adjacency matrices, incidence matrices, database schemas, and associative arrays," in *International Parallel and Distributed Processing Symposium Workshops (IPDPSW)*, IEEE, 2014.

[15] S. Maleki, G. Evans, and D. Padua, "Linear algebra operator extensions for performance tuning of graph algorithms," in *International Parallel and Distributed Processing Symposium Workshops (IPDPSW)*, IEEE, 2014.

[16] A. Buluç, G. Ballard, J. Demmel, J. Gilbert, L. Grigori, B. Lipshitz, A. Lugowski, O. Schwartz, E. Solomonik, and S. Toledo, "Communication-avoiding linear-algebraic primitives for graph analytics," in *International Parallel and Distributed Processing Symposium Workshops (IPDPSW)*, IEEE, 2014.

[17] A. Lumsdaine, "Standards: Lessons learned," in *International Parallel and Distributed Processing Symposium Workshops (IPDPSW)*, IEEE, 2014.

[18] D. J. Power, "A history of microcomputer spreadsheets," *Communications of the Association for Information Systems*, vol. 4, no. 1, p. 9, 2000.

[19] R. Mattessich, "Budgeting models and system simulation," *The Accounting Review*, vol. 36, no. 3, pp. 384–397, 1961.

[20] R. K. Pardo and R. Landau, "Process and apparatus for converting a source program into an object program," Aug. 9 1983. US Patent 4,398,249.

[21] D. Bricklin and B. Frankston, *Reference Manual: VisiCalc Computer Software Program*. Personal Software, Inc., 1979.

[22] R. Ramsdell, "The power of VisiCalc," *BYTE Magazine*, vol. 5, no. 11, p. 190, 1980.

[23] G. Williams, "Lotus Develoment Corporations's 1-2-3," *BYTE Magazine*, vol. 7, no. 12, p. 182, 1982.

[24] J. Pournell, "Computing at chaos manner," *BYTE Magazine*, vol. 10, no. 9, p. 347, 1985.

[25] IBM, *IBM FORTRAN Program Products for OS and the CMS Component of VM/370 General Information*. IBM Corporation, 1972.

[26] K. L. Berg, T. Seymour, and R. Goel, "History of databases," *International Journal of Management & Information Systems (Online)*, vol. 17, no. 1, p. 29, 2013.

[27] G. O'Regan, "History of databases," in *Introduction to the History of Computing*, pp. 275–283, Springer, 2016.

[28] E. F. Codd, "A relational model of data for large shared data banks," *Communications of the ACM*, vol. 13, no. 6, pp. 377–387, 1970.

[29] M. Stonebraker, G. Held, E. Wong, and P. Kreps, "The design and implementation of INGRES," *ACM Transactions on Database Systems*, vol. 1, no. 3, pp. 189–222, 1976.

[30] F. Chang, J. Dean, S. Ghemawat, W. C. Hsieh, D. A. Wallach, M. Burrows, T. Chandra, A. Fikes, and R. E. Gruber, "Bigtable: A distributed storage system for structured data," *ACM Transactions on Computer Systems*, vol. 26, no. 2, p. 4, 2008.

[31] G. DeCandia, D. Hastorun, M. Jampani, G. Kakulapati, A. Lakshman, A. Pilchin, S. Sivasubramanian, P. Vosshall, and W. Vogels, "Dynamo: Amazon's highly available key-value store," *ACM SIGOPS Operating Systems Review*, vol. 41, no. 6, pp. 205–220, 2007.

[32] A. Lakshman and P. Malik, "Cassandra: a decentralized structured storage system," *ACM SIGOPS Operating Systems Review*, vol. 44, no. 2, pp. 35–40, 2010.

[33] L. George, *HBase: The Definitive Guide: Random Access to Your Planet-Size Data*. O'Reilly Media, Inc., 2011.

[34] M. Stonebraker, D. J. Abadi, A. Batkin, X. Chen, M. Cherniack, M. Ferreira, E. Lau, A. Lin, S. Madden, E. O'Neil, P. O'Neil, A. Rasin, N. Tran, and S. Zdonik, "C-Store: A column-oriented DBMS," in *Proceedings of the 31st International Conference on Very Large Data Bases*, pp. 553–564, VLDB Endowment, 2005.

[35] R. Kallman, H. Kimura, J. Natkins, A. Pavlo, A. Rasin, S. Zdonik, E. P. Jones, S. Madden, M. Stonebraker, Y. Zhang, J. Hugg, and D. Abadi, "H-store: A high-performance, distributed main memory transaction processing system," *Proceedings of the VLDB Endowment*, vol. 1, no. 2, pp. 1496–1499, 2008.

[36] A. Lamb, M. Fuller, R. Varadarajan, N. Tran, B. Vandiver, L. Doshi, and C. Bear, "The Vertica analytic database: C-Store 7 years later," *Proceedings of the VLDB Endowment*, vol. 5, no. 12, pp. 1790–1801, 2012.

[37] M. Stonebraker and A. Weisberg, "The VoltDB main memory DBMS," *IEEE Data Engineering Bulletin*, vol. 36, no. 2, pp. 21–27, 2013.

[38] T. J. Green, G. Karvounarakis, and V. Tannen, "Provenance semirings," in *Proceedings of the Twenty-Sixth ACM SIGMOD-SIGACT-SIGART Symposium on Principles of Database Systems*, pp. 31–40, ACM, 2007.

[39] T. H. Cormen, C. E. Leiserson, R. L. Rivest, and C. Stein, *Introduction to Algorithms*. Cambridge: MIT Press, 2009.

[40] S. Brin and L. Page, "The anatomy of a large-scale hypertextual web search engine," *Computer Networks and ISDN Systems*, vol. 30, no. 1, pp. 107–117, 1998.

[41] J. Anderson and B. Reiser, "The LISP tutor," *BYTE Magazine*, vol. 10, no. 4, pp. 159–175, 1985.

[42] J. Dean and S. Ghemawat, "MapReduce: Simplified data processing on large clusters," *Communications of the ACM*, vol. 51, no. 1, pp. 107–113, 2008.

[43] D. Hutchison, J. Kepner, V. Gadepally, and A. Fuchs, "Graphulo implementation of server-side sparse matrix multiply in the Accumulo database," in *High Performance Extreme Computing Conference (HPEC)*, IEEE, 2015.

[44] N. Carriero and D. Gelernter, "Linda in context," *Communications of the ACM*, vol. 32, no. 4, pp. 444–458, 1989.

[45] O. Lassila and R. R. Swick, "Resource description framework (RDF) model and syntax specification," *W3C Recommendation*, vol. REC-rdf-syntax-19990222, 1999.

[46] E. Prud'hommeaux and A. Seaborne, "SPARQL query language for RDF," *W3C Recommendation*, Jan. 15 2008.

[47] M. J. Crawley, *The R Book*. John Wiley & Sons, 2012.

[48] M. Zaharia, M. Chowdhury, M. J. Franklin, S. Shenker, and I. Stoica, "Spark: Cluster computing with working sets," *HotCloud*, vol. 10, no. 10-10, p. 95, 2010.

[49] L. Wall, T. Christiansen, and J. Orwant, *Programming Perl*. O'Reilly Media, Inc., 2000.

[50] M. Lutz, *Learning Python*. O'Reilly Media, Inc., 2013.

[51] D. Flanagan, *Java in a Nutshell*. O'Reilly Media, Inc., 2005.

[52] D. J. Farber, R. E. Griswold, and I. P. Polonsky, "SNOBOL, a string manipulation language," *Journal of the ACM (JACM)*, vol. 11, no. 1, pp. 21–30, 1964.

[53] J. Bowie and G. O. Barnett, "MUMPS–an economical and efficient time-sharing system for information management," *Computer Programs in Biomedicine*, vol. 6, no. 1, pp. 11–22, 1976.

[54] M. F. Cowlishaw, "The design of the REXX language," *IBM Systems Journal*, vol. 23, no. 4, pp. 326–335, 1984.

[55] R. Ierusalimschy, W. Celes, L. de Figueiredo, and R. de Souza, "Lua: Uma linguagem para customizaçao de aplicaçoes," in *VII Simpósio Brasileiro de Engenharia de Software–Caderno de Ferramentas*, p. 55, 1993.

3 Dynamic Distributed Dimensional Data Model

Summary

The Dynamic Distributed Dimensional Data Model (D4M) software (d4m.mit.edu) is the first practical implementation of the full mathematics of associative arrays that successfully bridges spreadsheets, databases, matrices, and graphs. Associative arrays allow D4M users to implement complex algorithms with less effort than required by standard approaches. D4M brings the power of matrix mathematics directly to bear on both string and numeric data stored in spreadsheets, databases, and graphs. Using D4M, it is possible to develop a general-purpose sparse matrix schema for representing diverse data, such as document citations, webpages, computer networks, genetic sequence information, and social media. A common schema for diverse data has enabled a new class of analytics to be developed that can be applied generically to diverse data in a domain-independent manner to correlate data sets, detect outliers, and filter high-value data from low-value data. D4M leverages the decades of work on high performance linear algebra to enable high performance database interfaces that have achieved record-breaking database performance. This chapter highlights the features and technical successes of D4M in order to motivate understanding associative arrays at a deeper mathematical level.

3.1 Background

The development of the mathematical concept of associative arrays began in 1992 when Visual Intelligence Corporation developed the first webpage with dynamic content [1]. This system used a beta release of the first popular web browser (Mosaic) developed at the University of Illinois National Center for Supercomputing Applications by Marc Andreessen. The Visual Intelligence web system allowed the user to invoke programs from the browser that constructed user-defined queries to a relational database. The database query results were then parsed by programs to generate webpages that were then displayed in the browser. Prior to this system, nearly all webpage content was static. At this time, it was apparent that a tabular data structure that could be passed between browser, programs, and databases would simplify both the design and construction of web-based data processing systems.

In 1998, the founders of Google (Sergey Brin and Larry Page) released their web search engine that used matrix mathematics to rank webpages [2]. They also laid the foundation for a new style of key-value store database for indexing webpages called Google Bigtable.

The design of Google Bigtable was released in 2006 [3]. Many other organizations used this design to construct their own key-value store databases [4]. The U.S. National Security Agency developed its own key-value store database and released it to MIT in 2009 [5, 6] and to the Apache Foundation in 2011 [7] under the name Apache Accumulo.

The duality between the canonical representation of graphs as abstract collections of vertices and edges and a sparse adjacency matrix representation has been a part of graph theory since its inception [8, 9]. Matrix algebra has been recognized as a useful tool in graph theory for nearly as long (see [10] and references therein, in particular [11–17]). However, matrices were not traditionally used for practical computing with graphs, in part because dense two-dimensional arrays are not an efficient representation of a sparse graph. With the growth of efficient data structures and algorithms for sparse arrays and matrices, it has become possible to develop practical array-based approaches to computation on large, sparse graphs [18].

In 2010, researchers at MIT recognized the need for an environment that allowed spreadsheets, databases, matrices, and graphs to work together seamlessly, and they developed the mathematical concept of the associative array. The D4M software that implemented the mathematics of associative arrays was developed in 2010 and publicly released in 2012 as part of the nine-lecture MIT Lincoln Laboratory course *Signal Processing on Databases* (techtv.mit.edu/collections/d4m-class).

The name associative array derives from the Perl programming language data structure that allowed a string to be used as an index into an array [19]. The concept of using a string as an index in an array can also be found in Python dictionaries, Java hashmaps, and many other programming languages. The D4M mathematical associative array extends this concept by adding matrix operations that are mathematically closed so that any operation on an associative array produces another associative array.

3.2 Design

The D4M software (d4m.mit.edu) [20, 21] is the first practical implementation of associative array mathematics that successfully bridges spreadsheets, databases, matrices, and graphs. Associative arrays allow D4M users to implement high performance complex algorithms with significantly less effort than when using other implementation approaches. In D4M, a user can read data from a spreadsheet, load the data into a variety of databases, correlate rows and columns with matrix operations, and visualize relationships among the data using graph operations. These operations correspond to the steps necessary to build an end-to-end data processing system (see Figure 3.1). Often, the majority of time needed to build a data processing system is spent in the interfaces between the various steps, which normally require a conversion from one mathematical perspective to another. By using a

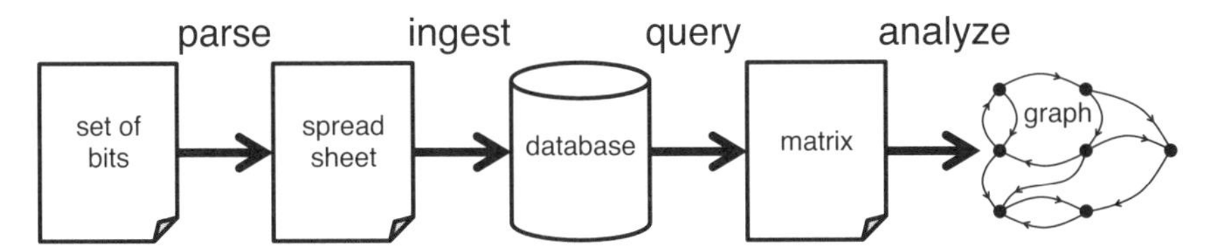

Figure 3.1
The standard steps in a data processing system often require different perspectives on the data. D4M associative arrays enable a common mathematical perspective to be used across all the steps.

common mathematical abstraction across all steps, the construction time of a data processing system is significantly reduced. The success of D4M in building real data processing systems has been a prime motivation for formalizing the mathematics of associative arrays. By making associative arrays mathematically rigorous, it becomes possible to implement associative arrays in a wide range of programming environments (not just in D4M).

The D4M software has provided a unified approach to these data structures that is based on intuitive ideas about what a user would expect when combining and manipulating data from all of these domains. In D4M, the unifying mathematical object is referred to as an associative array. Associative arrays have enabled D4M users to achieve a number of breakthroughs that are described in the subsequent sections.

3.3 Matrix Mathematics

D4M enables the power of matrix mathematics to be applied to both numeric and string data stored in spreadsheets, databases, and graphs. D4M allows diverse data to be easily represented in an intuitive manner. D4M enables complex algorithms to be represented succinctly with as much as a 50x reduction in code volume compared to traditional environments [21, 22].

To illustrate the use of D4M, consider the associative array **A** consisting of rows of documents and columns of terms shown in Figure 3.2. A common search operation on documents is to start with a term and then provide suggestions for additional terms. One approach to generating these suggestions is facet search. In this context, a facet search selects the subset of documents containing a set of terms and then computes the histogram of all the terms in this document subset. A facet search is particularly useful in helping a user build searches by providing guidance as to the most popular remaining terms as the search narrows. Facet search is a highly dynamic query because it is not possible to compute the histograms of all the subsets of keywords in advance.

Facet search in D4M begins with defining set of terms k in the table

$$k = \text{'UN Carl '}$$

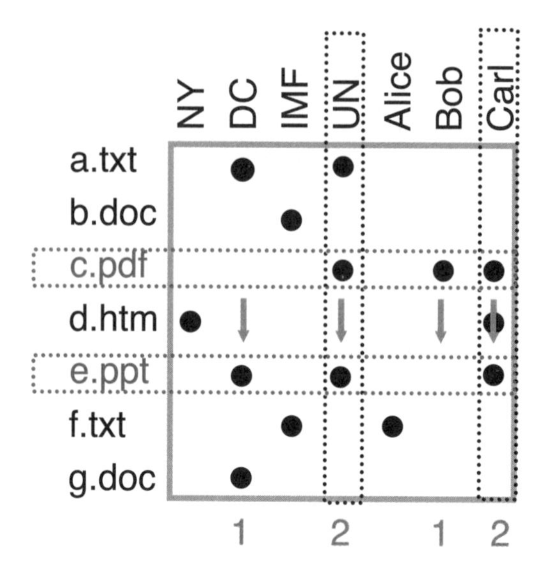

Figure 3.2
Facet search in D4M. Associative array **A** stores a list of documents as rows and their terms as columns. Selecting terms UN and Carl indicates that the documents c.pdf and e.ppt contain both. Selecting the documents c.pdf and e.ppt and summing the occurrences of their terms retrieves the facets DC and Bob. ©2012 IEEE. Reprinted, with permission, from [21]

Next, all documents that contain these terms are found by selecting the corresponding columns in **A**, summing the columns together, finding the resulting rows with more than one entry, and assigning these rows to a new associative array **B**. All of these operations can be carried out in a single line of D4M code

$$\mathbf{B} = \text{sum}(\mathbf{A}(:,k),2) > 1$$

Finally, the histogram of terms in the subset of documents that contains all the terms in k is computed via array multiplication of the transpose of the column vector **B** with **A**

$$\mathbf{F} = \text{transpose}(\mathbf{B}) * \mathbf{A}$$

This complex query can be performed efficiently in just two lines of D4M code that perform two database queries (one column query and one row query). Implementing a similar query using standard programming environments such as Java and SQL takes hundreds of lines of code and hides the inherent mathematical properties of the operation.

3.4 Common SQL, NoSQL, NewSQL Interface

Relational or SQL (Structured Query Language) databases have been the de facto interface to databases since the 1980s and are the bedrock of electronic transactions around the world. More recently key-value stores (NoSQL databases) have been developed for representing large sparse tables to aid in the analysis of data for Internet search. As a result, the majority of the data on the Internet are now analyzed using key-value stores. In response to the data analysis challenges, the relational database community has developed a new class of array store (NewSQL) databases to provide the features of relational databases while also scaling to very large data sets.

This diversity of databases has created a need to interoperate between them. Associative arrays can provide an abstraction that works with all of these classes of databases. D4M has demonstrated this capability [23, 24]. One example is in the field of medicine, where a SQL database might be used for patient records, a NoSQL database for analyzing the medical literature, and a NewSQL database for analyzing patient sensor data.

3.5 Key-Value Store Database Schema

Using D4M, it was possible to create a general sparse matrix database schema [25] for representing diverse data such as documents [22], citation graphs [26], computer networks [27–29], genetic sequence information [30, 31], and social media [32]. The D4M schema is now widely used throughout the Apache Accumulo database community that supports a wide range of government applications.

The D4M schema is best explained in the context of a specific example. Micro-blogs (such as Twitter) allow their users to globally post short messages. Micro-blogs are used by many humans and machines. Each micro-blog entry consists of a message and metadata. The simplest way to view an entry in a key-value store table is as a triple of strings consisting of a row key, a column key, and a value that corresponds to the entries of a sparse matrix. In the case of a micro-blog entry, a triple might be

(31963172416000001,user|getuki,1)

The above triple denotes that the micro-blog entry with row key 31963172416000001 was from the user called getuki. As is often the case in the D4M schema, the value of 1 is used to simply denote the existence of the relationship, and the value itself has no additional meaning.

Figure 3.3 shows how a four-table D4M schema can be applied to micro-blog data. The message text is stored in one column in the $\mathbf{T}_{\mathrm{edgeTxt}}$ table. All the message metadata such as its status (stat|), the message author (user|), the time the message was sent (time|), and the parsed text of the message (word|) are stored in the $\mathbf{T}_{\mathrm{edge}}$ table such that each column|value pair is a unique column. Combining the column and the value makes the resulting $\mathbf{T}_{\mathrm{edge}}$ table very sparse. Storing the transpose of the metadata in the $\mathbf{T}^{\mathsf{T}}_{\mathrm{edge}}$ table indexes every unique string in the data set and allows it to be looked up quickly. The sums of the unique column|value pairs are stored using an accumulator column labeled Degree in the $\mathbf{T}_{\mathrm{edgeDeg}}$ table. The sum table enables efficient query planning by allowing queries to

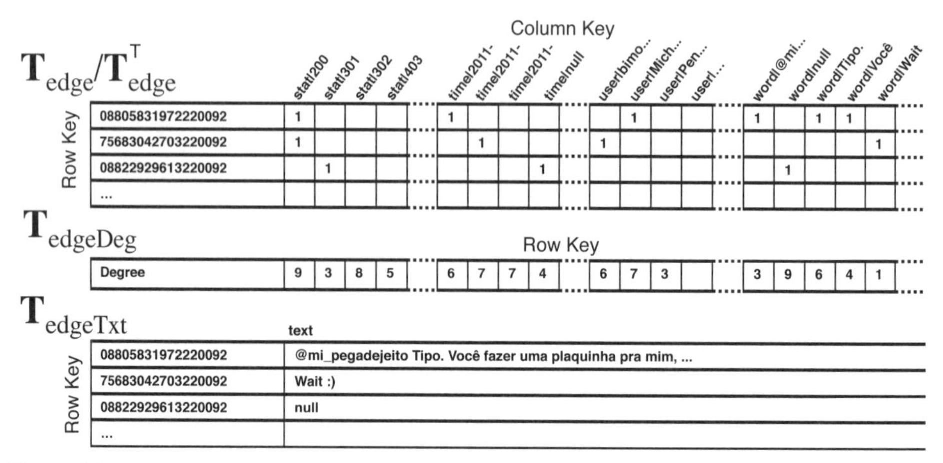

Figure 3.3

The D4M schema applied to micro-blog data consists of four tables. The text of the message is stored as one column in the $\mathbf{T}_{\text{edgeTxt}}$ table. All the metadata (stat|, user|, time|) and the parsed text (word|) are stored in $\mathbf{T}_{\text{edge}}$ such that each column|value pair is a unique column. Note: some of the column names have been truncated in the figure for display purposes. Storing the transpose of the metadata in $\mathbf{T}^{\mathsf{T}}_{\text{edge}}$ creates an index to every unique string in the data set and allows it to be selected efficiently. The sums of the unique column|value pairs are stored using an accumulator column labeled Degree in the $\mathbf{T}_{\text{edgeDeg}}$ table. The sum table allows query planning by enabling queries to estimate the sizes of their results prior to executing queries. Flipping the row keys allows for efficient load balancing while the table grows and in turn splits across multiple servers. ©2013 IEEE. Reprinted, with permission, from [25]

estimate the sizes of results prior to executing queries. The row keys are stored in a flipped format to allow for efficient load balancing as the table grows and is split (or sharded) across multiple database servers.

The D4M schema is designed to exploit a number of specific features found in key-value stores. These include

Row Store — Most key-value stores are row-oriented, so any row key identifier, such as 31963172416000001, can be looked up quickly. However, looking up a column, such as user|getuki), or value, such as 1, requires a complete scan of the table and can take a long time if there are a lot of data in the table. The D4M schema addresses this limitation by storing both the table ($\mathbf{T}_{\text{edge}}$) and its transpose ($\mathbf{T}^{\mathsf{T}}_{\text{edge}}$), allowing any row or column to be looked up quickly.

Sparse — Key-value storage is sparse. Only non-empty columns are stored in a row. This arrangement is critical because many of the data sets that key-value stores are used on are naturally represented as extremely sparse tables.

Unlimited Columns — Key-value stores can add new columns with no penalty. This critical capability of key-value stores is heavily exploited by the D4M schema. It is often the case that there are more unique columns than rows.

Arbitrary Text — Rows, columns, and values in a key-value store can often be arbitrary byte strings. This data type is very useful for storing numeric data, such as counts, or multilingual data, such as unicode. For example, consider the following micro-blog entry

identifier	stat	time	user	text
10000061427136913	200	2011-01-31 06:33:08	getuki	ハスなう

In a traditional database, the above entry would be represented by one row in a four-column table. In the D4M schema this entry is represented in the $\mathbf{T}_{\mathrm{edge}}$ table by the following four triples with flipped row keys:

(31963172416000001,stat|200,1)
(31963172416000001,time|2011-01-31 06:33:08,1)
(31963172416000001,user|getuki,1)
(31963172416000001,word|ハスなう,1)

Collective Updates — Key-value stores can perform simultaneous updates to tables that can update many triples at the same time. It is often optimal to have many triples in a single update.

Accumulators — Key-value stores can modify values at insert time. For example, if the triple (word|ハスなう,Degree,1) were inserted into the $\mathbf{T}_{\mathrm{edgeDeg}}$ table and the table entry already had a value of (word|ハスなう,Degree,16), then the key-value store can be instructed that any such collision on the column Degree should be handled by converting the strings 16 and 1 to numeric values, adding them, and then converting them back to a string to be stored as (word|ハスなう,Degree,17). An accumulator column is used to create the $\mathbf{T}_{\mathrm{edgeDeg}}$ column sum table in the D4M schema. The $\mathbf{T}_{\mathrm{edgeDeg}}$ sum table provides several benefits. First, the sum table allows tally queries like "How many tweets have a specific word?" to be answered trivially. Second, the sum table provides effective query planning. For example, to find all tweets containing two words, one first queries the sum table to determine which word is the least popular before proceeding to query the transpose table ($\mathbf{T}^{\mathsf{T}}_{\mathrm{edge}}$). Scanning for the least popular word first significantly reduces the amount of data that would need to be processed.

Parallel — Key-value stores are highly parallel. At any given time, it is possible to have many processes inserting and querying the database.

Distributed — Key-value stores use distributed storage. As the tables become large, they are broken up into pieces that can be stored on different servers.

Partitions — Tables in key-value stores are partitioned (or sharded) into different pieces at specific row keys that are called splits. As a table increases in size, a key-value store will automatically pick splits that keep the pieces approximately equal. If the row key has a time-like element to it (as does the micro-blog identifier), then it is important to convert it to a flipped format so that the most rapidly changing digits are first. This process will cause inserts to be spread across all the servers.

The power of the D4M schema is that it can be applied to almost any type of data that can be represented in a tabular fashion. In addition, new columns or new data sets can be added without making any changes to the schema. The resulting schema allows any unique string to be found quickly.

3.6 Data-Independent Analytics

The last decade has seen amazing progress on the primary data challenges: volume, velocity, variety, and veracity. Key-value stores and the map-reduce parallel programming model have enabled enormous volumes of data to be stored and processed. NoSQL databases allow these data to be ingested, indexed, and queried at high velocity. Graph analytics and machine learning provide mechanisms for analyzing and extracting value from a wide variety of data. Finally, new cryptographic approaches are starting to address ever-increasing security (aka veracity) concerns.

In the context of this extraordinary progress, the core issue of data preparation is now emerging as a major hurdle. The general recognition that a significant fraction of raw data records have known flaws that need to be addressed by data preparation has led to many professionals spending the majority of their time cleaning up data [33]. The challenge of developing tools and technologies to address data cleaning can be significantly aided by placing data onto a common mathematical framework [34].

A similar hurdle occurred early on in the digital sensor era. At the onset of digitization, digital data from sensors or communication systems were treated in a highly application-specific manner. The recognition that large classes of digital data could be represented as a regularly spaced time-series of digitally encoded numbers provided a common framework for processing data across a range of applications. More specifically, the concept that time series data could be represented mathematically by a vector allowed the field of digital signal processing to rigorously formalize a range of general-purpose concepts for preparing, cleaning up, and analyzing noisy data from a wide range of applications. The mathematical foundations of digital signal processing have stood the test of time and remain a cornerstone of modern communication and sensor processing systems.

A common schema that handles diverse data has enabled the development of analytics that can be applied generically to a variety of data in a domain-independent manner to allow correlating data sets, detecting outliers, and selecting high-value data and separating them from low-value data [32].

Comparing two data sets to determine how they are the same is one of the most common data processing operations. Identifying genetic sequence information from biological organisms is one common application of this technique. Associative arrays provide a natural

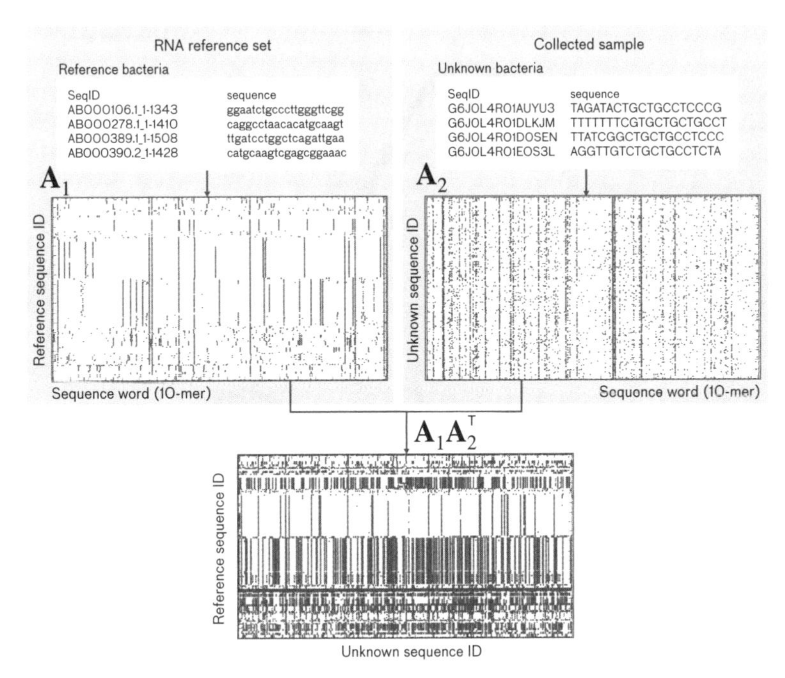

Figure 3.4
Sequence alignment via associative array multiplication. Genetic sequences hashed into words (10-mers) can be readily expressed as sparse associative matrices in which each row represents the name of the sequence and each column represents each unique genetic word. A reference set of known genetic data (left) can be compared with a set of unknown genetic data (right) by multiplying together their corresponding associative arrays. The result (bottom) is another associative array that has rows of known genetic sequence that are found in the columns of the unknown genetic sequence.

mechanism for both representing and correlating genetic data. Figure 3.4 [30, 31] shows two sets of genetic data (RNA) that are represented as associative arrays. One data set consists of reference sequences while the other is an unknown set of sample sequences. The rows of the associative arrays correspond to the labels of each sequence. The columns correspond to each unique combination of 10 RNA bases (a 10-mer). A value of 1 in the associative array indicates that a particular 10-mer was found in a specific labeled sequence. The correlation of the two sequences can be accomplished by a simple associative array multiplication.

This same correlation technique can be applied across many domains and can be used to correlate words across documents, diseases across patients, and domain names across computer network packets.

Correlating data is often done to find the best match between two data sets. In Figure 3.4, the goal is to determine the species of the organisms in the sample by comparing the sample sequence with a reference species sequence. A common feature of these data is that there

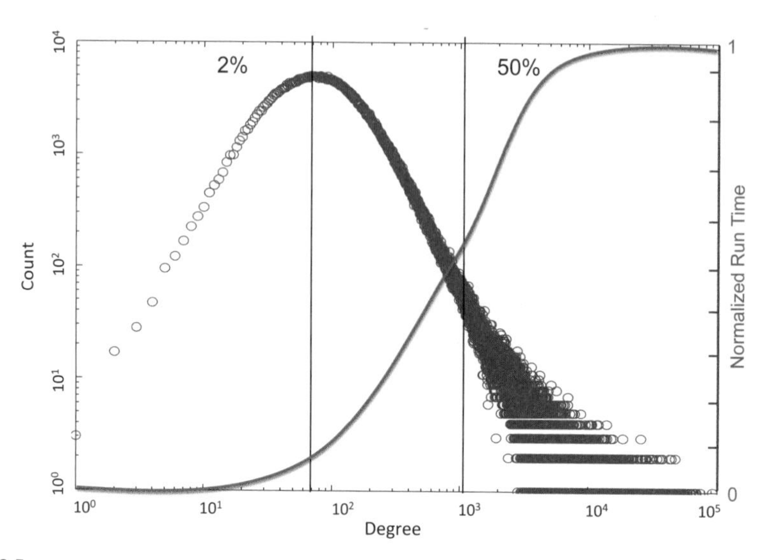

Figure 3.5
Degree distribution of the unique 10-mers in the unique bacterial RNA sequences. Vertical lines indicate the percentage of data. The data is dominated by a few common 10-mers. Increasing the percentage of data used drastically increases the run time as shown by the red curve. ©2015 IEEE. Reprinted, with permission, from [35]

are some sequences that are common and appear in many species and some sequences that are rare and appear in a few species. If the goal is to determine the best match, then the common sequences are not useful and can be eliminated from the comparison in advance. Figure 3.5 [35] shows the statistical distribution of 10-mers in a large reference database. It also shows the computing time required to compare against different fractions of the database. Interestingly, a good match can still be found with a fraction of the total data.

This same data sampling technique can be applied to a wide variety of applications. It is often the case that the most common words in a document are not very important in comparing two documents. In fact, there are standard sets of "stop" words that are often eliminated from document analysis in order to improve comparisons. Likewise, for computer networks, certain sites can be ignored because they are so common that they indicate little about the patterns in the network traffic. In all of these cases, the approach of eliminating the most common occurrences can be used to accelerate the algorithm or improve the quality of the analysis.

A common distribution that is observed in many social and computer network data sets is the power law distribution [36]. In a graph, a power law distribution means that a small number of vertices have a large number of edges and a large number of vertices have a small number of edges. For example, a small number of websites have many other websites that link to them, while a large number of websites have only a few websites that link to them [37]. This principle has also been referred to as the Pareto principle [38], Zipf's Law [39],

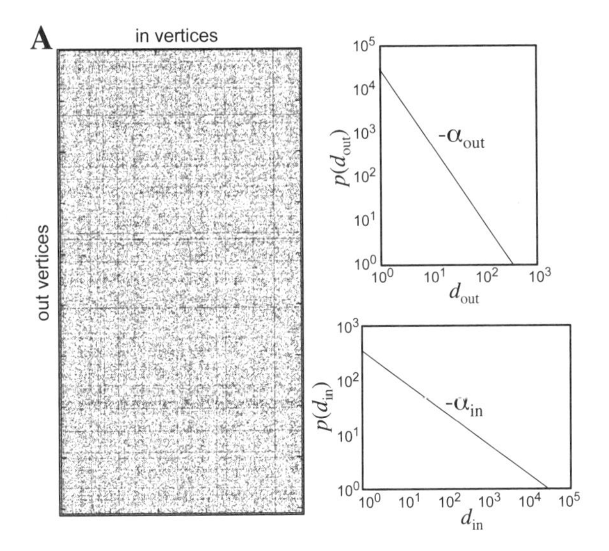

Figure 3.6
Adjacency array for a power law distributed random graph with corresponding in and out degree distributions.

or the 80-20 rule [40]. A power law distribution for a random variable d, is defined to be

$$p(d) \propto d^{-\alpha}$$

When d represents the degree of the vertex in a graph which is the number of edges connected to the vertex, then $p(d)$ is proportional to the number of vertices in the graph that have degree d. An illustrative example of 10,000 randomly generated points drawn from a power law distribution with exponent $\alpha = 1.8$ is shown in Figure 3.6. The graph consists of m_{out} vertices connected to n_{in} vertices. The adjacency array of the graph shown on the left has m_{out} rows and n_{in} columns. The out degree d_{out} is the sum of nonzero entries in each row. The in degree d_{in} is the sum of the nonzero entries in each column. The distributions $p(d_{\text{out}})$ and $p(d_{\text{in}})$ are shown on the top right and bottom right of Figure 3.6. In both cases, these distributions obey a power law, meaning that there are a few vertices with many edges and many vertices with a few edges.

Power law distributions are found in a wide range of data sets. Modeling data as a power law can be a useful tool for predicting the size and distribution of a data set and for identifying unusual data elements [41, 42].

The process of putting data into a sparse associative array using the D4M schema often reveals structures in the data that can be used to understand the data independent of their domain [32]. Figure 3.7 shows the sparse associative array representation of a business process database with 2.5 million entries. There are approximately 50 columns in the original database table. Following the D4M schema, each column and each value of the

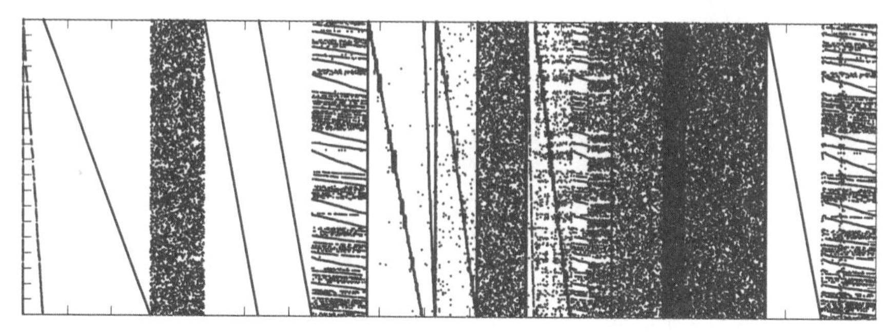

Figure 3.7
Sparse associative array representation of using the D4M schema of 2.5 million entries from a business process database. There are approximately 50 columns in the original database table that are visible as different patterns of columns in the associative array.

original table has been appended to create a new column in the sparse associative array **E**. Each of the original columns can be seen in the different column patterns in Figure 3.7. The entire associative array **E** has $\underline{m}$ non-empty rows, $\underline{n}$ non-empty columns, and nnz nonzero entries. [Note: underlined quantities refer to properties of associative arrays that are restricted to the non-empty rows and non-empty columns of the associative array.] Each sub-associative array $\mathbf{E}_i$ corresponds to a column i in the original database table and will have $\underline{m}_i$ non-empty rows, $\underline{n}_i$ non-empty columns, and nnz_i nonzero entries. These quantities are linked by several formulas. The entire associative array can be constructed by combining the sub-associative arrays of different sizes via element-wise associative array addition

$$\mathbf{E} = \bigoplus_i \mathbf{E}_i$$

The total number of non-empty rows is less than or equal to the sum of the non-empty rows of the sub-associative arrays

$$\underline{m} \leq \sum_i \underline{m}_i$$

The total number of non-empty columns is the sum of the non-empty columns of the sub-associative arrays

$$\underline{n} = \sum_i \underline{n}_i$$

The total number of nonzero entries is the sum of the nonzero entries of the sub-associative arrays

$$\text{nnz} = \sum_i \text{nnz}_i$$

The distinction between $\bigoplus$ and $\sum$ is defined precisely in Chapter 10.11. For now, it is sufficient to know that $\bigoplus$ allows associative arrays of different sizes to be combined together.

The different patterns of data in each $\mathbf{E}_i$ can be deduced by examining $\underline{m}_i$, $\underline{n}_i$, and nnz_i and comparing these values to $\underline{m}$, $\underline{n}$, and nnz. There are several characteristic patterns of data. Once the pattern has been identified, a model for the data can be constructed that can be compared with the real data to reveal outliers or omissions in the data. The most readily apparent pattern is the identity pattern (see Figure 1.8) whereby each row in $\mathbf{E}_i$ corresponds to exactly one column in $\mathbf{E}_i$. In an exact identity pattern, the following relations hold

$$\underline{m} = \underline{m}_i = \underline{n}_i = \text{nnz}_i$$

In other words, the number of non-empty rows, non-empty columns, and nonzero entries in $\mathbf{E}_i$ are equal to the number of non-empty rows in $\mathbf{E}$. A more interesting situation occurs when the data exhibit a pattern that is near the identity pattern

$$\underline{m} \approx \underline{m}_i \approx \underline{n}_i \approx \text{nnz}_i$$

If $\mathbf{E}_i$ is found to be close to the identity pattern, then the ways that $\mathbf{E}_i$ differs from an exact identity can often be very informative. For example, if there are more non-empty rows in $\mathbf{E}$ than in $\mathbf{E}_i$

$$\underline{m} > \underline{m}_i$$

then it is likely that data are missing from $\mathbf{E}_i$. If there are fewer non-empty rows in $\mathbf{E}_i$ than non-empty columns in $\mathbf{E}_i$

$$\underline{m}_i < \underline{n}_i$$

then it is likely that data were added inadvertently. Finally, if there are more non-empty rows in $\mathbf{E}_i$ than non-empty columns in $\mathbf{E}_i$

$$\underline{m}_i > \underline{n}_i$$

then it is likely that there are duplicate rows. Such pattern analysis is a powerful method for detecting unusual data entries in a way that is independent of the specifics of the application.

3.7 Parallel Performance

D4M utilizes the many years of effort on optimizing linear algebra performance to enable database interfaces that have achieved world records in database performance [43–45].

Key-value store databases are designed to handle unstructured data of the type found in document analysis, health records, bioinformatics, social media, computer networks, and computer logs. Often these data are represented as large graphs of nodes and edges. The Graph500 benchmark [46] is designed to test a computer's ability to process graph data. Graph500 contains a high performance, scalable graph generator that efficiently generates large power law graphs, which are graphs with a few nodes with many edges and many

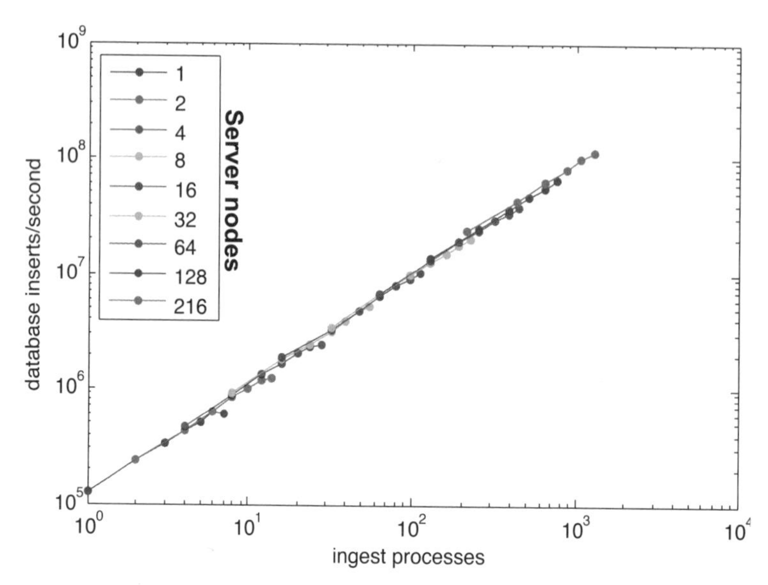

Figure 3.8
Ingest performance versus number of ingest processors for different instances of a key-value store database with different numbers of servers (see legend), demonstrating linear performance scaling. ©2011 IEEE. Reprinted, with permission, from [45]

nodes with a few edges. Figure 3.8 shows the near-perfect scalability achieved by D4M on inserting Graph500 data into a high performance key-value store database.

Achieving full performance on a key-value store database requires exploiting its ability to run well on parallel computers. Good parallel performance requires ensuring that there is sufficient parallelism in the application, load balancing the application across different parts of the system, and minimizing communication between processors. The techniques for achieving high performance on a key-value store are similar to the techniques for achieving high performance on other parallel computing applications.

D4M works seamlessly with the pMatlab [47, 48] parallel computing environment, which allows high performance parallel applications to be constructed with just a few lines of code. pMatlab uses a single-program-multiple-data (SPMD) parallel programming model and sits on top of a message passing interface (MPI) communication layer. SPMD and MPI are the primary tools used in much of the parallel computing world to achieve the highest levels of performance on the world's largest systems (see hpcchallenge.org). These tools can also be used for achieving high performance with a key-value store database.

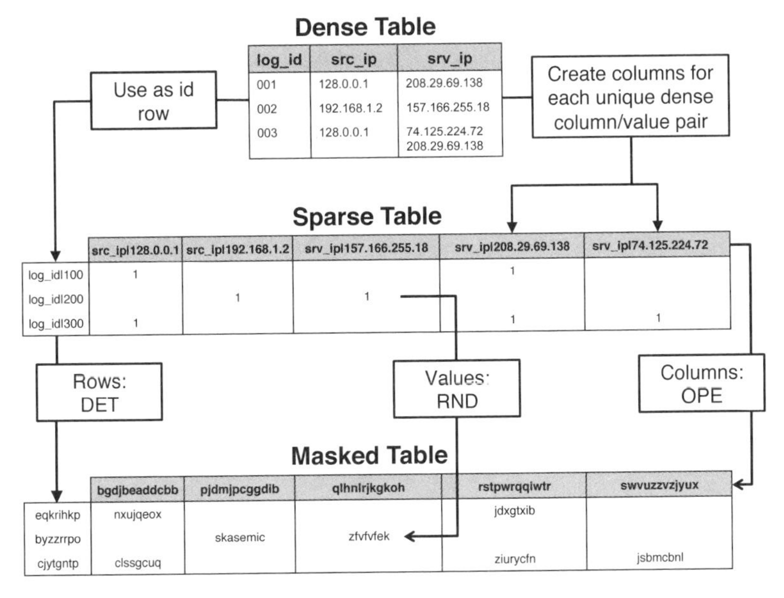

Figure 3.9
Masking network data records. Dense data are made sparse by using the D4M schema that is widely used in the key-value store database community. Dense table column names and values are appended to make columns in the sparse table, which moves most of the semantic content into the rows and columns. The sparse table is then masked by using a variety of encryption schemes depending upon the desired application. In this figure, the rows are masked using DET, the columns are masked using OPE, and the values are masked using RND. ©2014 IEEE. Reprinted, with permission, from [49]

3.8 Computing on Masked Data

Increasingly, data processing systems must address the confidentiality, integrity, and availability of their data. Data-centric protections are particularly useful for preserving the confidentiality of the data. Typical defenses of this type include encrypting the communication links between users and the data processing system, encrypting the communication links between the data sources and the data processing system, encrypting the data in the file system, and encrypting data in the database. These approaches are all significant steps forward in improving the confidentiality of a data processing system. However, all of these approaches require that the data be decrypted to be used inside the data processing system. Decryption requires that the passwords to the data be available to the data processing system, thus exposing the data to any attacker able to breach the boundaries of the system.

One vision for a secure data processing system is to have data sources encrypt data prior to transmittal to the system, have the data processing system operate on the data in encrypted form, and only allow authorized users the passwords to decrypt the answer for their results. Such a system makes the underlying big data technologies oblivious to the details of the data. As a result, the data and processing can reside in an untrusted environment while still enhancing the confidentiality of the data and results.

Computing on Masked Data (CMD) takes a step toward this vision by allowing basic computations on encrypted data. CMD combines efficient cryptographic encryption methods with an associative array representation of data. This low-computation cost approach permits both computation and query while revealing only a small amount of information about the underlying data. The overhead of CMD is sufficiently low (2x) to make it feasible for big data systems. Currently, many big data systems must operate on their data in the clear. CMD raises the bar by enabling some important computations on encrypted data while not dramatically increasing the computing resources required to perform those operations.

Several cryptographic tools could be used to build a system like CMD. First, fully homomorphic encryption (FHE) allows for arbitrary analytic computations to be performed on encrypted data without decryption and while preserving semantic security of the data so that no information about the data is leaked other than its length. FHE has been an active topic of research since its discovery [50]. Nevertheless, the best currently available schemes [51] have an overhead of 10^5 or more, making them too slow for use in practical big data systems.

If it is feasible to allow a limited amount of information about the encrypted data to be revealed, a much more efficient alternative to using FHE is to design protocols that leverage less secure cryptographic techniques to carry out queries on encrypted data. One example is CryptDB [52], which constructs a practical database system capable of handling most types of SQL queries on encrypted data. It uses deterministic encryption (DET), which always encrypts the same data to the same ciphertext, to enable equality queries; order-preserving encryption (OPE), which encrypts data in a way that preserves the original order of the data to enable range queries; and additively homomorphic encryption (HOM+), which enables summing values directly on encrypted data to perform basic analytics. Several other protocols for achieving alternative trade-offs between leakage and efficiency have also been proposed [53–55]. Additional solutions include using techniques for secure multi-party computation [56, 57], but these techniques require further improvement to achieve the required performance.

By leveraging the computational difficulty of unpermuting a sparse matrix, D4M can provide a new secure computing technique that allows data to be analyzed while they are in encrypted form [49]. The standard CMD use case is as follows (see Figure 3.9). First, users transform their data into associative arrays following the D4M schema. Then, the components of the associative array's rows, columns, and values are masked using different encryption schemes; this process induces a permutation on rows and columns as they are restructured in lexicographic order by their masks. At this point, the masked data structure can optionally be distributed to a system in the encrypted form. Next, algebraic operations are performed on the masked associative arrays. Finally, the results are collected by the user and unmasked.

Figure 3.9 demonstrates some of the masks that can be used in CMD: DET for the rows (since range queries on rows are not required), OPE for the columns (which allows for range queries), and RND (a semantically secure encryption scheme) for the values. Another option would be to use an additively homomorphic encryption scheme (HOM+) if the values require summing.

3.9 Conclusions, Exercises, and References

D4M is the first practical instantiation of the fully developed mathematics of associative arrays. Using D4M it is possible to accelerate the development of data processing systems. D4M can be used to analyze a variety of data, create high performance database systems, protect information, and even reason about the incompleteness of a data set. The effectiveness of associative array algebra as implemented in D4M indicates that spreadsheets, databases, matrices, and graphs can be linked at a much deeper mathematical level. The later chapters of the book will formalize the algebra of associative arrays by extending the ideas of sparse matrix algebra to more abstract mathematical concepts, such as semirings and tropical algebras (see [58] and references therein). This formalization will show how these diverse data representations are all different facets of a single unifying algebra. The next chapters will continue to explore associative arrays in a variety of contexts to illustrate the many varied and useful properties of associative arrays. This exploration will build intuition that will be invaluable for understanding the formalized mathematics of associative arrays.

Exercises

Exercise 3.1 — Download and install the D4M software package (d4m.mit.edu) by following the instructions in the `README.txt` file.

Exercise 3.2 — Run the example programs in

```
examples/1Intro/1AssocIntro/
```

Exercise 3.3 — Run the example programs in

```
examples/1Intro/2EdgeArt/
```

Exercise 3.4 — Run the example programs in

```
examples/2Apps/1EntityAnalysis/
```

References

[1] J. Kepner, "Summary of bugs database," 1993. Visual Intelligence Corporation.

[2] S. Brin and L. Page, "The anatomy of a large-scale hypertextual web search engine," *Computer Networks and ISDN Systems*, vol. 30, no. 1, pp. 107–117, 1998.

[3] F. Chang, J. Dean, S. Ghemawat, W. C. Hsieh, D. A. Wallach, M. Burrows, T. Chandra, A. Fikes, and R. E. Gruber, "Bigtable: A distributed storage system for structured data," *ACM Transactions on Computer Systems*, vol. 26, no. 2, p. 4, 2008.

[4] A. Khetrapal and V. Ganesh, "Hbase and Hypertable for large scale distributed storage systems," *Dept. of Computer Science, Purdue University*, pp. 22–28, 2006.

[5] D. Sherrill, J. Kurz, and C. McNally, "Toward a scalable knowledge space on the cloud," in *High Performance Embedded Computing Workshop (HPEC)*, MIT Lincoln Laboratory, 2010.

[6] G. Condon, J. Hepp, B. Landon, D. Sherrill, M. Yee, J. Allen, Y. Cao, B. Corwin, R. Delanoy, and N. Edst, "Taking advantage of big data," *R&D Magazine R&D 100 Awards*, 2013.

[7] S. Patil, M. Polte, K. Ren, W. Tantisiriroj, L. Xiao, J. López, G. Gibson, A. Fuchs, and B. Rinaldi, "YCSB++: benchmarking and performance debugging advanced features in scalable table stores," in *Proceedings of the 2nd ACM Symposium on Cloud Computing*, p. 9, ACM, 2011.

[8] D. König, "Graphen und matrizen (graphs and matrices)," *Mat. Fiz. Lapok*, vol. 38, no. 1931, pp. 116–119, 1931.

[9] D. König, *Theorie der endlichen und unendlichen Graphen: Kombinatorische Topologie der Streckenkomplexe*, vol. 16. Akademische Verlagsgesellschaft mbh, 1936.

[10] F. Harary, *Graph Theory*. Reading, MA: Addison-Wesley, 1969.

[11] G. Sabidussi, "Graph multiplication," *Mathematische Zeitschrift*, vol. 72, no. 1, pp. 446–457, 1959.

[12] P. M. Weichsel, "The Kronecker product of graphs," *Proceedings of the American Mathematical Society*, vol. 13, no. 1, pp. 47–52, 1962.

[13] M. McAndrew, "On the product of directed graphs," *Proceedings of the American Mathematical Society*, vol. 14, no. 4, pp. 600–606, 1963.

[14] H. Teh and H. Yap, "Some construction problems of homogeneous graphs," *Bulletin of the Mathematical Society of Nanying University*, vol. 1964, pp. 164–196, 1964.

[15] A. Hoffman and M. McAndrew, "The polynomial of a directed graph," *Proceedings of the American Mathematical Society*, vol. 16, no. 2, pp. 303–309, 1965.

[16] F. Harary and C. A. Trauth, Jr, "Connectedness of products of two directed graphs," *SIAM Journal on Applied Mathematics*, vol. 14, no. 2, pp. 250–254, 1966.

[17] R. A. Brualdi, "Kronecker products of fully indecomposable matrices and of ultrastrong digraphs," *Journal of Combinatorial Theory*, vol. 2, no. 2, pp. 135–139, 1967.

[18] J. Kepner and J. Gilbert, *Graph Algorithms in the Language of Linear Algebra*. SIAM, 2011.

[19] L. Wall, T. Christiansen, and J. Orwant, *Programming Perl*. O'Reilly Media, Inc., 2000.

[20] J. V. Kepner, "Multidimensional associative array database," Jan. 14 2014. US Patent 8,631,031.

[21] J. Kepner, W. Arcand, W. Bergeron, N. Bliss, R. Bond, C. Byun, G. Condon, K. Gregson, M. Hubbell, J. Kurz, A. McCabe, P. Michaleas, A. Prout, A. Reuther, A. Rosa, and C. Yee, "Dynamic Distributed Dimensional Data Model (D4M) database and computation system," in *2012 IEEE International Conference on Acoustics, Speech and Signal Processing (ICASSP)*, pp. 5349–5352, IEEE, 2012.

[22] J. Kepner, W. Arcand, W. Bergeron, C. Byun, M. Hubbell, B. Landon, A. McCabe, P. Michaleas, A. Prout, T. Rosa, D. Sherrill, A. Reuther, and C. Yee, "Massive database analysis on the cloud with D4M," in *High Performance Extreme Computing Conference (HPEC)*, IEEE, 2012.

[23] S. Wu, V. Gadepally, A. Whitaker, J. Kepner, B. Howe, M. Balazinska, and S. Madden, "Mimicviz: Enabling visualization of medical big data," *Intel Science & Technology Center Retreat, Portland, OR*, 2014.

[24] V. Gadepally, J. Kepner, W. Arcand, D. Bestor, B. Bergeron, C. Byun, L. Edwards, M. Hubbell, P. Michaleas, J. Mullen, A. Prout, A. Rosa, C. Yee, and A. Reuther, "D4M: Bringing associative arrays to database engines," in *High Performance Extreme Computing Conference (HPEC)*, IEEE, 2015.

[25] J. Kepner, C. Anderson, W. Arcand, D. Bestor, B. Bergeron, C. Byun, M. Hubbell, P. Michaleas, J. Mullen, D. O'Gwynn, A. Prout, A. Reuther, A. Rosa, and C. Yee, "D4M 2.0 schema: A general purpose high performance schema for the Accumulo database," in *High Performance Extreme Computing Conference (HPEC)*, IEEE, 2013.

[26] B. A. Miller, N. Arcolano, M. S. Beard, J. Kepner, M. C. Schmidt, N. T. Bliss, and P. J. Wolfe, "A scalable signal processing architecture for massive graph analysis," in *Acoustics, Speech and Signal Processing (ICASSP), 2012 IEEE International Conference on*, pp. 5329–5332, IEEE, 2012.

[27] M. Hubbell and J. Kepner, "Large scale network situational awareness via 3d gaming technology," in *High Performance Extreme Computing Conference (HPEC)*, IEEE, 2012.

[28] S. M. Sawyer, B. D. O'Gwynn, A. Tran, and T. Yu, "Understanding query performance in Accumulo," in *High Performance Extreme Computing Conference (HPEC)*, IEEE, 2013.

[29] S. M. Sawyer and B. D. O'Gwynn, "Evaluating Accumulo performance for a scalable cyber data processing pipeline," in *High Performance Extreme Computing Conference (HPEC)*, IEEE, 2014.

[30] J. Kepner, D. Ricke, and D. Hutchinson, "Taming biological big data with D4M," *Lincoln Laboratory Journal*, vol. 20, no. 1, 2013.

[31] S. Dodson, D. O. Ricke, and J. Kepner, "Genetic sequence matching using D4M big data approaches," in *High Performance Extreme Computing Conference (HPEC)*, IEEE, 2014.

[32] V. Gadepally and J. Kepner, "Big data dimensional analysis," in *High Performance Extreme Computing Conference (HPEC)*, IEEE, 2014.

[33] M. Soderholm, "Big data's dirty little secret," *Datanami*, July 2 2015.

[34] M. Soderholm, "Five steps to fix the data feedback loop and rescue analysis from 'bad' data," *Datanami*, Aug. 17 2015.

[35] S. Dodson, D. O. Ricke, J. Kepner, N. Chiu, and A. Shcherbina, "Rapid sequence identification of potential pathogens using techniques from sparse linear algebra," in *Symposium on Technologies for Homeland Security*, IEEE, 2015.

[36] A.-L. Barabási and R. Albert, "Emergence of scaling in random networks," *Science*, vol. 286, no. 5439, pp. 509–512, 1999.

[37] M. Faloutsos, P. Faloutsos, and C. Faloutsos, "On power-law relationships of the internet topology," in *ACM SIGCOMM Computer Communication Review*, vol. 29.4, pp. 251–262, ACM, 1999.

[38] V. Pareto, *Manuale di Economia Politica*, vol. 13. Societa Editrice, 1906.

[39] G. K. Zipf, *The Psycho-Biology of Language*. Houghton-Mifflin, 1935.

[40] F. M. Gryna, J. M. Juran, and L. A. Seder, *Quality Control Handbook*. McGraw-Hill, 1962.

[41] V. Gadepally and J. Kepner, "Using a power law distribution to describe big data," in *High Performance Extreme Computing Conference (HPEC)*, IEEE, 2015.

[42] J. Kepner, "Perfect power law graphs: Generation, sampling, construction and fitting," in *SIAM Annual Meeting*, 2012.

[43] C. Byun, W. Arcand, D. Bestor, B. Bergeron, M. Hubbell, J. Kepner, A. McCabe, P. Michaleas, J. Mullen, D. O'Gwynn, A. Prout, A. Reuther, A. Rosa, and C. Yee, "Driving big data with big compute," in *High Performance Extreme Computing Conference (HPEC)*, IEEE, 2012.

[44] R. Sen, A. Farris, and P. Guerra, "Benchmarking Apache Accumulo bigdata distributed table store using its continuous test suite," in *2013 IEEE International Congress on Big Data*, pp. 334–341, IEEE, 2013.

[45] J. Kepner, W. Arcand, D. Bestor, B. Bergeron, C. Byun, V. Gadepally, M. Hubbell, P. Michaleas, J. Mullen, A. Prout, A. Reuther, A. Rosa, and C. Yee, "Achieving 100,000,000 database inserts per second using Accumulo and D4M," in *High Performance Extreme Computing Conference (HPEC)*, IEEE, 2014.

[46] D. Bader, K. Madduri, J. Gilbert, V. Shah, J. Kepner, T. Meuse, and A. Krishnamurthy, "Designing scalable synthetic compact applications for benchmarking high productivity computing systems," *Cyberinfrastructure Technology Watch*, vol. 2, pp. 1–10, 2006.

[47] N. Travinin Bliss and J. Kepner, "pMATLAB Parallel MATLAB Library," *The International Journal of High Performance Computing Applications*, vol. 21, no. 3, pp. 336–359, 2007.

[48] J. Kepner, *Parallel MATLAB for Multicore and Multinode Computers*. SIAM, 2009.

[49] J. Kepner, V. Gadepally, P. Michaleas, N. Schear, M. Varia, A. Yerukhimovich, and R. K. Cunningham, "Computing on masked data: A high performance method for improving big data veracity," in *High Performance Extreme Computing Conference (HPEC)*, IEEE, 2014.

[50] C. Gentry, "Fully homomorphic encryption using ideal lattices," in *STOC*, vol. 9, pp. 169–178, 2009.

[51] H. Perl, M. Brenner, and M. Smith, "Poster: An implementation of the fully homomorphic Smart-Vercauteren crypto-system," in *Proceedings of the 18th ACM Conference on Computer and Communications Security*, pp. 837–840, ACM, 2011.

[52] R. A. Popa, C. Redfield, N. Zeldovich, and H. Balakrishnan, "Cryptdb: Protecting confidentiality with encrypted query processing," in *Proceedings of the Twenty-Third ACM Symposium on Operating Systems Principles*, pp. 85–100, ACM, 2011.

[53] D. Cash, S. Jarecki, C. Jutla, H. Krawczyk, M.-C. Roşu, and M. Steiner, "Highly-scalable searchable symmetric encryption with support for boolean queries," in *Advances in Cryptology–CRYPTO 2013*, pp. 353–373, Springer, 2013.

[54] M. Raykova, A. Cui, B. Vo, B. Liu, T. Malkin, S. M. Bellovin, and S. J. Stolfo, "Usable, secure, private search," *IEEE Security & Privacy*, vol. 10, no. 5, pp. 53–60, 2012.

[55] P. Pal, G. Lauer, J. Khoury, N. Hoff, and J. Loyall, "P3s: A privacy preserving publish-subscribe middleware," in *ACM/IFIP/USENIX International Conference on Distributed Systems Platforms and Open Distributed Processing*, pp. 476–495, Springer, 2012.

[56] A. C. Yao, "Protocols for secure computations," in *23rd Annual Symposium on Foundations of Computer Science*, pp. 160–164, IEEE, 1982.

[57] M. Ben-Or, S. Goldwasser, and A. Wigderson, "Completeness theorems for non-cryptographic fault-tolerant distributed computation," in *Proceedings of the Twentieth Annual ACM Symposium on Theory of computing*, pp. 1–10, ACM, 1988.

[58] B. De Schutter and B. De Moor, "The QR decomposition and the singular value decomposition in the symmetrized max-plus algebra revisited," *SIAM review*, vol. 44, no. 3, pp. 417–454, 2002.

4 Associative Arrays and Musical Metadata

Summary

Musical databases provide a ready data source for illustrating associative array concepts. In a dense representation, an associative array looks like a standard spreadsheet or database table that lists each song along with its properties. Sub-arrays can be selected using associative array indexing, array multiplication, and element-wise multiplication. Sub-arrays can be combined using addition and intersected using element-wise multiplication. A sparse representation of the data is useful for correlating different properties of the music, such as the song writers and the song genres. This chapter illustrates the practical application of associative array operations in the specific context of tabular text data drawn from a musical database.

4.1 Data and Metadata

An important concept in categorizing information is the distinction between data and metadata. The term *metadata* was first coined by Philip Bagley [1] to refer to data that are used to describe data. One of the clearest examples of the concept of data and metadata can be found in music. In a musical context, musical data usually refers to the lyrics, score, or recording of a performance and may consist of many words, notes, or sounds. Musical metadata typically refers to information associated with the music such as its title, author, year of composition, dedication, inspiration, performer, and producer.

Musical data has been written down for millennia. Hurrian Hymn No. 6 discovered in Syria [2] is believed to be over 3,000 years old. Metadata in the form of song titles, dedications, or instructions on performance are likely just as old. The Book of Psalms found in the Hebrew Bible and the Christian Old Testament contains 150 songs (or psalms) that are thought to be over 2,000 years old. Over 100 of the psalms begin with metadata indicating musical direction, the type of instrument to be used, type of composition, and dedications. Even in this early work, the challenge of keeping metadata consistent and connected with its data is apparent as much of the psalms metadata is incomplete [3].

As musical collections have increased over human history, the need for more organized musical metadata has become essential to allow performers and listeners to find the music they wish to perform and hear. The process of formalization of music metadata took place

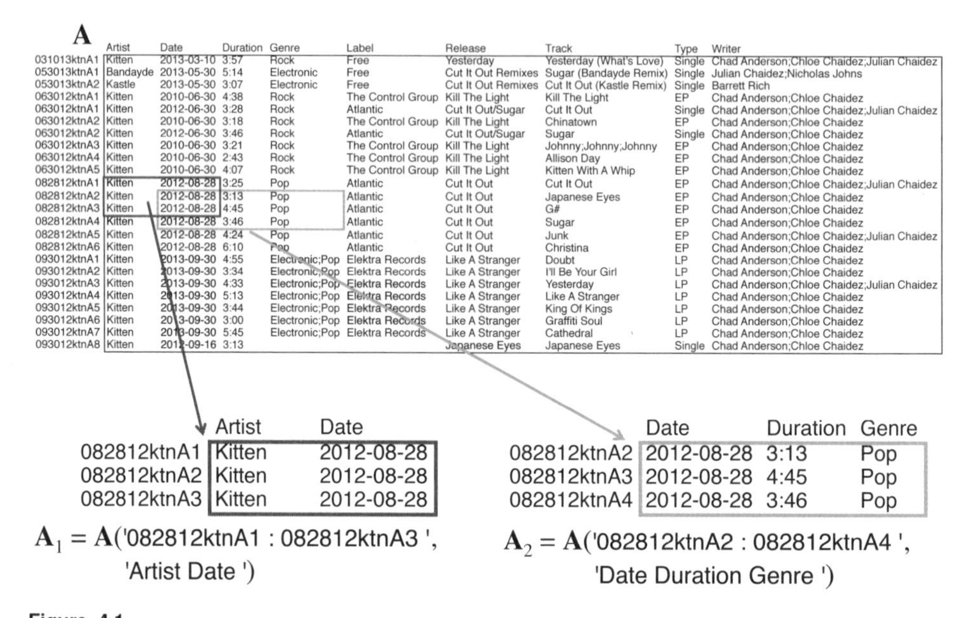

A

	Artist	Date	Duration	Genre	Label	Release	Track	Type	Writer
031013ktnA1	Kitten	2013-03-10	3:57	Rock	Free	Yesterday	Yesterday (What's Love)	Single	Chad Anderson;Chloe Chaidez;Julian Chaidez
053013ktnA1	Bandayde	2013-05-30	5:14	Electronic	Free	Cut It Out Remixes	Sugar (Bandayde Remix)	Single	Julian Chaidez;Nicholas Johns
053013ktnA2	Kastle	2013-05-30	3:07	Electronic	Free	Cut It Out Remixes	Cut It Out (Kastle Remix)	Single	Barrett Rich
063012ktnA1	Kitten	2010-06-30	4:38	Rock	The Control Group	Kill The Light	Kill The Light	EP	Chad Anderson;Chloe Chaidez
063012ktnA1	Kitten	2012-06-30	3:28	Rock	Atlantic	Cut It Out/Sugar	Cut It Out	Single	Chad Anderson;Chloe Chaidez;Julian Chaidez
063012ktnA2	Kitten	2010-06-30	3:18	Rock	The Control Group	Kill The Light	Chinatown	EP	Chad Anderson;Chloe Chaidez
063012ktnA2	Kitten	2012-06-30	3:46	Rock	Atlantic	Cut It Out/Sugar	Sugar	Single	Chad Anderson;Chloe Chaidez
063012ktnA3	Kitten	2010-06-30	3:21	Rock	The Control Group	Kill The Light	Johnny;Johnny;Johnny	EP	Chad Anderson;Chloe Chaidez
063012ktnA4	Kitten	2010-06-30	2:43	Rock	The Control Group	Kill The Light	Allison Day	EP	Chad Anderson;Chloe Chaidez
063012ktnA5	Kitten	2010-06-30	4:07	Rock	The Control Group	Kill The Light	Kitten With A Whip	EP	Chad Anderson;Chloe Chaidez
082812ktnA1	Kitten	2012-08-28	3:25	Pop	Atlantic	Cut It Out	Cut It Out	EP	Chad Anderson;Chloe Chaidez;Julian Chaidez
082812ktnA2	Kitten	2012-08-28	3:13	Pop	Atlantic	Cut It Out	Japanese Eyes	EP	Chad Anderson;Chloe Chaidez
082812ktnA3	Kitten	2012-08-28	4:45	Pop	Atlantic	Cut It Out	G#	EP	Chad Anderson;Chloe Chaidez
082812ktnA4	Kitten	2012-08-28	3:46	Pop	Atlantic	Cut It Out	Sugar	EP	Chad Anderson;Chloe Chaidez
082812ktnA5	Kitten	2012-08-28	4:24	Pop	Atlantic	Cut It Out	Junk	EP	Chad Anderson;Chloe Chaidez;Julian Chaidez
082812ktnA6	Kitten	2012-08-28	6:10	Pop	Atlantic	Cut It Out	Christina	EP	Chad Anderson;Chloe Chaidez
093012ktnA1	Kitten	2013-09-30	4:55	Electronic;Pop	Elektra Records	Like A Stranger	Doubt	LP	Chad Anderson;Chloe Chaidez
093012ktnA2	Kitten	2013-09-30	3:34	Electronic;Pop	Elektra Records	Like A Stranger	I'll Be Your Girl	LP	Chad Anderson;Chloe Chaidez
093012ktnA3	Kitten	2013-09-30	4:33	Electronic;Pop	Elektra Records	Like A Stranger	Yesterday	LP	Chad Anderson;Chloe Chaidez;Julian Chaidez
093012ktnA4	Kitten	2013-09-30	5:13	Electronic;Pop	Elektra Records	Like A Stranger	Like A Stranger	LP	Chad Anderson;Chloe Chaidez
093012ktnA5	Kitten	2013-09-30	3:44	Electronic;Pop	Elektra Records	Like A Stranger	King Of Kings	LP	Chad Anderson;Chloe Chaidez
093012ktnA6	Kitten	2013-09-30	3:00	Electronic;Pop	Elektra Records	Like A Stranger	Graffiti Soul	LP	Chad Anderson;Chloe Chaidez
093012ktnA7	Kitten	2013-09-30	5:45	Electronic;Pop	Elektra Records	Like A Stranger	Cathedral	LP	Chad Anderson;Chloe Chaidez
093012ktnA8	Kitten	2012-09-16	3:13			Japanese Eyes	Japanese Eyes	Single	Chad Anderson;Chloe Chaidez

	Artist	Date
082812ktnA1	Kitten	2012-08-28
082812ktnA2	Kitten	2012-08-28
082812ktnA3	Kitten	2012-08-28

$\mathbf{A}_1 = \mathbf{A}(\text{'082812ktnA1 : 082812ktnA3 ', 'Artist Date '})$

	Date	Duration	Genre
082812ktnA2	2012-08-28	3:13	Pop
082812ktnA3	2012-08-28	4:45	Pop
082812ktnA4	2012-08-28	3:46	Pop

$\mathbf{A}_2 = \mathbf{A}(\text{'082812ktnA2 : 082812ktnA4 ', 'Date Duration Genre '})$

Figure 4.1

D4M associative array **A** representation of a dense table of data from a music database. Row keys are an ordered set of music track identifiers. The column keys are an ordered set of the fields in the database. Sub-arrays $\mathbf{A}_1$ and $\mathbf{A}_2$ are selected using MATLAB-style notation with ranges of row keys and sets of column keys.

at the great public libraries of the late 1800s. In America, these libraries included Brooklyn Public Library, the Boston Public Library, the Peabody Institute, the Lenox Branch of the New York Public Library (NYPL), the Newberry Library, Harvard University, Forbes Library, and the Library of Congress [4]. The work of these early music librarians standardized the categorization of music and musical metadata and is the basis of most modern musical databases. More recently, machine learning has greatly enhanced musical metadata by enabling automatic labeling of music to identify genres [5], instruments [6], and even the emotions experienced by the listener [7].

Metadata often receives the most attention in data analysis because it is the basis for finding and retrieving information and is the most readily understandable by humans. Thus, many of the examples in this text deal with metadata. However, is is worth emphasizing, that the mathematics of associative arrays applies to both data and metadata without distinction.

4.2 Dense Data

Associative arrays in the Dynamic Distributed Dimensional Data Model (D4M) define their operations based on what feels intuitive to users. D4M is often used on large data

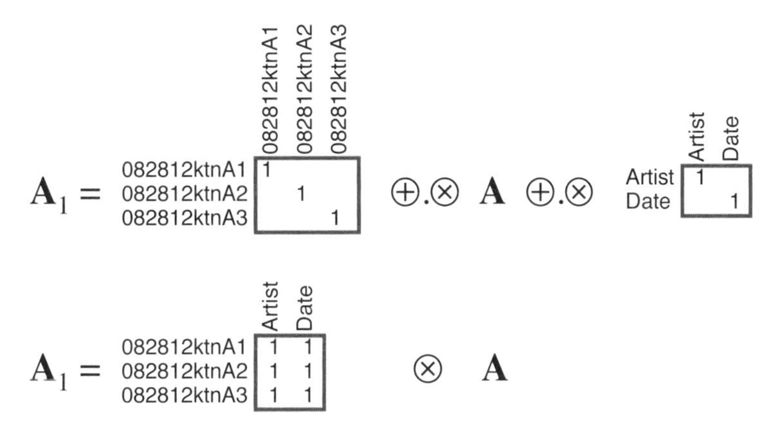

Figure 4.2
Selection of a sub-array $\mathbf{A}_1$ from $\mathbf{A}$ shown in Figure 4.1 via array multiplication (top) by the appropriate diagonal arrays and via element-wise multiplication (bottom) with an array of all 1's.

sets stored in tables with many rows and columns. However, for illustrative purposes, it is sufficient to understand the behavior of associative arrays in the context of a small example. Consider the small 25×10 associative array $\mathbf{A}$ of music data shown in Figure 4.1, where song identification strings are the leftmost column (or row key) and the corresponding artist, length, and genre data are stored in the other columns. This table could easily be stored in a spreadsheet or a database.

The first behavior of $\mathbf{A}$ that makes it an associative array is that each row and column is labeled with a string called a key. An entry in $\mathbf{A}$ is defined by a triple consisting of a row key, a column key, and value. For example, the first entry in $\mathbf{A}$ is defined as

$$\mathbf{A}(\text{'031013ktnA1 '},\text{'Artist '}) = \text{'Kitten '}$$

Associative arrays are similar to matrices in that each entry is defined by a row index, column index, and a value, except that in an associative array the row and the column are not limited to positive integers and can be strings of characters. Likewise, the values can be numbers or strings or even sets. As a matter of practice, the rows and columns are always represented in some sorted order such as lexigraphic ordering. This ordering is not a strict requirement of an associative array, but it is necessary as a practical matter to make retrieval of information efficient. Thus, associative array row keys and column keys each need to be orderable sets.

D4M users have found having sorted row and column keys a very intuitive way to think about tabular data. This approach directly maps to NoSQL databases and some NewSQL databases that employ a tuple in a similar way. SQL database tables also have a row key and column key, but the row key is often hidden from the user. However, since underlying relational algebra in SQL is based on sets, SQL tables map into associative arrays quite

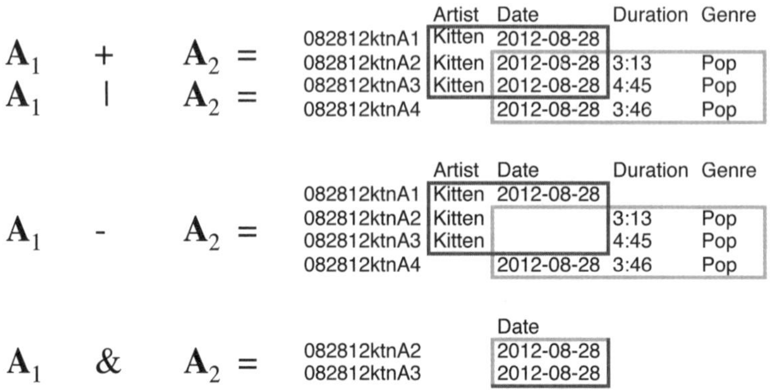

$\mathbf{A}_1 \; + \; \mathbf{A}_2 =$
$\mathbf{A}_1 \; | \; \mathbf{A}_2 =$

	Artist	Date	Duration	Genre
082812ktnA1	Kitten	2012-08-28		
082812ktnA2	Kitten	2012-08-28	3:13	Pop
082812ktnA3	Kitten	2012-08-28	4:45	Pop
082812ktnA4		2012-08-28	3:46	Pop

$\mathbf{A}_1 \; - \; \mathbf{A}_2 =$

	Artist	Date	Duration	Genre
082812ktnA1	Kitten	2012-08-28		
082812ktnA2	Kitten		3:13	Pop
082812ktnA3	Kitten		4:45	Pop
082812ktnA4		2012-08-28	3:46	Pop

$\mathbf{A}_1 \; \& \; \mathbf{A}_2 =$

	Date
082812ktnA2	2012-08-28
082812ktnA3	2012-08-28

Figure 4.3
D4M operations performed on the sub-arrays $\mathbf{A}_1$ and $\mathbf{A}_2$ as defined in Figure 4.1. From top to bottom, the operations are addition +, logical or |, subtraction −, and logical and &. The +, |, and − operations result in a new associative array that is the union of the row and column keys, while the & operation produces an associative array that is the intersection.

well. Spreadsheets often have their own row and column labeling systems, such as A1 or R1C1, but these labels can be confusing, and users just as often prefer to have table entries identified by user-defined row and column keys. Finally, as will be discussed in later sections, graphs are often defined by labeled vertices with labeled edges, and so associative arrays naturally allow this information to be incorporated into adjacency array representations of graphs (see Figure 4.6).

4.3 Dense Operations

Perhaps the most widely used operations on spreadsheets, databases, graphs, and matrices are extracting and inserting sub-spreadsheets, sub-databases, sub-graphs, and sub-matrices. One of the most interesting properties of an associative array is how sub-arrays are handled. In Figure 4.1, the sub-arrays $\mathbf{A}_1$ and $\mathbf{A}_2$ are extracted from $\mathbf{A}$ with MATLAB-style index notation for ranges and sets of keys. In this example, rows are selected by a range of keys, and the columns are selected by a distinct set of keys. As expected, the row keys and column keys are carried along into the sub-array. In addition, associative arrays allow the same sub-array selection to be performed via array multiplication or element-wise multiplication (see Figure 4.2). The duality between array selection and array multiplication allows this essential operation to be manipulated in the same manner as other algebraic operations.

Inserting, or assigning, values to an associative array can also be carried out via simple addition. For example, given the associative array

$$\mathbf{A}_3(\text{'093012ktnA8 ','093012ktnA8 '}) = \text{'New Wave '}$$

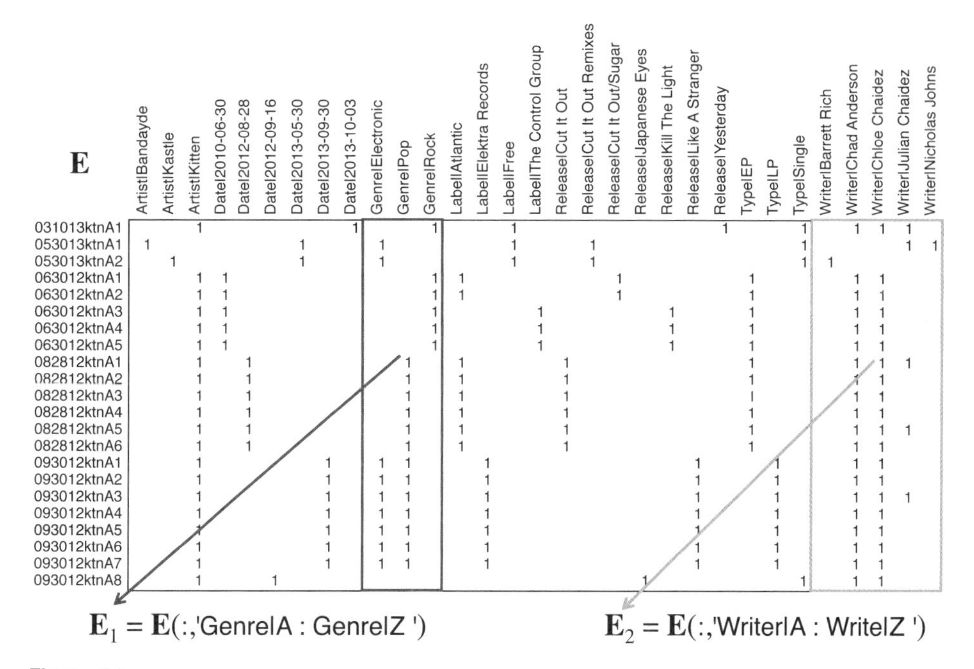

Figure 4.4
D4M sparse associative array $\mathbf{E}$ representation of a table of data from a music database. The column key and the value are concatenated with a separator symbol (in this case |) resulting in every unique pair of column and value having its own column in the sparse view. The new value is usually 1 to denote the existence of an entry. Column keys are an ordered set of database fields. Sub-arrays $\mathbf{E}_1$ and $\mathbf{E}_2$ are selected with MATLAB-style notation to denote all of the row keys and ranges of column keys.

this value can be assigned or inserted into $\mathbf{A}$ via

$$\mathbf{A} = \mathbf{A} + \mathbf{A}_3$$

Thus, it is simple to pull out pieces of an associative array and put them back where they were taken from because their global keys are preserved. In addition, these operations can be performed both algebraically and by standard sub-array referencing techniques. Associative array row and column keys relieve the user of the burden of having to keep track of both the relative position and the meaning of their rows and columns. Users of D4M have found this behavior of associative array row and column keys both intuitive and desirable.

There are many intuitive ways to combine the associative arrays. Figure 4.3 illustrates combining associative arrays $\mathbf{A}_1$ and $\mathbf{A}_2$ from Figure 4.1. From top to bottom, the operations are addition +, logical or |, subtraction −, and logical and &. In this particular example, the +, |, − operations produce a new associative array whose set of row and columns keys is the union of the sets of row and column keys found in $\mathbf{A}_1$ and $\mathbf{A}_2$. In short,

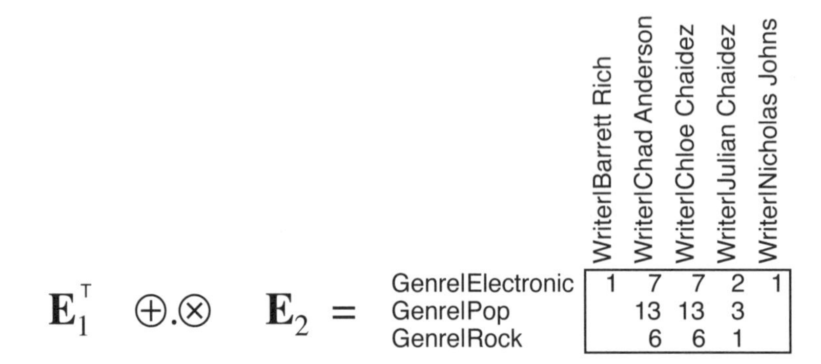

Figure 4.5
Correlating the music writers with the music genres can be accomplished by multiplying $\mathbf{E}_1$ and $\mathbf{E}_2$ as defined in Figure 4.4. This correlation is performed using the transpose operation ^{T} and the array multiplication operation $\oplus.\otimes$. The resulting associative array has row keys taken from the column keys of $\mathbf{E}_1$ and column keys taken from the column keys of $\mathbf{E}_2$. The values represent the correlation, or the number of common music tracks, between the input arrays.

associative arrays lift the constraint that exists in matrix algebra that all matrices have apparently equal numbers of rows and columns in order for them to be combined. The & operation results in an associative array that is the intersection of the row and column keys found in $\mathbf{A}_1$ and $\mathbf{A}_2$ and illustrates a fundamental property of associative arrays: empty rows or empty columns are not stored. Empty rows and empty columns are removed from the result and only non-empty rows and non-empty columns are kept.

All the examples in Figure 4.3 involve two entries that overlap

$$\mathbf{A}(\text{'08281ktnA2 '},\text{'Date '}) = \text{'2012-08-28 '}$$
$$\mathbf{A}(\text{'08281ktnA3 '},\text{'Date '}) = \text{'2012-08-28 '}$$

How these values are combined is determined by the rules of the value set. Traditionally, linear algebra is defined on the standard numeric field where the values are real (or complex) numbers and the operations are standard arithmetic addition and arithmetic multiplication. Associative arrays provide a broader selection of potential sets to choose from, the details of which are at the heart of this book and will be discussed in due course.

4.4 Sparse Data

Figure 4.1 shows a standard dense view of data that is commonly used in spreadsheets and SQL databases. Figure 4.4 shows a sparse view of the data that is commonly used in NoSQL databases, NewSQL databases, graphs, and sparse linear algebra [8]. In the sparse view, the column key and the value are concatenated with a separate symbol (in this case |)

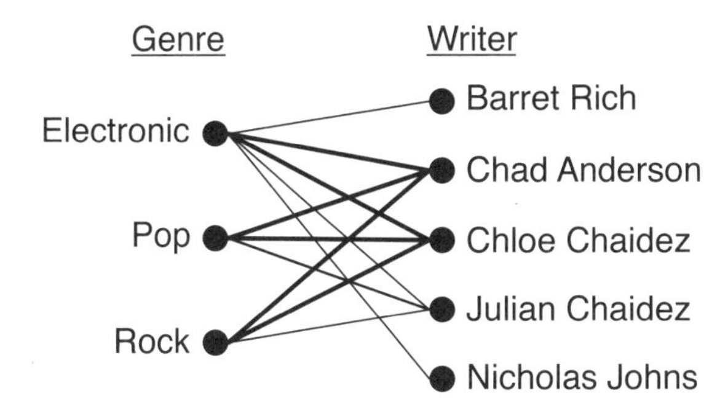

Figure 4.6
Graph representation of the result of the associative array multiplication shown in Figure 4.5. The bipartite graph shows the strengths of connections between sets of music genre vertices and music writer vertices. The width of the edge between vertices is proportional to the number of music tracks of a particular genre and writer.

so that every unique pair of column and value has its own column in the sparse view. The new value is usually 1 to denote the existence of an entry.

4.5 Sparse Operations

The power of the sparse view can be seen in a several ways. First, the structure of the data set becomes readily apparent from visual inspection. Both rare and popular columns are easy to spot. Second, and perhaps more importantly, correlating the data or constructing graphs from the data can be accomplished via array multiplication. For example, in Figure 4.4, the columns corresponding to music genre are placed in the sub-array $\mathbf{E}_1$ and the columns corresponding to music writer are placed in the sub-array $\mathbf{E}_2$. The correlation between music genre and music writer is then computed by array multiplication of the two associative arrays (see Figure 4.5). The result is a new associative array in which the row keys are the genre, the columns keys are the writer, and the value shows how many tracks there are that correspond to a particular genre and writer.

The resulting associative array can also be viewed as a graph from a set of vertices corresponding to genres to a set of vertices corresponding to writers where the thickness of the lines (or edge weights) is proportional to the number of tracks between vertices (see Figure 4.6). The graph in Figure 4.6 also illustrates the concept of a bipartite graph that connects two sets of vertices (genres and writers) with no edges within these sets. In this context, $\mathbf{E}$ can be viewed as a graph incidence array whereby every hyper-edge in the larger graph is represented by a row. The result of multiplying the two sub-arrays $\mathbf{E}_1$ and $\mathbf{E}_2$ is an adjacency array of the sub-graph corresponding to the genre vertices and writer vertices represented by the sub-arrays.

Associative arrays allow a variety of array multiplication operations. By returning to the dense view of the data shown in Figure 4.1, it is possible to define an array multiplication

$$\mathbf{A}_1^{\mathsf{T}} \quad \cup.\cup \quad \mathbf{A}_2 =$$

	Date	Duration	Genre
Artist	Kitten;2012-08-28;Kitten;2012-08-28;	Kitten;3:46;Kitten;4:24;	Kitten;Pop;Kitten;Pop;
Date	2012-08-28;2012-08-28;2012-08-28;2012-08-28;	2012-08-28;3:46;2012-08-28;4:24;	2012-08-28;Pop;2012-08-28;Pop;

Figure 4.7
Transpose ^{T} and array multiply $\cup.\cup$ performed on the sub-arrays $\mathbf{A}_1$ and $\mathbf{A}_2$ as defined in Figure 4.1. The output is an associative array whose row keys are taken from the column keys of $\mathbf{A}_1$ and whose column keys are taken from the column keys of $\mathbf{A}_2$. The values represent the concatenation (or union) of the values in the input arrays where a semicolon (;) is used to separate the values.

$\cup.\cup$ whereby the output values are sets representing the union (or concatenation) of the values of the input arrays (see Figure 4.7).

Even more exotic array multiplications can be constructed. In Figure 4.5, the values correspond to the number of music tracks with a specific genre and writer. Another very useful array multiplication is one in which the resulting values are sets containing the corresponding intersecting row and column keys (see Figure 4.8). Such an array multiplication not only provides information on the strongest correlations, but makes it possible to look up the precise music tracks that produced the correlation. This behavior is sometimes referred to as *pedigree-preserving* computation because information on the source of the information is preserved through the computation. This computation is in contrast to a standard array multiplication in which the source information is lost.

All of the above array multiplications represent specific choices of the element-wise addition operation $\oplus$ and the element-wise multiplication operation $\otimes$ over a value set V. The general version of this array multiplication can be written as

$$\mathbf{C} = \mathbf{AB} = \mathbf{A} \oplus.\otimes \mathbf{B}$$

or more specifically

$$\mathbf{C}(k_1,k_2) = \bigoplus_{k_3} \mathbf{A}(k_1,k_3) \otimes \mathbf{B}(k_3,k_2)$$

where the row and column keys k_1, k_2, and k_3 are elements of the sets

$$k_1 \in K_1$$
$$k_2 \in K_2$$
$$k_3 \in K_3$$

and $\mathbf{A}$, $\mathbf{B}$, and $\mathbf{C}$ are associative arrays that map from pairs of keys to values in the set V

$$\mathbf{A} : K_1 \times K_3 \to V$$
$$\mathbf{B} : K_3 \times K_2 \to V$$
$$\mathbf{C} : K_1 \times K_2 \to V$$

$$\mathbf{E}_1^{\mathsf{T}} \ \cup.\otimes \ \mathbf{E}_2(:,\text{'WriterJ : WriterZ '}) =$$

	WriterIJulian Chaidez	WriterINicholas Johns
GenreIElectronic	053013ktnA1;093012ktnA3;	053013ktnA1;
GenreIPop	082812ktnA1;082812ktnA5;093012ktnA3;	
GenreIRock	031013ktnA1;	

Figure 4.8
Transpose ^{T} and array multiplication $\cup.\otimes$ performed on the array $\mathbf{E}_1$ and a sub-array of $\mathbf{E}_2$ as defined in Figure 4.4. The resulting array has row keys taken from the column keys of $\mathbf{E}_1$ and column keys taken from $\mathbf{E}_2$. The values of each entry hold the intersecting keys from the array multiplication.

Array multiplication is very similar to traditional matrix multiplication because matrix multiplication is a special case of array multiplication in which $\oplus$ is traditional arithmetic addition $+$, $\otimes$ is traditional arithmetic multiplication $\times$ over the real numbers $\mathbb{R}$, and the row and columns keys are sets of integers

$$K_1 = \{1,\ldots,m\}$$
$$K_2 = \{1,\ldots,n\}$$
$$K_3 = \{1,\ldots,\ell\}$$

In this case, traditional matrix multiplication can be written as the more familiar notation

$$\mathbf{C} = \mathbf{AB} = \mathbf{A} \ +.\times \ \mathbf{B}$$

or more specifically

$$\mathbf{C}(i,j) = \sum_k \mathbf{A}(i,k) \times \mathbf{B}(k,j)$$

where the row and column indices i, j, and k are elements of the sets

$$i \in \{1,\ldots,m\}$$
$$j \in \{1,\ldots,n\}$$
$$k \in \{1,\ldots,\ell\}$$

4.6 Conclusions, Exercises, and References

Databases of music provide a rich source of data for illustrating associative array concepts. There are two predominant ways to represent data in associative arrays: dense and sparse. In the dense representation, an associative array is very similar to a standard spreadsheet or database table. Each row is a record and each column is a field. Selecting sub-arrays can be be accomplished with associative array indexing, array multiplication, and element-wise multiplication. Sub-arrays can be combined using addition. Sub-arrays can be intersected using element-wise multiplication. In a sparse representation of the data, every unique field and value in the record is a unique column in the associative array. The sparse view is very useful for correlating different musical properties, such as the song writers with the song genres. These correlations can also be depicted as graphs from one set of vertices (genres) to another set of vertices (writers).

Exercises

Exercise 4.1 — Refer to Figure 4.1 and for each array $\mathbf{A}$, $\mathbf{A}_1$, and $\mathbf{A}_2$

(a) Compute the number of rows m and number of columns n in the arrays.

(b) Compute the total number of entries mn.

(c) Compute how many empty entries there are.

(d) Compute how many non-empty entries there are.

(e) What are the values of $\mathbf{A}$('063012ktnA2 ','Track '), $\mathbf{A}_1(1,1)$, and $\mathbf{A}_2(3,3)$?

Exercise 4.2 — Refer to Figure 4.3 and for the top, middle, and bottom arrays depicted

(a) Compute the number of rows m and number of columns n in the arrays.

(b) Compute the total number of entries mn.

(c) Compute how many empty entries there are.

(d) Compute how many non-empty entries there are.

Exercise 4.3 — Refer to Figure 4.4 and for each array $\mathbf{E}$, $\mathbf{E}_1$, and $\mathbf{E}_2$

(a) Compute the number of rows m and number of columns n in the arrays.

(b) Compute the total number of entries mn.

(c) Compute how many empty entries there are.

(d) Compute how many non-empty entries there are.

(e) Which genre(s) and artist(s) are affiliated with the most tracks?

Exercise 4.4 — Refer to Figure 4.6.

(a) Compute the number of edges in the graph.

(b) What is the maximum possible number of edges between the genres and the writers?

(c) How many of the possible edges are not in the graph?

Check your answers by referring to Figure 4.5.

Exercise 4.5 — Write down a dense associative array representing some of the properties of your favorite music.

Exercise 4.6 — Select two sub-arrays using array indexing, array multiplication, and element-wise multiplication.

Exercise 4.7 — Combine two sub-arrays using addition.

Exercise 4.8 — Combine two sub-arrays using element-wise multiplication.

Exercise 4.9 — Write down a sparse associative array representing some of the properties of your favorite music.

Exercise 4.10 — Correlate two properties of the music using array multiplication of sparse sub-arrays.

References

[1] P. R. Bagley, "Extension of programming language concepts," tech. rep., University City Science Center, 1968.

[2] E. Laroche, "Notes sur le panthéon hourrite de ras shamra," *Journal of the American Oriental Society*, pp. 148–150, 1968.

[3] A. Pietersma, "David in the Greek psalms," *Vetus Testamentum*, vol. 30, no. Fasc. 2, pp. 213–226, 1980.

[4] C. J. Bradley, "Classifying and cataloguing music in American libraries: A historical overview," *Cataloging & Classification Quarterly*, vol. 35, no. 3-4, pp. 467–481, 2003.

[5] T. Li, M. Ogihara, and Q. Li, "A comparative study on content-based music genre classification," in *Proceedings of the 26th Annual International ACM SIGIR Conference on Research and Development in Information Retrieval*, pp. 282–289, ACM, 2003.

[6] P. Herrera-Boyer, G. Peeters, and S. Dubnov, "Automatic classification of musical instrument sounds," *Journal of New Music Research*, vol. 32, no. 1, pp. 3–21, 2003.

[7] K. Trohidis, G. Tsoumakas, G. Kalliris, and I. P. Vlahavas, "Multi-label classification of music into emotions," in *ISMIR*, vol. 8, pp. 325–330, 2008.

[8] J. Kepner, C. Anderson, W. Arcand, D. Bestor, B. Bergeron, C. Byun, M. Hubbell, P. Michaleas, J. Mullen, D. O'Gwynn, A. Prout, A. Reuther, A. Rosa, and C. Yee, "D4M 2.0 schema: A general purpose high performance schema for the Accumulo database," in *High Performance Extreme Computing Conference (HPEC)*, IEEE, 2013.

5 Associative Arrays and Abstract Art

Summary

An essential element to data representation is the ability to abstract important elements and enable viewing data at different scales. Associative arrays are well suited for depicting data in summary or in detail. The vertices and edges of a graph can be represented in an associative array. The two primary representations of graphs are adjacency arrays and incidence arrays. Adjacency arrays can be computed from incidence arrays via array multiplication. This chapter depicts these important visual representational properties of associative arrays using a work of abstract art as an example.

5.1 Visual Abstraction

A common step in analyzing the relationships among data is abstracting the data into a visual form that preserves key concepts and eliminates unimportant items. Graphs consisting of vertices and edges are a powerful tool for enabling visual abstraction. Among the most well-known early formal applications of graph theory is Leonhard Euler's famous "Seven Bridges of Konigsberg" problem [1], which asks if there is path across all seven bridges of the City of Konigsberg that crosses each bridge only once. The first step in solving the problem is to visually abstract the islands, river banks, and bridges of the city into a simplified graph of vertices and edges.

The process of reducing a real-world visual scene to a simplified composition of vertices is easily appreciated in abstract art [2]. Reducing a real-world environment to its essential edges built on earlier Cubist geometric approaches initiated by Pablo Picasso and Georges Braque, but provided the necessary break with the physical world to create completely abstract representations [3, 4]. This view of abstraction was championed by the Dutch De Stijl artists and its most noted member, Piet Mondrian, who believed

> " ... through horizontal and vertical lines constructed with awareness, but not with calculation, led by high intuition, and brought to harmony and rhythm, these basic forms of beauty, supplemented if necessary by other direct lines or curves, can become a work of art, as strong as it is true ... " [5, 6]

In many respects, the above quotation also eloquently captures how a mathematician tries to create abstract mathematical models of the world. The relationships that can be depicted in a graph go beyond what can be depicted by abstracting a view of the real world.

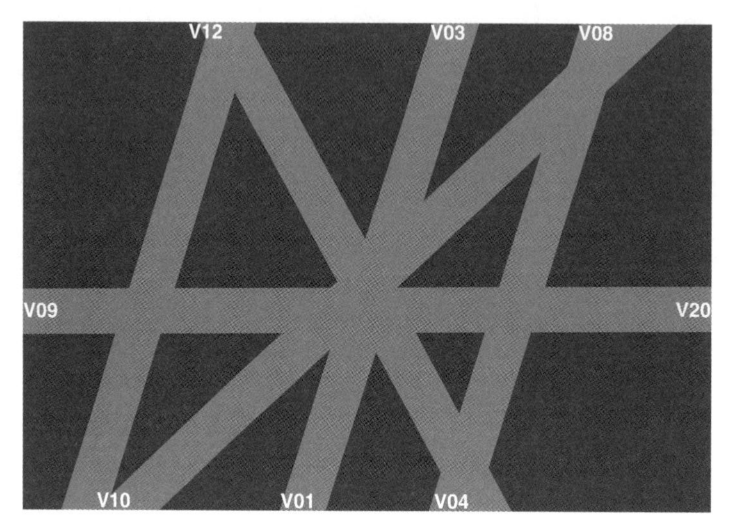

Figure 5.1
Simplified two-toned depiction of the abstract line-art painting (*XCRS* by Ann Pibal). In this view, there appear to be six edges with eight termination vertices labeled V01, V03, V04, V08, V09, V10, V12, and V20.

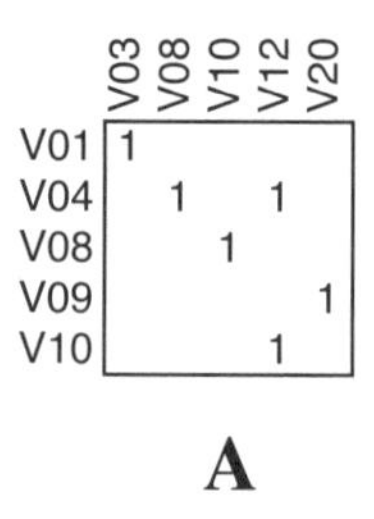

	V03	V08	V10	V12	V20
V01	1				
V04		1		1	
V08			1		
V09					1
V10				1	

A

Figure 5.2
A minimal associative array representation **A** of the edges depicted in Figure 5.1.

Fortunately, artists have continued to extend abstraction to include inspirations from architecture, graphic design, and landscape [7]. These modern artworks can depict the fuller range of multi-hyper-directed-weighted graphs that are representable by associative arrays; multi- in that the graph can have multiple identical edges; hyper- in that an edge can connect more than two vertices; directed- because the edges can have direction; and weighted- because the edges can have a weight or label.

An important aspect of an effective abstraction is the ability to support data exploration at different scales. One effective approach to data exploration can be summed up in the phrase "overview first, zoom and filter, then details on demand" [8]. Essential to this approach is the ability to view data at different scales. From a distance, the details of the data should appear blurry, but the overall picture should be clear. Likewise, when the data are viewed

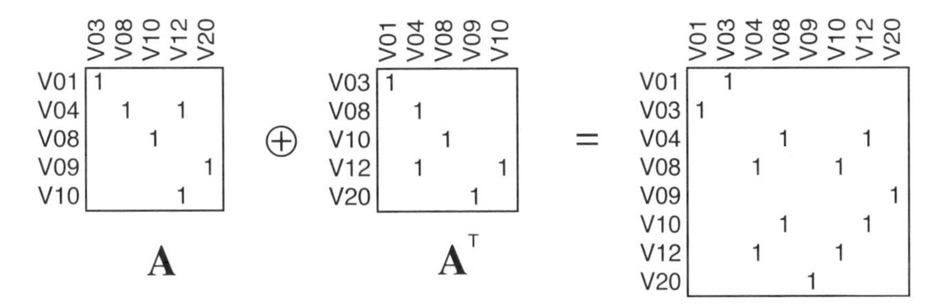

Figure 5.3
Calculation of the square-symmetric representation of the edges from Figure 5.1 using associative array addition.

up close, additional details should jump out. Visually, this scenario is analogous to moving from a fuzzy black-and-white depiction of a picture to a clear multicolor depiction of the same picture. It is important that the mathematics of associative arrays have this same property.

5.2 Minimal Adjacency Array

Figure 5.1 depicts a simple, two-toned version of an abstract line painting. This illustration is analogous to viewing data in its lowest-resolution form. Figure 5.1 can be approximated as a graph with six edges with eight termination vertices V01, V03, V04, V08, V09, V10, V12, and V20. The edges can be represented as pairs of vertices (V01, V03), (V04, V08), (V04, V12), (V08, V10), (V09, V20), and (V10, V12). Likewise, these vertices can be the rows and columns of an associative array.

Figure 5.1 can be represented as a 5×5 associative array (see Figure 5.2) where the existence of an edge between two vertices is denoted by a 1 in the corresponding row and column. This representation of a graph is referred to as an adjacency array. Because the edges in the picture have no particular direction, each edge can be represented as one of two possible pairs of vertices. For example, the first edge could be listed as either (V01, V03) or (V03, V01). This type of graph is referred to as undirected because the edge has no particular direction, and the starting vertex and the ending vertex of the edge are unspecified. Furthermore, the only thing that is known about the edge is that it exists, and so the only values that can be assigned to the edge are either 0 or 1. Such a graph is said to be unweighted because there is no particular weight on the edges. All edges that exist are the same. Combined, these two attributes of the graph would cause it to be categorized as an undirected-unweighted graph. Such a graph is among the simplest graphs and is extensively used in a variety of graph algorithms [9].

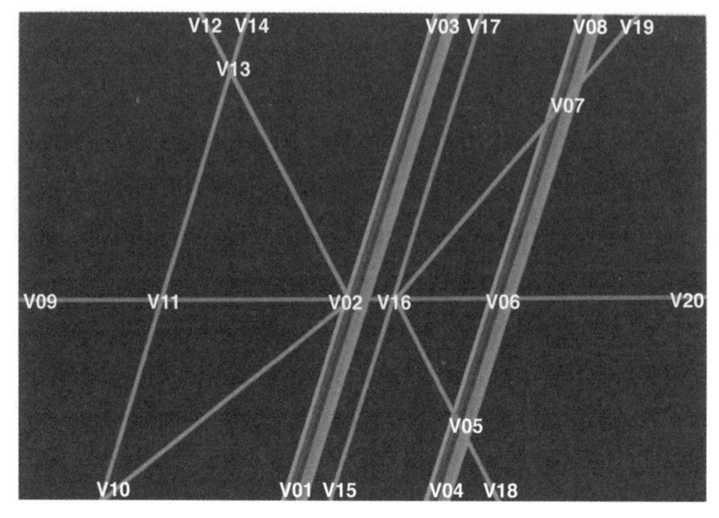

Figure 5.4
Multitone depiction of an abstract edge painting (*XCRS* by Ann Pibal) showing various edges. The intersections and terminations of the edges are labeled with vertices (V01,...,V20) that have been superimposed onto the painting with white letters.

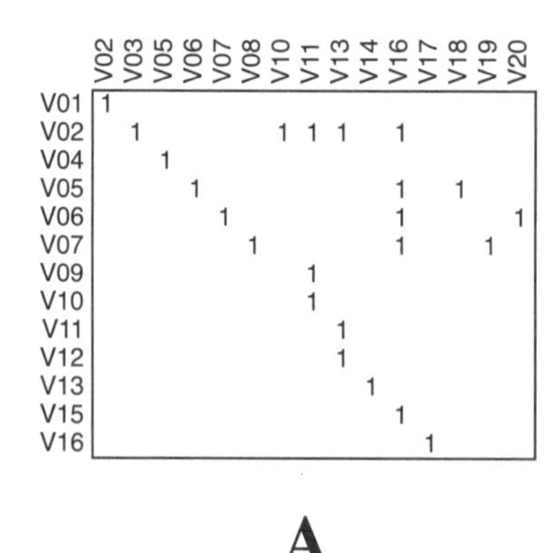

	V02	V03	V05	V06	V07	V08	V10	V11	V13	V14	V16	V17	V18	V19	V20
V01	1														
V02		1					1	1	1		1				
V04			1												
V05				1							1		1		
V06					1						1				1
V07						1					1			1	
V09								1							
V10								1							
V11									1						
V12									1						
V13										1					
V15											1				
V16												1			

A

Figure 5.5
A minimal associative array representation **A** of the edges depicted in Figure 5.4.

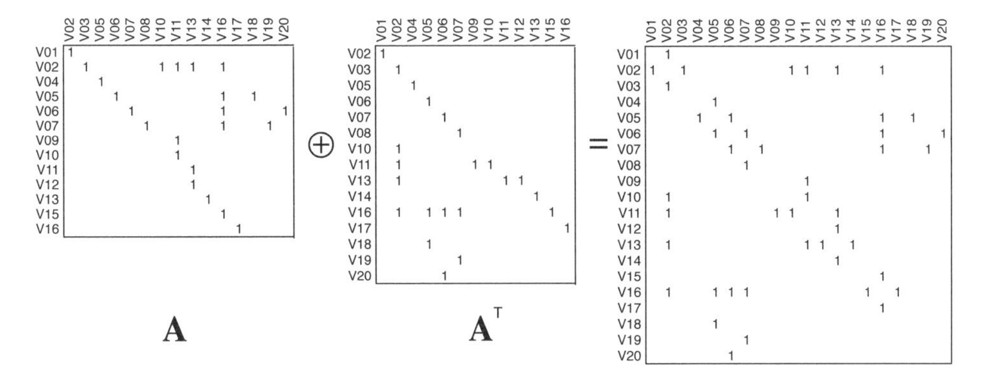

Figure 5.6
Construction of a square-symmetric representation of the edges depicted in Figure 5.4 via associative array addition.

Figure 5.2 captures all that is known about the vertices and edges in Figure 5.1, but it is not necessarily the most convenient form to work with. It is often desirable that the associative array representation of an undirected graph be square so that it has equal numbers of row and columns, and be symmetric so that entries for any possible vertex pairs could represent an edge. Computing the square-symmetric representation of an undirected graph can be accomplished by simply adding the associative array to its transpose. For example, if the associative array in Figure 5.1 is denoted $\mathbf{A}$ and its transpose is denoted $\mathbf{A}^\mathsf{T}$, then the square-symmetric representation of the graph can be computed via

$$\mathbf{A} \oplus \mathbf{A}^\mathsf{T}$$

This operation is depicted in Figure 5.3 and results in an 8×8 associative array.

5.3 Symmetric Adjacency Array

As data become higher resolution with more gradations, additional details become apparent. Features that were blurred together begin to separate into distinct elements. Figure 5.4 depicts the multitoned painting shown in Figure 5.2.
Many of the vertices in Figure 5.2 split into the multiple vertices in Figure 5.4

V01 → {V01, V15}
V03 → {V03, V17}
V04 → {V04, V05, V18}
V08 → {V07, V08, V19}
V12 → {V12, V13, V14}

The increased clarity also makes apparent five new vertices V02, V06, V11, V13, and V16. Simply connecting these vertices results in 23 undirected edges. A 13×15 associative array representation of Figure 5.4 is shown in Figure 5.5.

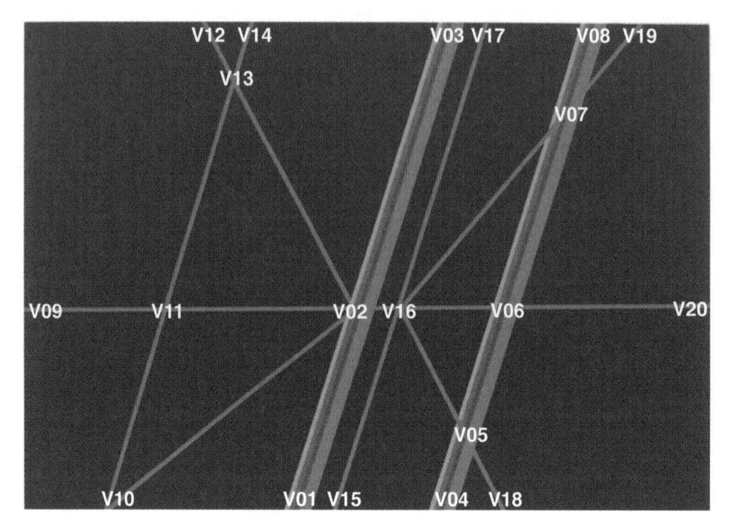

Figure 5.7
Abstract line painting (*XCRS* by Ann Pibal) showing various colored lines. The intersections and terminations of the lines are labeled vertices (V1,...,V20) and have been superimposed onto the painting in white letters.

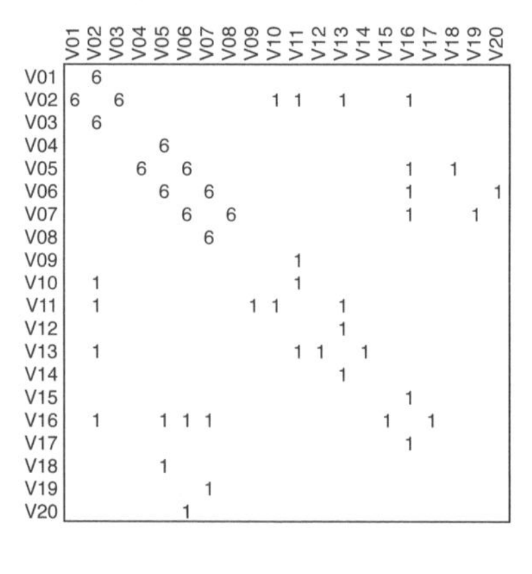

	V01	V02	V03	V04	V05	V06	V07	V08	V09	V10	V11	V12	V13	V14	V15	V16	V17	V18	V19	V20
V01		6																		
V02	6		6							1	1		1			1				
V03		6																		
V04					6															
V05				6		6										1		1		
V06					6		6									1				1
V07						6		6								1			1	
V08							6													
V09											1									
V10		1									1									
V11		1							1	1			1							
V12													1							
V13		1									1	1		1						
V14													1							
V15																1				
V16		1			1	1	1								1		1			
V17																1				
V18					1															
V19							1													
V20						1														

A

Figure 5.8
Square-symmetric associative array representation **A** of the edges depicted in Figure 5.7. Each value represents the number of edges connecting each pair of vertices.

The square-symmetric representation of the edges is constructed in the same manner as in the prior section by adding the 13×15 associative array to its 15×13 transpose to produce a 20×20 associative array with 46 non-empty entries (see Figure 5.6).

5.4 Weighted Adjacency Array

A color representation of the picture shows that there are more details in Figure 5.7 than just edges connecting the vertices. For example, some edges appear to have multiple vertices associated with them.

The square-symmetric associative array representation of the edges can be augmented to capture some of this information. In Figure 5.8, each value of the associative array represents the number of edges going between each pair of vertices. In this case, there are six pairs of vertices that all have six edges between them

(V01, V02), (V02, V03), (V04, V05), (V05, V06), (V06, V07), (V07, V08)

This value is referred to as the edge weight, and the corresponding graph is described as a weighted-undirected graph. If the edge weight specifically represents the number of edges between two vertices, then the graph is a multi-graph.

5.5 Incidence Array

The edges in Figure 5.7 have additional properties, including specific colors for each edge: blue, silver, green, orange, and pink. Note: some of the color differences are subtle, but for the purposes of this discussion the colors can be thought of as arbitrary labels on the edges. There are also multiple edges going between the same vertices. Some of the edges appear to lie on top of each other and so there is an order to the edges. All of these properties should be captured in the corresponding associative array representation. Including this information in the associative array requires additional labels. Figure 5.9 adds a label based on the edge color to each of the straight edges in the picture.

The additional information expressed about the edges cannot be represented in an adjacency array in which the rows and columns are vertices. In particular, hyper-edges that connect more than two vertices cannot be represented with an adjacency array. A different representation is required in which each row represents an edge and the columns represent the different possible attributes of an edge. This representation is referred to as an incidence array. Figure 5.10 is an associative array representation of an incidence array with hyper-edges.

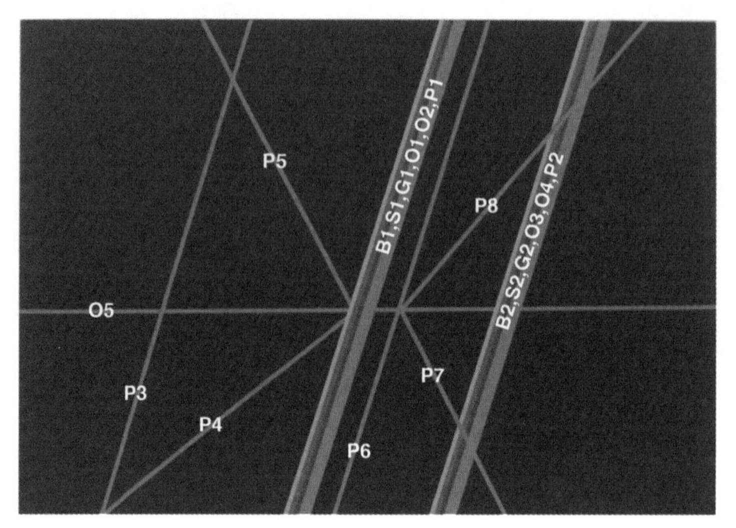

Figure 5.9
Edge labeling of each straight edge based on the color of the edge.

	Color	Order	V01	V02	V03	V04	V05	V06	V07	V08	V09	V10	V11	V12	V13	V14	V15	V16	V17	V18	V19	V20
B1	blue	2	1	1	1																	
B2	blue	2				1	1	1	1	1												
G1	green	2	1	1	1																	
G2	green	2				1	1	1	1	1												
O1	orange	1	1	1	1																	
O2	orange	1	1	1	1																	
O3	orange	1				1	1	1	1	1												
O4	orange	1				1	1	1	1	1												
O5	orange	1		1				1			1		1					1				1
P1	pink	3	1	1	1																	
P2	pink	3				1	1	1	1	1												
P3	pink	3										1	1		1	1						
P4	pink	3		1								1										
P5	pink	3		1										1	1							
P6	pink	3															1	1	1			
P7	pink	3					1											1		1		
P8	pink	3							1									1			1	
S1	silver	2	1	1	1																	
S2	silver	2				1	1	1	1	1												

E

Figure 5.10
Associative array representation of the incidence array **E** of the hyper-edges depicted in Figure 5.9.

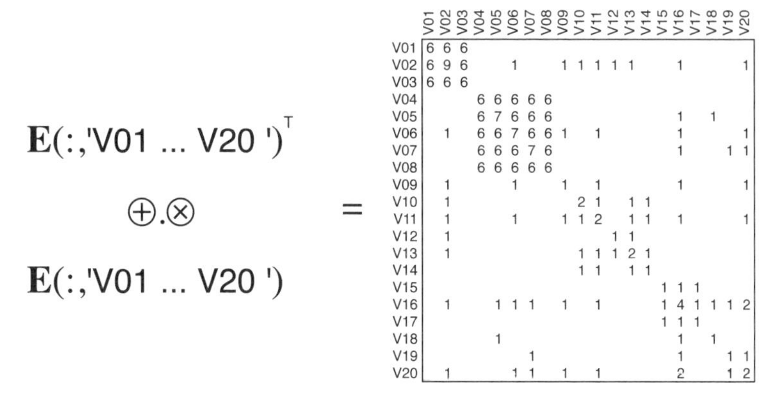

Figure 5.11
Computation of the vertex adjacency array via array multiplication of the incidence array **E** shown in Figure 5.10. For this computation, only the columns associated with the vertices are selected. The values indicate the number of times a given pair of vertices shares the same edge. The values differ from those shown in Figure 5.8 because the edges are hyper-edges.

Edges that connect many vertices are called hyper-edges. A graph with both multi-edges and hyper-edges is referred to as a multi-hyper-graph (or hyper-multi-graph). Associative array representations of incidence arrays can capture the full range of properties found in most any graph. Adjacency arrays can be constructed from incidence arrays via array multiplication. For example, if the incidence array depicted by the associative array in Figure 5.9 is denoted **E** and its transpose is denoted $\mathbf{E}^\mathsf{T}$, then the adjacency array of the vertices can be computed via

$$\mathbf{E}^\mathsf{T} \oplus.\otimes \mathbf{E}$$

This operation is depicted in Figure 5.11 and results in a 20×20 square-symmetric associative array in which each value is the number of times a pair of vertices (denoted by a row and a column) share the same edge.

For a graph with hyper-edges, the vertex adjacency array is not the only adjacency array that can be computed from the incidence array. The edge adjacency array can also be computed via

$$\mathbf{E} \oplus.\otimes \mathbf{E}^\mathsf{T}$$

This operation is depicted in Figure 5.12 and results in a 19×19 square-symmetric associative array in which each value stores the number of times a given pair of edges (denoted by a row and a column) shares the same vertex. The values differ from those shown in Figure 5.8 because the edges in Figure 5.10 are hyper-edges so that one edge can connect more that two vertices.

$$\mathbf{E}(:,'\text{V01} \ldots \text{V20 }') \oplus.\otimes \mathbf{E}(:,'\text{V01} \ldots \text{V20 }')^{\mathsf{T}} =$$

	B1	B2	G1	G2	O1	O2	O3	O4	O5	P1	P2	P3	P4	P5	P6	P7	P8	S1	S2
B1	3		3		3	3			1	3			1	1				3	
B2		5		5			5	5	1		5					1	1		5
G1	3		3		3	3			1	3			1	1				3	
G2		5		5			5	5	1		5					1	1		5
O1	3		3		3	3			1	3			1	1				3	
O2	3		3		3	3			1	3			1	1				3	
O3		5		5			5	5	1		5					1	1		5
O4		5		5			5	5	1		5					1	1		5
O5	1	1	1	1	1	1	1	1	6	1	1	1	1	1	1	1	2	1	1
P1	3		3		3	3			1	3			1	1				3	
P2		5		5			5	5	1		5					1	1		5
P3									1			4	1	1					
P4	1		1		1	1			1	1		1	2	1				1	
P5	1		1		1	1			1	1		1	1	3				1	
P6									1						3	1	1		
P7		1		1			1	1	1		1				1	3	1		1
P8		1		1			1	1	2		1				1	1	4		1
S1	3		3		3	3			1	3			1	1				3	
S2		5		5			5	5	1		5					1	1		5

Figure 5.12

Calculation of the edge adjacency array using array multiplication of the incidence array **E** shown in Figure 5.10. For this computation, only the columns associated with the vertices are selected. The values indicate the number of times a given pair of edges shares the same vertex.

5.6 Conclusions, Exercises, and References

Associative arrays can be used for showing both a summary or the details of data. A graph consisting of unweighted, undirected edges can be easily captured with a minimal adjacency array. The full symmetric adjacency array can be computed by adding the minimal adjacency array with its transpose. A more detailed graph might have multiple edges between vertices and can be summarized by setting to the adjacency array values that are the number of edges between any two vertices. An even more detailed description of a graph can be constructed by accounting for hyper-edges between vertices. An incidence array representation of a graph is well suited for capturing all of these details. The adjacency array can be calculated from the incidence array via array multiplication of the incidence array with its transpose.

Exercises

Exercise 5.1 — Refer to Figure 5.1.

(a) What type of graph is this? Directed/undirected, weighted/unweighted, hyper, multi and why?

(b) Compute the number of vertices and edges in the graph.

(c) What is the maximum possible number of edges between the vertices (not including self-edges)?

(d) How many of the possible edges are not in the graph?

Exercise 5.2 — Refer to Figure 5.3 and for each array $\mathbf{A}$, $\mathbf{A}^{\mathsf{T}}$, and $\mathbf{A} \oplus \mathbf{A}^{\mathsf{T}}$

(a) Compute the number of rows m and number of columns n in the arrays.

(b) Compute the total number of entries mn.

(c) Compute how many zero entries there are.

(d) Compute how many non-empty entries there are, and explain how the values relate to each other.

Exercise 5.3 — Refer to Figure 5.6 and for each array $\mathbf{A}$, $\mathbf{A}^\mathsf{T}$, and $\mathbf{A} \oplus \mathbf{A}^\mathsf{T}$

(a) Compute the number of rows m and number of columns n in the arrays.

(b) Compute the total number of entries mn.

(c) Compute how many empty entries there are.

(d) Compute how many non-empty entries there are, and explain how the values relate to each other.

Exercise 5.4 — Refer to Figure 5.11 and explain how the (a) 3×3, (b) 5×5, and (c) 3×3 dense blocks of non-empty values along the diagonal of the array relate back to vertices and edges in the paintings in Figure 5.7 and 5.9.

Exercise 5.5 — Refer to Figure 5.12 and explain how the dense row and column corresponding to edge O5 of the array relates back to the vertices and edges in the paintings in Figure 5.7 and 5.9.

Exercise 5.6 — Write down an associative array representing an incidence array of only the pink edges from Figure 5.10. Compute the vertex adjacency array and the edge adjacency array from this incidence array.

Exercise 5.7 — Select a picture of your own choosing. Label the vertices and the edges. Write down an associative array representing the incidence array of the picture. Compute the vertex adjacency array and the edge adjacency array from this incidence array.

References

[1] L. Euler, "Solutio problematis ad geometriam situs pertinentis," *Commentarii Academiae Scientiarum Petropolitanae*, vol. 8, pp. 128–140, 1736.

[2] R. Zimmer, "Abstraction in art with implications for perception," *Philosophical Transactions of the Royal Society of London B: Biological Sciences*, vol. 358, no. 1435, pp. 1285–1291, 2003.

[3] A. H. Barr Jr, *Cubism and Abstract Art, exhibition catalog*. New York: Museum of Modern Art, 1936. Jacket cover: diagram of stylistic evolution from 1890 until 1935.

[4] A. Schmidt Burkhardt, "Shaping modernism: Alfred Barr's genealogy of art," *Word & Image*, vol. 16, no. 4, pp. 387–400, 2000.

[5] P. Mondrian, 1914. Letter written to H. Bremmer.

[6] P. Mondrian, H. Holtzman, and M. S. James, *The New Art–The New Life: The Collected Writings of Piet Mondrian*. Thames and Hudson, 1987.

[7] A. Pibal, 2016. Personal communication.

[8] B. Shneiderman, "The eyes have it: A task by data type taxonomy for information visualizations," in *Proceedings of IEEE Symposium on Visual Languages*, pp. 336–343, IEEE, 1996.

[9] T. H. Cormen, C. E. Leiserson, R. L. Rivest, and C. Stein, *Introduction to Algorithms*. Cambridge: MIT Press, 2009.

6 Manipulating Graphs with Matrices

Summary

Associative arrays have many properties that are similar to matrices. Understanding how matrices can be used to manipulate graphs is a natural starting point for understanding the mathematics of associative arrays. Graphs represent connections between vertices with edges. Matrices can represent a wide range of graphs using adjacency matrices or incidence matrices. Adjacency matrices are often easier to analyze while incidence matrices are often better for representing data. Fortunately, the two are easily connected by matrix multiplication. A key feature of matrix mathematics is that a very small number of matrix operations is sufficient to manipulate a very wide range of graphs. Furthermore, matrix mathematics enables graph algorithms to be written as linear systems, enabling many graph operations to be reordered without changing the result. This chapter provides specific formulas and examples for the manipulation of graphs via matrices and lays the ground work for extending these operations into the broader domain of associative arrays.

6.1 Introduction

Matrix-based computational approaches have emerged as one of the most useful tools for analyzing and understanding graphs [2]. Likewise, because graphs are also often used to study non-numeric data such as documents, graphs have a strong connection to associative arrays. Thus, matrix-based approaches to graphs are a good starting point for introducing matrix mathematics concepts that are important to the deeper mathematics of associative arrays. The foundational concepts for matrix-based graph analysis are the adjacency matrix and incidence matrix representations of graphs. From these concepts, a more formal definition of a matrix can be constructed. How such a matrix can be manipulated depends on the types of values the matrix holds and the operations allowed on those values. More importantly, the mathematical properties of the operations on the values of the matrix determine the operations that can be performed on the whole matrix.

The mathematical development of graphs and matrices, along with their corresponding operations to represent larger collections of data, is a continuation of a process that dates to the earliest days of computing. The initial connection of bits (0's and 1's) with logical

---

operations such as and, or, and xor was laid out by George Boole [3, 4] and inspired Claude Shannon's early work on using digital circuits for controlling switches in communication networks [5]. Charles Babbage's concept for a general-purpose Analytical Engine [6] went beyond simple logical operations and envisioned computations on integers and rational numbers using +, ×, −, and /. Babbage's work inspired other computing pioneers, such as Percy Ludgate [7], Leonardo Torres y Quevedo [8], and Vannevar Bush [9], to design practical systems to implement these data representations and mathematical operations that could be built using electrical technologies [10].

The ability to digitize analog signals allowed sensor data to be processed by a computer. Claude Shannon's pioneering work on information theory [11] enabled the lossless communication of digital data over a lossy network. The representation of a signal as a vector of bits naturally led to the definition of mathematical operations on those vectors. Among these vector operations was the idea of filtering one vector in with another vector (or matrix), and this concept became the foundation of the field of digital signal processing [12].

The power of the rigorous pairing of data and operations in the form of a general abstract data type was first fully described by Barbara Liskov [13] and has become the basis of most modern programming languages.

Computational Benefits

Graphs are among the most important abstract data structures in computer science, and the algorithms that operate on them are critical to applications in bioinformatics, computer networks, and social media [14–18]. Graphs have been shown to be powerful tools for modeling complex problems because of their simplicity and generality [19, 20]. For this reason, the field of graph algorithms has become one of the pillars of theoretical computer science, informing research in such diverse areas as combinatorial optimization, complexity theory, and topology. Graph algorithms have been adapted and implemented by the military, commercial industry, and researchers in academia, and have become essential in controlling the power grid, telephone systems, and, of course, computer networks.

As the size of graphs increases, the time to analyze them can become prohibitive. One approach to addressing these computation challenges is to implement graph algorithms on a parallel computer. However, parallel graph algorithms are difficult to implement and optimize [21–26]. Irregular data access patterns and high communication found in graph algorithms mean that even the best algorithms will have parallel efficiencies that decrease as the number of processors is increased [27, 28]. Recent work on communication-avoiding algorithms, and their applications to graph computations [29, 30], might defer but cannnot completely eliminate the parallel computing bottleneck. Consequently, novel hardware architectures will also be required [31–33]. A common matrix-based graph processing

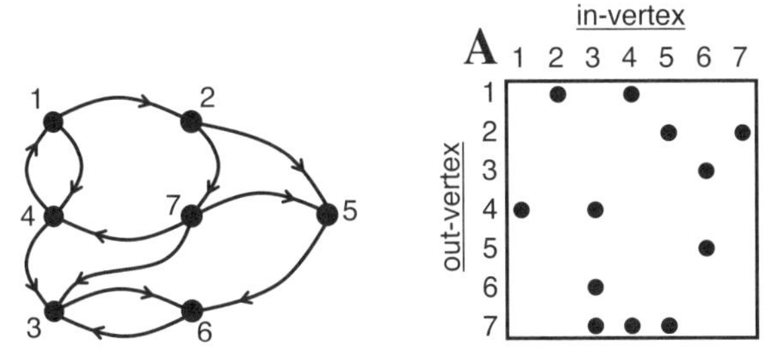

Figure 6.1
(Left) seven-vertex graph with 12 edges. Each vertex is labeled with an integer. (Right) 7×7 adjacency matrix **A** representation of the graph. **A** has 12 nonzero entries corresponding to the edges in the graph.

interface provides a useful tool for optimizing both software and hardware to provide high performance graph applications.

The duality between the canonical representation of graphs as abstract collections of vertices and edges and a matrix representation has been a part of graph theory since its inception [34, 35]. Matrix algebra has been recognized as a useful tool in graph theory for nearly as long (see [36] and references therein, in particular [37–43]). Likewise, graph-based approaches have also been very useful in matrix calculations [44–46]. The modern description of the duality between graph algorithms and matrix mathematics (or sparse linear algebra) has been extensively covered in the literature and is summarized in the cited text [2]. This text has further spawned the development of the GraphBLAS math library standard (GraphBLAS.org) [47] that has been developed in a series of proceedings [48–55] and implementations [56–62].

Graphs and matrices extend the aforementioned computing concepts far beyond individual bits and allow vast collections of data to be represented and manipulated in the same manner and with the same rigor. In this context, the most useful representation of graphs will be as adjacency matrices, typically denoted **A**, and as incidence (or edge) matrices, typically denoted **E**.

Adjacency Matrix: Undirected Graphs, Directed Graphs, and Weighted Graphs

Given an adjacency matrix **A**, if

$$\mathbf{A}(i,j) = 1$$

then there exists an edge going from vertex i to vertex j (see Figure 6.1). Likewise, if

$$\mathbf{A}(i,j) = 0$$

then there is no edge from i to j. Adjacency matrices have direction, which means that $\mathbf{A}(i,j)$ is not necessarily the same as $\mathbf{A}(j,i)$. Adjacency matrices can also have edge weights. If

$$\mathbf{A}(i,j) = v \neq 0$$

then the edge going from i to j is said to have weight v. Adjacency matrices provide a simple way to represent the connections between vertices in a graph. Adjacency matrices are often square, and both out-vertices (rows) and in-vertices (columns) are the same set of vertices. Adjacency matrices can be rectangular, in which case the out-vertices (rows) and the in-vertices (columns) are different sets of vertices. If there is no overlap between the out-vertices and in-vertices, then such graphs are often called bipartite graphs. In summary, adjacency matrices can represent many graphs, including any graph with any set of the following properties: directed, weighted, and/or bipartite.

Incidence Matrix: Multi-graphs and Hyper-graphs

An incidence, or edge matrix $\mathbf{E}$, uses the rows to represent every edge in the graph, and the columns represent every vertex. There are a number of conventions for denoting an edge in an incidence matrix. One such convention is to use two incidence matrices

$$\mathbf{E}_{\text{out}}(k,i) = 1 \quad \text{and} \quad \mathbf{E}_{\text{in}}(k,j) = 1$$

to indicate that edge k is a connection from i to j (see Figure 6.2). Incidence matrices are useful because they can easily represent multi-graphs and hyper-graphs. These complex graphs are difficult to capture with an adjacency matrix. A multi-graph has multiple edges between the same vertices. If there was another edge, k', from i to j, this relationship can be captured in an incidence matrix by setting

$$\mathbf{E}_{\text{out}}(k',i) = 1 \quad \text{and} \quad \mathbf{E}_{\text{in}}(k',j) = 1$$

(see Figure 6.3). In a hyper-graph, one edge can connect more than two vertices. For example, to denote edge k has a connection from i to j and j' can be accomplished by also setting (see Figure 6.3)

$$\mathbf{E}_{\text{in}}(k,j') = 1$$

Thus, an incidence matrix can be used to represent a graph with any set of the following graph properties: directed, weighted, multi-edge, and/or hyper-edge.

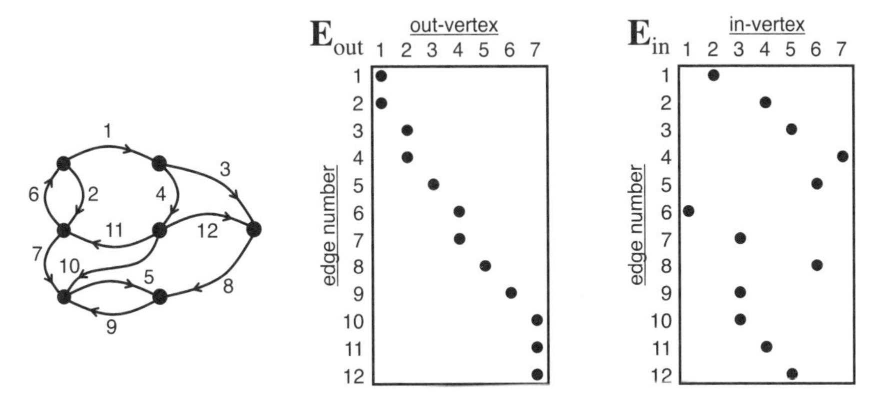

Figure 6.2
(Left) seven-vertex graph with 12 edges. Each edge is labeled with an integer; the vertex labels are the same as in Figure 6.1. (Middle) 12×7 incidence matrix $\mathbf{E}_{\text{out}}$ representing the out-vertices of the graph edges. (Right) 12×7 incidence matrix $\mathbf{E}_{\text{in}}$ representing the in-vertices of the graph edges. Both $\mathbf{E}_{\text{start}}$ and $\mathbf{E}_{\text{in}}$ have 12 nonzero entries corresponding to the edges in the graph.

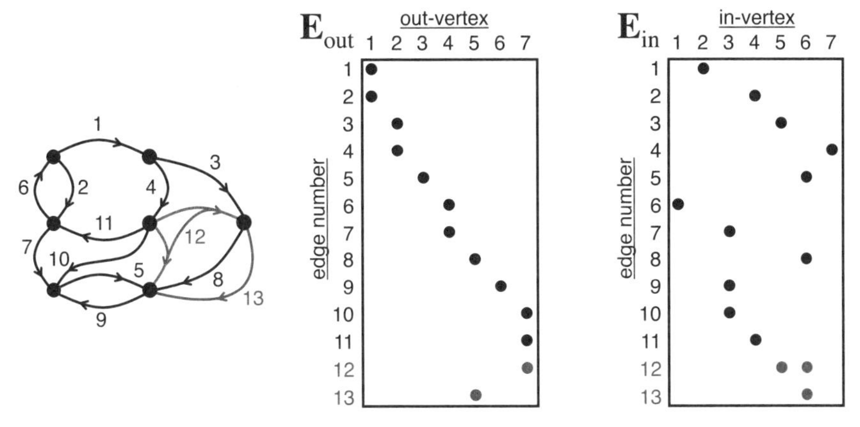

Figure 6.3
Graph and incidence matrices from Figure 6.2 with a hyper-edge (edge 12) and a multi-edge (edge 13). The graph is a hyper-graph because edge 12 has more than one in-vertex. The graph is a multi-graph because edge 8 and edge 13 have the same out- and in-vertex.

6.2 Matrix Indices and Values

A typical matrix has m rows and n columns of real numbers. Such a matrix can be denoted as

$$\mathbf{A} \in \mathbb{R}^{m \times n}$$

The row and column indices of the matrix $\mathbf{A}$ are

$$i \in I = \{1, \ldots, m\}$$

and

$$j \in J = \{1, \ldots, n\}$$

so that any particular value $\mathbf{A}$ can be denoted as $\mathbf{A}(i, j)$. The row and column indices of matrices are sets of natural numbers $I, J \subset \mathbb{N}$. [Note: a specific *implementation* of these matrices might use IEEE 64-bit double-precision floating-point numbers to represent real numbers, 64-bit unsigned integers to represent row and column indices, and the compressed sparse rows (CSR) format or the compressed sparse columns (CSC) format to store the nonzero values inside the sparse matrix.]

Matrices can be defined over a wide range of scalars. Some common classes of scalar are as follows. A matrix of complex numbers

$$\mathbb{C} = \{x + y\sqrt{\text{-}1} \mid x, y \in \mathbb{R}\}$$

is denoted

$$\mathbf{A} \in \mathbb{C}^{m \times n}$$

A matrix of integers

$$\mathbb{Z} = \{\ldots, \text{-}1, 0, 1, \ldots\}$$

is denoted

$$\mathbf{A} \in \mathbb{Z}^{m \times n}$$

A matrix of natural numbers

$$\mathbb{N} = \{1, 2, 3, \ldots\}$$

is denoted

$$\mathbf{A} \in \mathbb{N}^{m \times n}$$

Using the above concepts, a matrix is defined as the following two-dimensional (2D) mapping

$$\mathbf{A} : I \times J \to \mathbb{S}$$

where the indices $I, J \subset \mathbb{Z}$ are finite sets of integers with m and n elements, respectively, and

$$\mathbb{S} \in \{\mathbb{R}, \mathbb{Z}, \mathbb{N}, \ldots\}$$

is a set of scalars. Without loss of generality, matrices over arbitrary scalars can be denoted

$$\mathbf{A} \in \mathbb{S}^{m \times n}$$

A *vector* is a matrix where either $m = 1$ or $n = 1$. A *column vector* is denoted

$$\mathbf{v} \in \mathbb{S}^{m \times 1}$$

A *row vector* is denoted

$$\mathbf{v} \in \mathbb{S}^{1 \times n}$$

A *scalar* is a single element of a set

$$s \in \mathbb{S}$$

and has no matrix dimensions.

Scalar Operations: Combining and Scaling Graph Edge Weights

Matrix operations are built on top of scalar operations. The primary scalar operations are standard arithmetic addition, such as

$$1 + 1 = 2$$

and arithmetic multiplication, such as

$$2 \times 2 = 4$$

These scalar operations of addition and multiplication can be defined to be a wide variety of functions. To prevent confusion with standard arithmetic addition and arithmetic multiplication, $\oplus$ will be used to denote scalar addition and $\otimes$ will be used to denote scalar multiplication. In this notation, standard arithmetic addition and arithmetic multiplication of real numbers

$$a, b, c \in \mathbb{R}$$

where

$$\oplus \equiv + \quad \text{and} \quad \otimes \equiv \times$$

results in

$$c = a \oplus b \quad \text{implies} \quad c = a + b$$

and

$$c = a \otimes b \quad \text{implies} \quad c = a \times b$$

Generalizing $\oplus$ and $\otimes$ to a variety of operations enables a wide range of algorithms on scalars of all different types (not just real or complex numbers).

Scalar Properties: Composable Graph Edge Weight Operations

Certain $\oplus$ and $\otimes$ combinations over certain sets of scalars are particularly useful because they preserve desirable mathematical properties of linear systems such as additive commutativity

$$a \oplus b = b \oplus a$$

multiplicative commutativity

$$a \otimes b = b \otimes a$$

additive associativity

$$(a \oplus b) \oplus c = a \oplus (b \oplus c)$$

multiplicative associativity

$$(a \otimes b) \otimes c = a \otimes (b \otimes c)$$

and the distributivity of multiplication over addition

$$a \otimes (b \oplus c) = (a \otimes b) \oplus (a \otimes c)$$

The properties of commutativity, associativity, and distributivity are *extremely* useful properties for building graph applications because they allow the builder to swap operations without changing the result. Example combinations of $\oplus$ and $\otimes$ that preserve scalar commutativity, associativity, and distributivity include (but are not limited to) standard arithmetic

$$\oplus \equiv + \quad \otimes \equiv \times \quad a,b,c \in \mathbb{R}$$

max-plus algebras

$$\oplus \equiv \max \quad \otimes \equiv + \quad a,b,c \in \mathbb{R} \cup \{-\infty\}$$

max-min algebras

$$\oplus \equiv \max \quad \otimes \equiv \min \quad a,b,c \in [0,\infty]$$

finite (Galois) fields such as GF(2)

$$\oplus \equiv \text{xor} \quad \otimes \equiv \text{and} \quad a,b,c \in \{0,1\}$$

and the power set of the set of integers

$$\oplus \equiv \cup \quad \otimes \equiv \cap \quad a,b,c \subset \mathbb{Z}$$

The above pairs of operation are all classes of mathematics ofter referred to as a semiring. Semirings are an important part of associative arrays and are discussed at length in subsequent chapters. Other functions that do not preserve the above properties (and are not semirings) can also be defined for $\oplus$ and $\otimes$. For example, it is often useful for $\oplus$ or $\otimes$ to pull in other data, such as vertex indices of a graph.

6.3 Composable Graph Operations and Linear Systems

Associativity, distributivity, and commutativity are very powerful properties that enable the construction of composable graph algorithms with all the properties of linear systems. Such linear operations can be reordered with the knowledge that the answers will remain unchanged. Composability makes it easy to build a wide range of graph algorithms with just a few functions. Given matrices

$$\mathbf{A}, \mathbf{B}, \mathbf{C} \in \mathbb{S}^{m \times n}$$

let their elements be specified by

$$a = \mathbf{A}(i, j)$$
$$b = \mathbf{B}(i, j)$$
$$c = \mathbf{C}(i, j)$$

Commutativity, associativity, and distributivity of scalar operations translates into similar properties on matrix operations in the following manner.

Additive Commutativity — Allows graphs to be swapped and combined via matrix element-wise addition (see Figure 6.4) without changing the result

$$a \oplus b = b \oplus a \quad \text{implies} \quad \mathbf{A} \oplus \mathbf{B} = \mathbf{B} \oplus \mathbf{A}$$

where matrix element-wise addition

$$\mathbf{C} = \mathbf{A} \oplus \mathbf{B}$$

is given by

$$\mathbf{C}(i, j) = \mathbf{A}(i, j) \oplus \mathbf{B}(i, j)$$

Multiplicative Commutativity — Allows graphs to be swapped, intersected, and scaled via matrix element-wise multiplication (see Figure 6.5) without changing the result

$$a \otimes b = b \otimes a \quad \text{implies} \quad \mathbf{A} \otimes \mathbf{B} = \mathbf{B} \otimes \mathbf{A}$$

where matrix element-wise (Hadamard) multiplication

$$\mathbf{C} = \mathbf{A} \otimes \mathbf{B}$$

is given by

$$\mathbf{C}(i, j) = \mathbf{A}(i, j) \otimes \mathbf{B}(i, j)$$

Additive Associativity — Allows graphs to be combined via matrix element-wise addition in any grouping without changing the result

$$(a \oplus b) \oplus c = a \oplus (b \oplus c) \quad \text{implies} \quad (\mathbf{A} \oplus \mathbf{B}) \oplus \mathbf{C} = \mathbf{A} \oplus (\mathbf{B} \oplus \mathbf{C})$$

Multiplicative Associativity — Allows graphs to be intersected and scaled via matrix element-wise multiplication in any grouping without changing the result

$$(a \otimes b) \otimes c = a \otimes (b \otimes c) \quad \text{implies} \quad (\mathbf{A} \otimes \mathbf{B}) \otimes \mathbf{C} = \mathbf{A} \otimes (\mathbf{B} \otimes \mathbf{C})$$

Element-Wise Distributivity — Allows graphs to be intersected and/or scaled and then combined or vice versa without changing the result

$$a \otimes (b \oplus c) = (a \otimes b) \oplus (a \otimes c) \quad \text{implies} \quad \mathbf{A} \otimes (\mathbf{B} \oplus \mathbf{C}) = (\mathbf{A} \otimes \mathbf{B}) \oplus (\mathbf{A} \otimes \mathbf{C})$$

Matrix Multiply Distributivity — Allows graphs to be transformed via matrix multiply and then combined or vice versa without changing the result

$$a \otimes (b \oplus c) = (a \otimes b) \oplus (a \otimes c) \quad \text{implies} \quad \mathbf{A}(\mathbf{B} \oplus \mathbf{C}) = (\mathbf{AB}) \oplus (\mathbf{AC})$$

where matrix multiply

$$\mathbf{C} = \mathbf{AB} = \mathbf{A} \ {\oplus.\otimes} \ \mathbf{B}$$

is given by

$$\mathbf{C}(i,j) = \bigoplus_{k=1}^{\ell} \mathbf{A}(i,k) \otimes \mathbf{B}(k,j)$$

for matrices with dimensions

$$\mathbf{A} \in \mathbb{S}^{m \times \ell}$$
$$\mathbf{B} \in \mathbb{S}^{\ell \times n}$$
$$\mathbf{C} \in \mathbb{S}^{m \times n}$$

Matrix Multiply Associativity — Another implication of scalar distributivity is that graphs can be transformed via matrix multiplication in any grouping without changing the result

$$a \otimes (b \oplus c) = (a \otimes b) \oplus (a \otimes c) \quad \text{implies} \quad (\mathbf{AB})\mathbf{C} = \mathbf{A}(\mathbf{BC})$$

Matrix Multiply Commutativity — In general, $\mathbf{AB} \neq \mathbf{BA}$. Some examples where $\mathbf{AB} = \mathbf{BA}$ include when either matrix is all zeros, either matrix matrix is the identity matrix, both matrices are diagonal matrices, or both matrices are rotation matrices with the same axis of rotation.

0-Element: No Graph Edge

Sparse matrices play an important role in graphs. Many implementations of sparse matrices reduce storage by not storing the 0 valued elements in the matrix. In adjacency matrices, the 0 element is equivalent to no edge from the vertex that is represented by the row to the vertex that is represented by the column. In incidence matrices, the 0 element is equivalent to the edge denoted by the row not including the vertex denoted by the column. In most

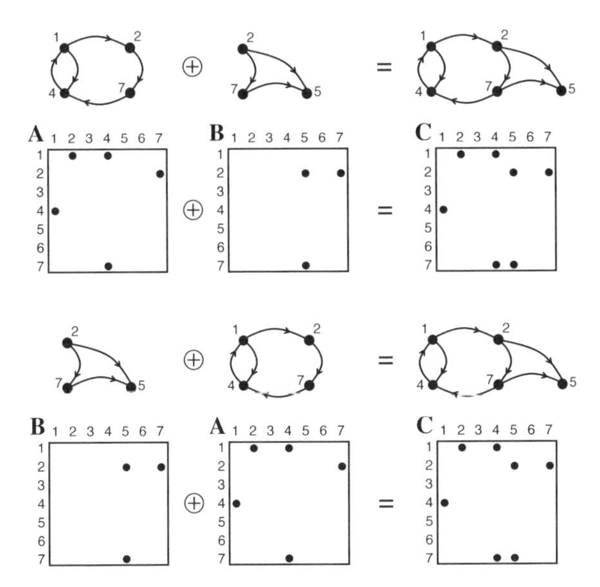

Figure 6.4
Illustration of the commutative property of the element-wise addition of two graphs and their corresponding adjacency matrix representations.

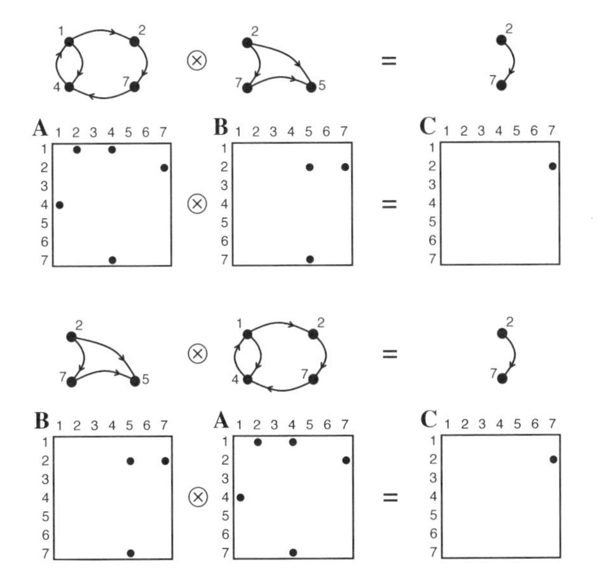

Figure 6.5
Depiction of the commutative property of the element-wise multiplication of two graphs along with their adjacency matrices.

cases, the 0 element is standard arithmetic 0, but in other cases it can be a different value. Nonstandard 0 values can be helpful when combined with different $\oplus$ and $\otimes$ operations. For example, in different contexts 0 might be $+\infty$, $-\infty$, or $\emptyset$. For any value of 0, if the 0 element has certain properties with respect to scalar $\oplus$ and $\otimes$, then the sparsity of matrix operations can be managed efficiently. These properties are the additive identity

$$a \oplus 0 = a$$

and the multiplicative annihilator

$$a \otimes 0 = 0$$

There are many combinations of $\oplus$ and $\otimes$ that exhibit the additive identity and multiplicative annihilator. Some of the more important combinations of these semirings that are used throughout the book are carefully described as follows.

Arithmetic on Real Numbers $(+.\times)$ — Given standard arithmetic over the real numbers

$$a \in \mathbb{R}$$

where addition is

$$\oplus \equiv +$$

multiplication is

$$\otimes \equiv \times$$

and zero is

$$0 \equiv 0$$

which results in additive identity

$$a \oplus 0 = a + 0 = a$$

and multiplicative annihilator

$$a \otimes 0 = a \times 0 = 0$$

Max-Plus Algebra (max.+) — Given real numbers with a minimal element

$$a \in \mathbb{R} \cup \{-\infty\}$$

where addition is

$$\oplus \equiv \max$$

multiplication is

$$\otimes \equiv +$$

and zero is

$$0 \equiv -\infty$$

which results in additive identity

$$a \oplus 0 = \max(a, -\infty) = a$$

and multiplicative annihilator

$$a \otimes 0 = a + -\infty = -\infty$$

Min-Plus Algebra (min.+) — Given real numbers with a maximal element

$$a \in \mathbb{R} \cup \{\infty\}$$

where addition is

$$\oplus \equiv \min$$

multiplication is

$$\otimes \equiv +$$

and zero is

$$0 \equiv \infty$$

which results in additive identity

$$a \oplus 0 = \min(a, \infty) = a$$

and multiplicative annihilator

$$a \otimes 0 = a + \infty = \infty$$

Max-Min Algebra (max.min) — Given non-negative real numbers

$$\mathbb{R}_{\geq 0} = \{a \in \mathbb{R} \mid 0 \leq a < \infty\} = [0, \infty)$$

where addition is

$$\oplus \equiv \max$$

multiplication is

$$\otimes \equiv \min$$

and zero is

$$0 \equiv 0$$

which results in additive identity

$$a \oplus 0 = \max(a, 0) = a$$

and multiplicative annihilator

$$a \otimes 0 = \min(a, 0) = 0$$

Min-Max Algebra (min.max) — Given non-positive real numbers

$$\mathbb{R}_{\leq 0} = \{a \in \mathbb{R} \mid -\infty < a \leq 0\} = (-\infty, 0]$$

where addition is

$$\oplus \equiv \min$$

multiplication is

$$\otimes \equiv \max$$

and zero is

$$\mathbb{0} \equiv 0$$

which results in additive identity

$$a \oplus \mathbb{0} = \min(a, 0) = a$$

and multiplicative annihilator

$$a \otimes \mathbb{0} = \max(a, 0) = 0$$

Galois Field (xor.and) — Given a set of two numbers

$$a \in \{0, 1\}$$

where addition is

$$\oplus \equiv \text{xor}$$

multiplication is

$$\otimes \equiv \text{and}$$

and zero is

$$\mathbb{0} \equiv 0$$

which results in additive identity

$$a \oplus \mathbb{0} = \text{xor}(a, 0) = a$$

and multiplicative annihilator

$$a \otimes \mathbb{0} = \text{and}(a, 0) = 0$$

Power Set ($\cup.\cap$) — Given any subset of integers

$$a \subset \mathbb{Z}$$

where addition is

$$\oplus \equiv \cup$$

multiplication is

$$\otimes \equiv \cap$$

and zero is

$$0 \equiv \emptyset$$

which results in additive identity

$$a \oplus 0 = a \cup \emptyset = a$$

and multiplicative annihilator

$$a \otimes 0 = a \cap \emptyset = \emptyset$$

The above examples are a small selection of the operators and sets that form semirings that are useful for building graph algorithms with linear systems properties. Many more are possible. The ability to change the scalar values and operators while preserving the overall behavior of the graph operations is one of the principal benefits of using matrices for graph algorithms. For example, relaxing the requirement that the multiplicative annihilator be the additive identity, as in the above examples, yields additional operations, such as

*Max-Max Algebra (*max.max*)* — Given non-positive real numbers with a minimal element

$$a \in \mathbb{R}_{\leq 0} \cup \{-\infty\}$$

where addition is

$$\oplus \equiv \max$$

multiplication is (also)

$$\otimes \equiv \max$$

and zero is

$$0 \equiv -\infty$$

which results in additive identity

$$a \oplus 0 = \max(a, -\infty) = a$$

*Min-Min Algebra (*min.max*)* — Given non-negative real numbers with a maximal element

$$a \in \mathbb{R}_{\geq 0} \cup \{\infty\}$$

where addition is

$$\oplus \equiv \min$$

multiplication is (also)

$$\otimes \equiv \min$$

and zero is

$$0 \equiv \infty$$

which results in additive identity

$$a \oplus 0 = \min(a, \infty) = a$$

6.4 Matrix Graph Operations Overview

The main benefit of a matrix approach to graphs is the ability to perform a wide range of graph operations on diverse types of graphs with a small number of matrix operations. The core set of matrix functions has been shown to be useful for implementing a wide range of graph algorithms. These matrix functions strike a balance between providing enough functions to be useful to application builders while being few enough that they can be implemented effectively. Some of these core matrix operations and some example graph operations they support are describe subsequently.

Building a Matrix: Edge List to Graph

A foundational matrix operation is to build a sparse matrix from row, column, and value triples. Constructing a matrix from triples is equivalent to graph construction from vectors of out-vertices $\mathbf{i}$, in-vertices $\mathbf{j}$, and edge weight values $\mathbf{v}$. Each triple

$$(\mathbf{i}(k), \mathbf{j}(k), \mathbf{v}(k))$$

corresponds to edge k in the graph. Directed graphs can be represented as triples of vectors $\mathbf{i}$, $\mathbf{j}$, and $\mathbf{v}$ corresponding to the nonzero elements in the sparse matrix. Constructing an $m \times n$ sparse matrix from vector triples can be denoted

$$\mathbf{C} \ = \ \mathbb{S}^{m \times n}(\mathbf{i}, \mathbf{j}, \mathbf{v}, \oplus)$$

where

$$\mathbf{i} \in I^{\ell}$$
$$\mathbf{j} \in J^{\ell}$$
$$\mathbf{v} \in \mathbb{S}^{\ell}$$

are all ℓ element vectors, and $\mathbb{S}$ denotes a set of scalar values. The optional $\oplus$ operation defines how multiple entries with the same row and column are handled. Other variants include replacing any or all of the vector inputs with single-element vectors. For example

$$\mathbf{C} = \mathbb{S}^{m \times n}(\mathbf{i}, \mathbf{j}, 1)$$

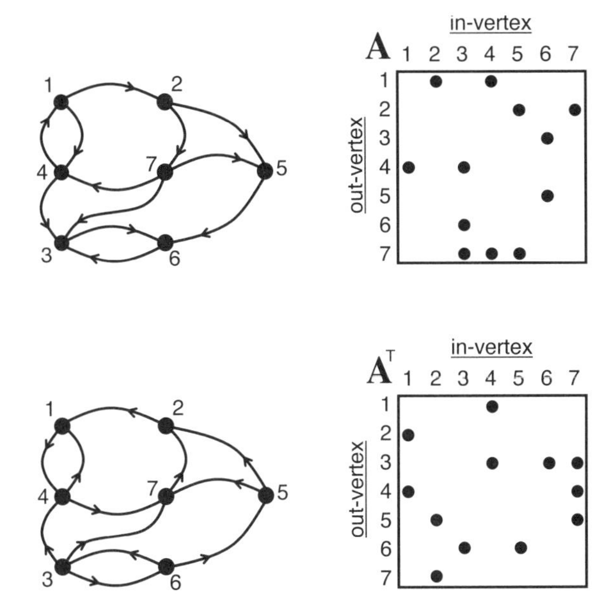

Figure 6.6
Transposing the adjacency matrix of a graph switches the directions of its edges.

would use the value of 1 for nonzero matrix values. Likewise, a row vector can be constructed using

$$\mathbf{C} = \mathbb{S}^{m \times n}(1, \mathbf{j}, \mathbf{v})$$

and a column vector can be constructed using

$$\mathbf{C} = \mathbb{S}^{m \times n}(\mathbf{i}, 1, \mathbf{v})$$

The scalar value type of the sparse matrix can be further specified using standard matrix notation. For example, a sparse matrix containing real numbers can be specified via

$$\mathbf{C} = \mathbb{R}^{m \times n}(\mathbf{i}, \mathbf{j}, \mathbf{v})$$

Extracting Tuples: Graph to Vertex List

It is just as important to be able to extract the row, column, and value triples corresponding to the nonzero elements in a sparse matrix, which is equivalent to listing the edges in a graph. Extracting the nonzero triples from a sparse matrix can be denoted mathematically as

$$(\mathbf{i}, \mathbf{j}, \mathbf{v}) = \mathbf{A}$$

Transpose: Swap Out-Vertices and In-Vertices

Transposing a matrix has the effect of swapping the out-vertices and the in-vertices of a graph. Swapping the rows and columns of a sparse matrix is a common tool for changing the direction of vertices in a graph (see Figure 6.6). The transpose is denoted as

$$\mathbf{C} = \mathbf{A}^{\mathsf{T}}$$

or more explicitly

$$\mathbf{C}(j,i) = \mathbf{A}(i,j)$$

where

$$\mathbf{A} \in \mathbb{S}^{m \times n}$$
$$\mathbf{C} \in \mathbb{S}^{n \times m}$$

Transpose also can be implemented using triples as follows

$$(\mathbf{i},\mathbf{j},\mathbf{v}) = \mathbf{A}$$
$$\mathbf{C} = \mathbb{S}^{n \times m}(\mathbf{j},\mathbf{i},\mathbf{v})$$

Matrix Multiplication: Breadth-First-Search and Adjacency Matrix Construction

Perhaps the most useful matrix operation is matrix multiplication, which can be used to perform a number of graph traversal operations, such as single-source breadth-first search, multisource breadth-first search, and weighted breadth-first search.

Matrix multiplication can be used to implement a wide range of graph algorithms. Examples include finding the nearest neighbors of a vertex (see Figure 6.7) and constructing an adjacency matrix from an incidence matrix (see Figure 6.8). In its most common form, matrix multiplication using standard arithmetic addition and multiplication is given by

$$\mathbf{C} = \mathbf{A}\mathbf{B}$$

or more explicitly

$$\mathbf{C}(i,j) = \sum_{k=1}^{\ell} \mathbf{A}(i,k)\mathbf{B}(k,j)$$

where

$$\mathbf{A} \in \mathbb{R}^{m \times \ell}$$
$$\mathbf{B} \in \mathbb{R}^{\ell \times n}$$
$$\mathbf{C} \in \mathbb{R}^{m \times n}$$

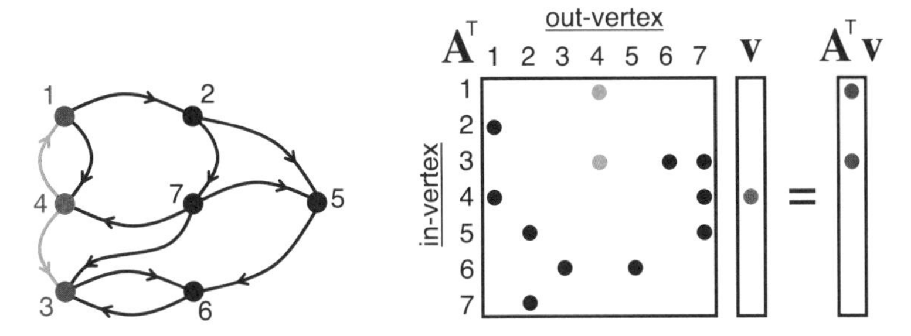

Figure 6.7
(Left) breadth-first-search of a graph starting at vertex 4 and traversing to vertices 1 and 3. (Right) matrix-vector multiplication of the adjacency matrix of a graph performs the equivalent operation.

Matrix multiplication has many important variants that include non-arithmetic addition and multiplication

$$\mathbf{C} = \mathbf{A} \oplus.\otimes \mathbf{B}$$

where

$$\mathbf{A} \in \mathbb{S}^{m \times \ell}$$
$$\mathbf{B} \in \mathbb{S}^{\ell \times n}$$
$$\mathbf{C} \in \mathbb{S}^{m \times n}$$

and the notation $\oplus.\otimes$ makes explicit that $\oplus$ and $\otimes$ can be other functions.

One of the most common uses of matrix multiplication is to construct an adjacency matrix from an incidence matrix representation of a graph. For a graph with out-vertex incidence matrix $\mathbf{E}_{\text{out}}$ and in-vertex incidence matrix $\mathbf{E}_{\text{in}}$, the corresponding adjacency matrix is

$$\mathbf{A} = \mathbf{E}_{\text{out}}^\mathsf{T} \mathbf{E}_{\text{in}}$$

The individual values in $\mathbf{A}$ can be computed via

$$\mathbf{A}(i, j) = \bigoplus_k \mathbf{E}_{\text{out}}^\mathsf{T}(i, k) \otimes \mathbf{E}_{\text{in}}(k, j)$$

Matrix Multiplication: Combining and Scaling Edges

Standard matrix multiplication on real numbers first performs scalar arithmetic multiplication on the elements and then performs scalar arithmetic addition on the results. The scalar operations of addition $\oplus$ and multiplication $\otimes$ can be replaced with other functions. This replacement can be formally denoted as

$$\mathbf{C} = \mathbf{A} \oplus.\otimes \mathbf{B}$$

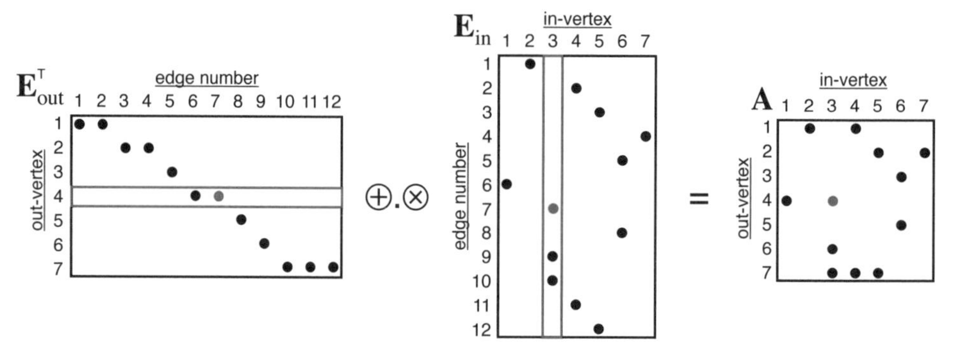

Figure 6.8
Construction of an adjacency matrix of a graph from its incidence matrices via matrix-matrix multiply. The entry $\mathbf{A}(4,3)$ is obtained by combining the row vector $\mathbf{E}_{\text{out}}^{\mathsf{T}}(4,k)$ with the column vector $\mathbf{E}_{\text{in}}(k,3)$ via matrix product $\mathbf{A}(4,3) = \bigoplus_{k=1}^{12} \mathbf{E}_{\text{out}}^{\mathsf{T}}(4,k) \otimes \mathbf{E}_{\text{in}}(k,3)$.

or more explicitly

$$\mathbf{C}(i,j) = \bigoplus_{k=1}^{\ell} \mathbf{A}(i,k) \otimes \mathbf{B}(k,j)$$

where

$$\mathbf{A} \in \mathbb{S}^{m \times \ell}$$
$$\mathbf{B} \in \mathbb{S}^{\ell \times n}$$
$$\mathbf{C} \in \mathbb{S}^{m \times n}$$

In this notation, standard matrix multiply can be written

$$\mathbf{C} = \mathbf{A} \ {+}.{\times} \ \mathbf{B}$$

where $\mathbb{S} \to \mathbb{R}$. Other matrix multiplications of interest include max-plus algebras

$$\mathbf{C} = \mathbf{A} \ \max.{+} \ \mathbf{B}$$

or more explicitly

$$\mathbf{C}(i,j) = \max_k \{\mathbf{A}(i,k) + \mathbf{B}(k,j)\}$$

where $\mathbb{S} = \mathbb{R} \cup \{\text{-}\infty\}$; min-max algebras

$$\mathbf{C} = \mathbf{A} \ \min.\max \ \mathbf{B}$$

or more explicitly

$$\mathbf{C}(i,j) = \min_k \{\max(\mathbf{A}(i,k), \mathbf{B}(k,j))\}$$

where $\mathbb{S} = [0, \infty)$; the Galois field of order 2

$$\mathbf{C} = \mathbf{A} \text{ xor.and } \mathbf{B}$$

or more explicitly

$$\mathbf{C}(i,j) = \underset{k}{\text{xor}}\{\text{and}(\mathbf{A}(i,k), \mathbf{B}(k,j))\}$$

where $\mathbb{S} = \{0, 1\}$; and power set of the set of integers

$$\mathbf{C} = \mathbf{A} \cup.\cap \mathbf{B}$$

or more explicitly

$$\mathbf{C}(i,j) = \bigcup_k \mathbf{A}(i,k) \cap \mathbf{B}(k,j)$$

where $\mathbb{S} = \mathcal{P}(\mathbb{Z})$.

Extract: Selecting Sub-Graphs

Extracting a sub-matrix from a larger matrix is equivalent to selecting a sub-graph from a larger graph. Selecting sub-graphs is a very common graph operation (see Figure 6.9). This operation is performed by selecting out-vertices (row) and in-vertices (columns) from a matrix $\mathbf{A} \in \mathbb{S}^{m \times n}$

$$\mathbf{C} = \mathbf{A}(\mathbf{i}, \mathbf{j})$$

or more explicitly

$$\mathbf{C}(i,j) = \mathbf{A}(\mathbf{i}(i), \mathbf{j}(j))$$

where

$$i \in \{1, ..., m_C\}$$
$$j \in \{1, ..., n_C\}$$
$$\mathbf{i} \in I^{m_C}$$
$$\mathbf{j} \in J^{m_C}$$

select specific sets of rows and columns in a specific order. The resulting matrix

$$\mathbf{C} \in \mathbb{S}^{m_C \times n_C}$$

can be larger or smaller than the input matrix $\mathbf{A}$. This operation can also be used to replicate and/or permute rows and columns in a matrix.

Extraction can also be implemented with matrix multiplication as

$$\mathbf{C} = \mathbf{S}(\mathbf{i}) \; \mathbf{A} \; \mathbf{S}^{\mathsf{T}}(\mathbf{j})$$

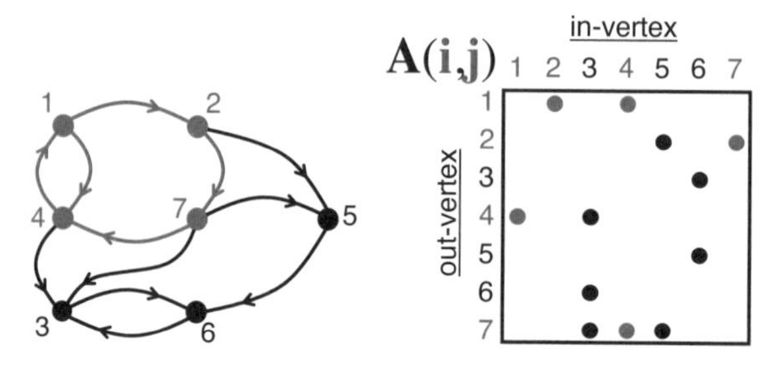

Figure 6.9
Selection of a 4-vertex sub-graph from the adjacency matrix via selecting subsets of rows and columns $\mathbf{i} = \mathbf{j} = (1,2,4,7)$.

where $\mathbf{S}(\mathbf{i})$ and $\mathbf{S}(\mathbf{j})$ are selection matrices given by

$$\mathbf{S}(\mathbf{i}) = \mathbb{S}^{m_C \times m}(\{1,...,m_C\}, \mathbf{i}, 1)$$

$$\mathbf{S}(\mathbf{j}) = \mathbb{S}^{n_C \times n}(\{1,...,n_C\}, \mathbf{j}, 1)$$

Assign: Modifying Sub-Graphs

Assigning a matrix to a set of indices in a larger matrix is equivalent to inserting or modifying a sub-graph into a graph. Modifying sub-graphs is a very common graph operation. This operation is performed by selecting out-vertices (row) and in-vertices (columns) from a matrix $\mathbf{C} \in \mathbb{S}^{m \times n}$ and assigning new values to them from another sparse matrix, $\mathbf{A} \in \mathbb{S}^{m_A \times n_A}$

$$\mathbf{C}(\mathbf{i},\mathbf{j}) = \mathbf{A}$$

or more explicitly

$$\mathbf{C}(\mathbf{i}(i),\mathbf{j}(j)) = \mathbf{A}(i,j)$$

where

$$i \in \{1,...,m_A\}$$
$$j \in \{1,...,n_A\}$$
$$\mathbf{i} \in I^{m_A}$$
$$\mathbf{j} \in J^{n_A}$$

select specific sets of rows and columns.

The additive form of this operation can be implemented with sparse matrix multiplication as

$$\mathbf{C} = \mathbf{S}^{\mathsf{T}}(\mathbf{i})\,\mathbf{A}\,\mathbf{S}(\mathbf{j})$$

where $\mathbf{S}(\mathbf{i})$ and $\mathbf{S}(\mathbf{j})$ are selection matrices given by

$$\mathbf{S}(\mathbf{i}) = \mathbb{S}^{m_A \times m}(\{1,...,m_A\}, \mathbf{i}, 1)$$

$$\mathbf{S}(\mathbf{j}) = \mathbb{S}^{n_A \times n}(\{1,...,n_A\}, \mathbf{j}, 1)$$

Element-Wise Addition and Element-Wise Multiplication: Combining Graphs, Intersecting Graphs, and Scaling Graphs

Element-wise addition of matrices and element-wise multiplication of matrices are among the main ways to combine and correlate matrices. These matrix operations are in many ways equivalent to graph union and intersection. Combining graphs along with adding their edge weights can be accomplished by adding together their sparse matrix representations

$$\mathbf{C} = \mathbf{A} \oplus \mathbf{B}$$

where

$$\mathbf{A}, \mathbf{B}, \mathbf{C} \in \mathbb{S}^{m \times n}$$

or more explicitly

$$\mathbf{C}(i,j) = \mathbf{A}(i,j) \oplus \mathbf{B}(i,j)$$

where $i \in \{1,...,m\}$, and $j \in \{1,...,n\}$.

Intersecting graphs along with scaling their edge weights can be accomplished by element-wise multiplication of their sparse matrix representations

$$\mathbf{C} = \mathbf{A} \otimes \mathbf{B}$$

where

$$\mathbf{A}, \mathbf{B}, \mathbf{C} \in \mathbb{S}^{m \times n}$$

or more explicitly

$$\mathbf{C}(i,j) = \mathbf{A}(i,j) \otimes \mathbf{B}(i,j)$$

where $i \in \{1,...,m\}$, and $j \in \{1,...,n\}$.

Kronecker: Graph Generation

Generating graphs is a common operation in a wide range of graph algorithms. Graph generation is used in testing graph algorithms, in creating graph templates to match against, and for comparing real graph data with models. The Kronecker product of two matrices is a convenient, well-defined matrix operation that can be used for generating a wide range of graphs from a few a parameters [63, 64].

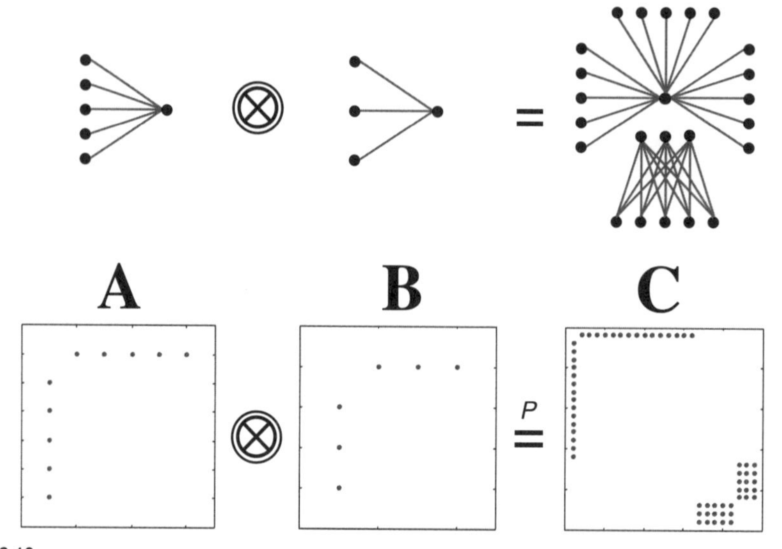

Figure 6.10
Kronecker product of the adjacency matrix of two bipartite graphs **A** and **B** results in a graph **C** with two bipartite sub-graphs. The $\stackrel{P}{=}$ notation is used to indicate that the adjacency matrix **C** has been permuted so that the two bipartite sub-graphs are more apparent.

The Kronecker product is defined as follows [65]

$$\mathbf{C} = \mathbf{A} \circledast \mathbf{B} = \begin{pmatrix} \mathbf{A}(1,1)\otimes\mathbf{B} & \mathbf{A}(1,2)\otimes\mathbf{B} & \ldots & \mathbf{A}(1,n_A)\otimes\mathbf{B} \\ \mathbf{A}(2,1)\otimes\mathbf{B} & \mathbf{A}(2,2)\otimes\mathbf{B} & \ldots & \mathbf{A}(2,n_A)\otimes\mathbf{B} \\ \vdots & \vdots & \ddots & \vdots \\ \mathbf{A}(m_A,1)\otimes\mathbf{B} & \mathbf{A}(m_A,2)\otimes\mathbf{B} & \ldots & \mathbf{A}(m_A,n_A)\otimes\mathbf{B} \end{pmatrix}$$

where

$$\mathbf{A} \in \mathbb{S}^{m_A \times n_A}$$
$$\mathbf{B} \in \mathbb{S}^{m_B \times n_B}$$
$$\mathbf{C} \in \mathbb{S}^{m_A m_B \times n_A n_B}$$

More explicitly, the Kronecker product can be written as

$$\mathbf{C}((i_A-1)m_B+i_B,(j_A-1)n_B+j_B) = \mathbf{A}(i_A,j_A)\otimes\mathbf{B}(i_B,j_B)$$

The element-wise multiply operation $\otimes$ can be user-defined so long as the resulting operation obeys the aforementioned rules on element-wise multiplication, such as the multiplicative annihilator. If element-wise multiplication and addition obey the conditions specified in in the previous sections, then the Kronecker product has many of the same desirable properties, such as associativity

$$(\mathbf{A} \circledast \mathbf{B}) \circledast \mathbf{C} = \mathbf{A} \circledast (\mathbf{B} \circledast \mathbf{C})$$

and element-wise distributivity over addition

$$\mathbf{A} \circledast (\mathbf{B} \oplus \mathbf{C}) = (\mathbf{A} \circledast \mathbf{B}) \oplus (\mathbf{A} \circledast \mathbf{C})$$

Finally, one unique feature of the Kronecker product is its relation to the matrix product. Specifically, the matrix product of two Kronecker products is equal to the Kronecker product of two matrix products

$$(\mathbf{A} \circledast \mathbf{B})(\mathbf{C} \circledast \mathbf{D}) = (\mathbf{AC}) \circledast (\mathbf{BD})$$

The relation of the Kronecker product to graphs is easily illustrated in the context of bipartite graphs. Bipartite graphs have two sets of vertices, and every vertex has an edge to the other set of vertices but no edges within its own set of vertices. The Kronecker product of such graphs was first looked at by Weischel [38], who observed that the Kronecker product of two bipartite graphs resulted in a new graph consisting of two bipartite subgraphs (see Figure 6.10).

6.5 Graph Algorithms and Diverse Semirings

The ability to change $\oplus$ and $\otimes$ operations allows different graph algorithms to be implemented using the same element-wise addition, element-wise multiplication, and matrix multiplication operations. Different semirings are best suited for certain classes of graph algorithms. The pattern of nonzero entries resulting from breadth-first-search illustrated in Figure 6.7 is generally preserved for various semirings. However, the nonzero values assigned to the edges and vertices can be very different and enable different graph algorithms.

Figure 6.11 illustrates performing a single-hop breadth-first-search using seven semirings (+.×, max.+, min.+, max.min, min.max, max.×, and min.×). For display convenience, some operator pairs that produce the same result *in this specific example* are stacked. In Figure 6.11, the starting vertex 4 is assigned a value of .5 and the edges to its vertex neighbors 1 and 3 are assigned values of .2 and .4. Empty values are assumed to be the corresponding 0 of the operator pair. In all cases, the pattern of nonzero entries of the results are the same. In each case, because there is only one path from vertex 4 to vertex 1 and from vertex 4 to vertex 3, the only effect of the $\oplus$ operator is to compare the nonzero output of the $\otimes$ operator with 0. Thus, the differences between the $\oplus$ operators have no impact in this specific example because for any values of a

$$a \oplus 0 = 0 \oplus a = a$$

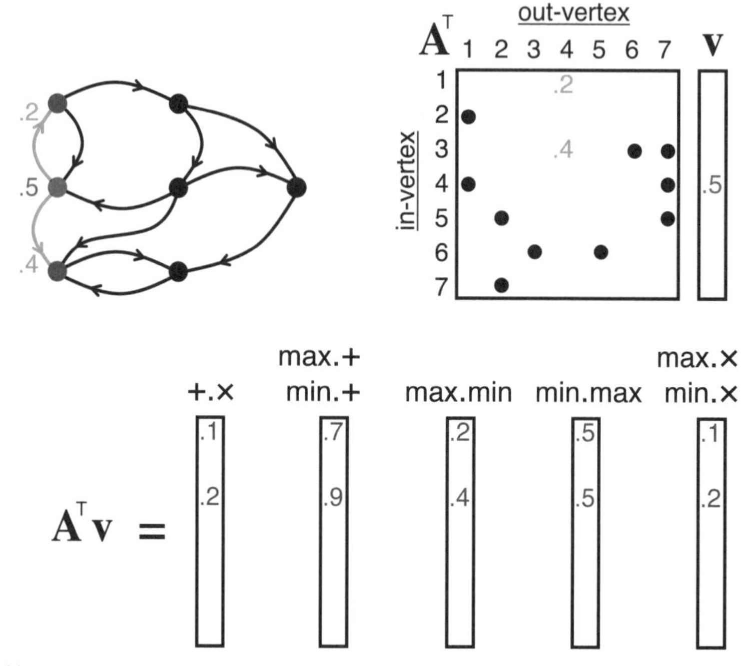

Figure 6.11
(Top left) one-hop breadth-first-search of a weighted graph starting at vertex 4 and traversing to vertices 1 and 3. (Top right) matrix representation of the weighted graph and vector representation of the starting vertex. (Bottom) matrix-vector multiplication of the adjacency matrix of a graph performs the equivalent operation. Different pairs of operations ⊕ and ⊗ produce different results. For display convenience, some operator pairs that produce the same values *in this specific example* are stacked.

The graph algorithm implications of different ⊕.⊗ operator pairs is more clearly seen in the two-hop breadth-first-search. Figure 6.12 illustrates graph traversal that starts at vertex 4, goes to vertices 1 and 3, and then continues on to vertices 2, 4, and 6. For simplicity, the additional edge weights are assigned values of .3. The first operator pair +.× provides the product of all the weights of all paths from the starting vertex to each ending vertex. The +.× semiring is valuable for determining the strengths of all the paths between the starting and ending vertices. In this example, there is only one path between the starting vertex and the ending vertices, so +.×, max.×, and min.× all produce the same results. If there were multiple paths between the start and end vertices, then ⊕ would operate on more than one nonzero value and the differences would be apparent. Specifically, +.× combines all paths while max.× and min.× select either the minimum or the maximum path. Thus, these different operator pairs represent different graph algorithms. One algorithm produces a value that combines all paths while the other algorithm produces a value that is derived from a single path.

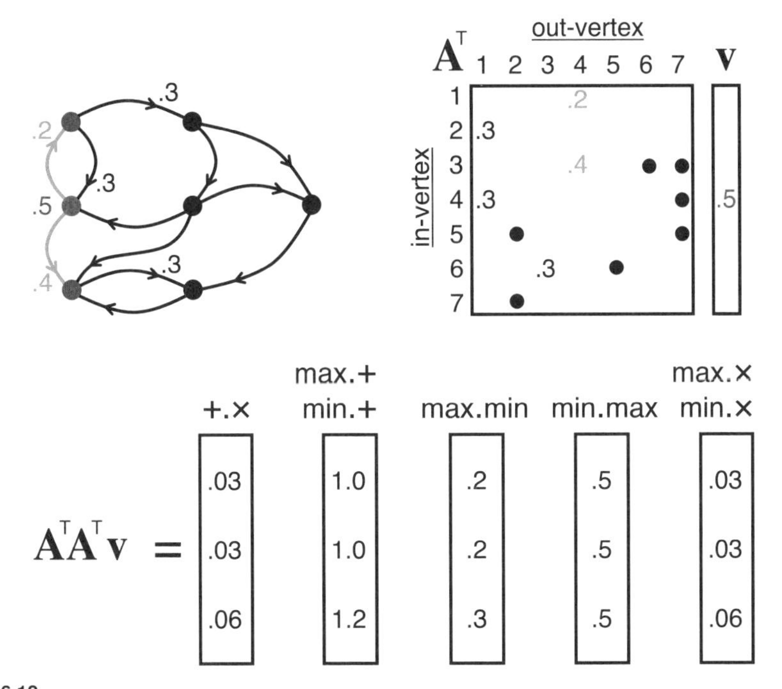

Figure 6.12
(Top left) two-hop breadth-first-search of a weighted graph starting at vertex 4 and traversing to vertices 1 and 3 and continuing on to vertices 2, 4, and 6. (Top right) matrix representation of the weighted graph and vector representation of the starting vertex. (Bottom) matrix-matrix-vector multiplication of the adjacency matrix of a graph performs the two-hop breadth-first-search operation. Different pairs of operations ⊕ and ⊗ produce different results. For display convenience, some operator pairs that produce the same result *in this specific example* are stacked.

A similar pattern can be seen among the other operator pairs. max.+ and min.+ compute the sum of the weights along each path from the starting vertex to each ending vertex and then select the largest (or smallest) weighted path. Likewise, max.min and min.max compute the minimum (or maximum) of the weights along each path and then select the largest (or smallest) weighted path.

A wide range of breadth-first-search weight computations can be performed via matrix multiplication with different operator pairs. A synopsis of the types of calculations illustrated in Figures 6.11 and 6.12 is as follows

+.× — Sum of products of weights along each path; computes the strength of all connections between the starting vertex and the ending vertices.

max.× — Maximum of products of weights along each path; computes the longest product of all of the connections between the starting vertex and the ending vertices.

min.× — Minimum of products of weights along each path; computes the shortest product of all of the connections between the starting vertex and the ending vertices.

max.+ — Maximum of sum of weights along each path; computes the longest sum of all of the connections between the starting vertex and the ending vertices.

min.+ — Minimum of sum of weights along each path; computes the shortest sum of all of the connections between the starting vertex and the ending vertices.

max.min — Maximum of minimum of weight along each path; computes the longest of all the shortest connections between the starting vertex and the ending vertices.

min.max — Minimum of maximum of weight along each path; computes the shortest of all the longest connections between the starting vertex and the ending vertices.

6.6 Conclusions, Exercises, and References

Matrices are a powerful tool for representing and manipulating graphs. Adjacency matrices represent directed-weighted-graphs with each row and column in the matrix representing a vertex and the values representing the weights of the edges. Incidence matrices represent directed-weighted-multi-hyper-graphs with each row representing an edge and each column representing a vertex. Perhaps the most important aspects of matrix-based graphs are the mathematical properties of commutativity, associativity, and distributivity. These properties allow a very small number of matrix operations to be used to construct a large number of graphs. These properties of the matrix are determined by the element-wise properties of addition and multiplication on the values in the matrix.

Exercises

Exercise 6.1 — Refer to Figure 6.1.

(a) What type of graph is this? Directed/undirected, weighted/unweighted, hyper, and/or multi?

(b) How is the type of graph apparent from the graph?

(c) How is the type of graph apparent from the adjacency matrix?

Exercise 6.2 — Refer to Figure 6.3.

(a) What type of graph is this? Directed/undirected, weighted/unweighted, hyper, and/or multi?

(b) How is the type of graph apparent from the graph?

(c) How is the type of graph apparent from the adjacency matrix?

Exercise 6.3 — How does Figure 6.4 illustrate the additive identity property?

Exercise 6.4 — How does Figure 6.5 illustrate the multiplicative annihilator property?

Exercise 6.5 — Pick non-commutative $\otimes$ and $\oplus$ functions on the real numbers and write an example for each, demonstrating that the functions are non-commutative.

Exercise 6.6 — Pick non-associative $\otimes$ and $\oplus$ functions on the real numbers and write an example for each, demonstrating that the functions are non-associative.

Exercise 6.7 — Pick non-distributive $\otimes$ and $\oplus$ functions on the real numbers and write an example demonstrating that the functions are non-distributive.

Exercise 6.8 — For semirings $\otimes$ distributes over $\oplus$

$$a \otimes (b \oplus c) = (a \otimes b) \oplus (a \otimes c)$$

For some values of a, b, c it is also true that $\oplus$ distributes over $\otimes$

$$a \oplus (b \otimes c) = (a \oplus b) \otimes (a \oplus c)$$

Show that these values are $a = 0$ or $a \oplus b \oplus c = 1$.

Exercise 6.9 — Take a graph from your own experience and write down its adjacency matrix (see Figure 6.1) and its incidence matrices (see Figure 6.2).

Exercise 6.10 — Using the adjacency matrix from the previous exercise, compute the nearest neighbors of a vertex by using vector matrix multiplication (see Figure 6.7).

Exercise 6.11 — Using the incidence matrices from the previous exercise, compute the adjacency matrix by using matrix multiplication (see Figure 6.8).

Exercise 6.12 — Compute the degree distribution of matrices **A**, **B**, and **C** in Figure 6.10.

Exercise 6.13 — For matrices, element-wise addition $\otimes$ distributes over element-wise multiplication $\oplus$

$$\mathbf{A} \otimes (\mathbf{B} \oplus \mathbf{C}) = (\mathbf{A} \otimes \mathbf{B}) \oplus (\mathbf{A} \otimes \mathbf{C})$$

For some values of $\mathbf{A}, \mathbf{B}, \mathbf{C}$ it is also true that $\oplus$ distributes over $\otimes$

$$\mathbf{A} \oplus (\mathbf{B} \otimes \mathbf{C}) = (\mathbf{A} \oplus \mathbf{B}) \otimes (\mathbf{A} \oplus \mathbf{C})$$

Show that these values are

$$\mathbf{A} = \mathbb{0} \quad \text{or} \quad \mathbf{A} \oplus \mathbf{B} \oplus \mathbf{C} = \mathbb{1}$$

where $\mathbb{0}$ is a matrix of all zeros and $\mathbb{1}$ is a matrix of all ones.

Exercise 6.14 — For matrices, element-wise addition $\otimes$ distributes over array multiplication $\oplus.\otimes$

$$\mathbf{A} \oplus.\otimes (\mathbf{B} \oplus \mathbf{C}) = (\mathbf{A} \oplus.\otimes \mathbf{B}) \oplus (\mathbf{A} \oplus.\otimes \mathbf{C})$$

For some values of $\mathbf{A}, \mathbf{B}, \mathbf{C}$ it is also true that $\oplus$ distributes over $\oplus.\otimes$

$$\mathbf{A} \oplus (\mathbf{B} \oplus.\otimes \mathbf{C}) = (\mathbf{A} \oplus \mathbf{B}) \oplus.\otimes (\mathbf{A} \oplus \mathbf{C})$$

Show that these values are

$$\mathbf{A} = \mathbb{0} \quad \text{or} \quad \mathbf{A} \oplus \mathbf{B} \oplus \mathbf{C} = \mathbb{I}$$

where $\mathbb{0}$ is a matrix of all zeros and $\mathbb{I}$ is the identity matrix.

References

[1] J. Kepner, P. Aaltonen, D. Bader, A. Buluç, F. Franchetti, J. Gilbert, D. Hutchison, M. Kumar, A. Lumsdaine, H. Meyerhenke, S. McMillan, J. Moreira, J. Owens, C. Yang, M. Zalewski, and T. Mattson, "Mathematical foundations of the GraphBLAS," in *High Performance Extreme Computing Conference (HPEC)*, IEEE, 2016.

[2] J. Kepner and J. Gilbert, *Graph Algorithms in the Language of Linear Algebra*. SIAM, 2011.

[3] G. Boole, *The Mathematical Analysis of Logic: Being an Essay Towards a Calculus of Deductive Reasoning*. Cambridge: MacMillan, Barclay, & MacMillan, 1847.

[4] G. Boole, *An Investigation of the Laws of Thought: On which are Founded the Mathematical Theories of Logic and Probabilities*. New York: Dover Publications, 1854.

[5] C. E. Shannon, "A symbolic analysis of relay and switching circuits," *Electrical Engineering*, vol. 57, no. 12, pp. 713–723, 1938.

[6] L. F. Menabrea and A. King, Countess of Lovelace, "Sketch of the analytical engine invented by charles babbage, esq," *Bibliotheque Universelle de Geneve*, vol. 82, 1842.

[7] P. Ludgate, "On a proposed analytical machine," *Scientific Proceedings of the Royal Dublin Society*, vol. 12, no. 9, pp. 77–91, 1909.

[8] T. y. Quevedo, "Ensayos sobre automatica-su definition. extension teorica de sus aplicaciones," *Revista de la Real Academia de Ciencias Exactas, Fisicas y Naturales*, vol. 12, pp. 391–418, 1913.

[9] V. Bush, "Instrumental analysis," *Bulletin of the American Mathematical Society*, vol. 42, no. 10, pp. 649–669, 1936.

[10] B. Randell, "From analytical engine to electronic digital computer: The contributions of Ludgate, Torres, and Bush," *Annals of the History of Computing*, vol. 4, no. 4, pp. 327–341, 1982.

[11] C. E. Shannon, "A mathematical theory of communication," *ACM SIGMOBILE Mobile Computing and Communications Review*, vol. 5, no. 1, pp. 3–55, 2001.

[12] B. Gold and C. M. Rader, *Digital Processing of Signals*. McGraw-Hill, 1969.

[13] B. Liskov and S. Zilles, "Programming with abstract data types," *ACM Sigplan Notices*, vol. 9, no. 4, pp. 50–59, 1974.

[14] B. Hendrickson and T. G. Kolda, "Graph partitioning models for parallel computing," *Parallel Computing*, vol. 26, no. 12, pp. 1519–1534, 2000.

[15] D. Ediger, K. Jiang, J. Riedy, and D. A. Bader, "Massive streaming data analytics: A case study with clustering coefficients," in *International Parallel and Distributed Processing Symposium Workshops (IPDPSW)*, pp. 1–8, IEEE, 2010.

[16] D. Ediger, J. Riedy, D. A. Bader, and H. Meyerhenke, "Tracking structure of streaming social networks," in *International Parallel and Distributed Processing Symposium Workshops (IPDPSW)*, pp. 1691–1699, IEEE, 2011.

[17] J. Riedy, D. A. Bader, and H. Meyerhenke, "Scalable multi-threaded community detection in social networks," in *International Parallel and Distributed Processing Symposium Workshops (IPDPSW)*, pp. 1619–1628, IEEE, 2012.

[18] J. Riedy and D. A. Bader, "Multithreaded community monitoring for massive streaming graph data," in *International Parallel and Distributed Processing Symposium Workshops (IPDPSW)*, pp. 1646–1655, IEEE, 2013.

[19] E. Bergamini, H. Meyerhenke, and C. L. Staudt, "Approximating betweenness centrality in large evolving networks," in *2015 Proceedings of the Seventeenth Workshop on Algorithm Engineering and Experiments (ALENEX)*, pp. 133–146, SIAM, 2015.

[20] E. Bergamini and H. Meyerhenke, "Approximating betweenness centrality in fully dynamic networks," *Internet Mathematics*, vol. 12, no. 5, pp. 281–314, 2016.

[21] D. Ediger, R. McColl, J. Riedy, and D. A. Bader, "Stinger: High performance data structure for streaming graphs," in *High Performance Extreme Computing Conference (HPEC)*, IEEE, 2012.

[22] D. Ediger and D. A. Bader, "Investigating graph algorithms in the BSP model on the Cray XMT," in *International Parallel and Distributed Processing Symposium Workshops (IPDPSW)*, pp. 1638–1645, IEEE, 2013.

[23] A. McLaughlin and D. A. Bader, "Revisiting edge and node parallelism for dynamic GPU graph analytics," in *International Parallel and Distributed Processing Symposium (IPDPS)*, pp. 1396–1406, IEEE, 2014.

[24] A. McLaughlin and D. A. Bader, "Scalable and high performance betweenness centrality on the GPU," in *Proceedings of the International Conference for High Performance Computing, Networking, Storage and Analysis*, pp. 572–583, IEEE Press, 2014.

[25] A. McLaughlin, J. Riedy, and D. A. Bader, "Optimizing energy consumption and parallel performance for static and dynamic betweenness centrality using GPUs," in *High Performance Extreme Computing Conference (HPEC)*, IEEE, 2014.

[26] C. L. Staudt and H. Meyerhenke, "Engineering parallel algorithms for community detection in massive networks," *IEEE Transactions on Parallel and Distributed Systems*, vol. 27, no. 1, pp. 171–184, 2016.

[27] A. Buluç and J. R. Gilbert, "Parallel sparse matrix-matrix multiplication and indexing: Implementation and experiments," *SIAM Journal on Scientific Computing*, vol. 34, no. 4, pp. C170–C191, 2012.

[28] A. Azad, G. Ballard, A. Buluc, J. Demmel, L. Grigori, O. Schwartz, S. Toledo, and S. Williams, "Exploiting multiple levels of parallelism in sparse matrix-matrix multiplication," *SIAM Journal on Scientific Computing*, vol. 38, no. 6, pp. C624–C651, 2016.

[29] G. Ballard, A. Buluc, J. Demmel, L. Grigori, B. Lipshitz, O. Schwartz, and S. Toledo, "Communication optimal parallel multiplication of sparse random matrices," in *Proceedings of the Twenty-Fifth Annual ACM Symposium on Parallelism in Algorithms and Architectures*, pp. 222–231, ACM, 2013.

[30] E. Solomonik, A. Buluc, and J. Demmel, "Minimizing communication in all-pairs shortest paths," in *International Parallel and Distributed Processing Symposium (IPDPS)*, pp. 548–559, IEEE, 2013.

[31] W. S. Song, J. Kepner, H. T. Nguyen, J. I. Kramer, V. Gleyzer, J. R. Mann, A. H. Horst, L. L. Retherford, R. A. Bond, N. T. Bliss, E. Robinson, S. Mohindra, and J. Mullen, "3-D graph processor," in *High Performance Embedded Computing Workshop (HPEC)*, MIT Lincoln Laboratory, 2010.

[32] W. S. Song, J. Kepner, V. Gleyzer, H. T. Nguyen, and J. I. Kramer, "Novel graph processor architecture," *Lincoln Laboratory Journal*, vol. 20, no. 1, pp. 92–104, 2013.

[33] W. S. Song, V. Gleyzer, A. Lomakin, and J. Kepner, "Novel graph processor architecture, prototype system, and results," in *High Performance Extreme Computing Conference (HPEC)*, IEEE, 2016.

[34] D. König, "Graphen und matrizen (graphs and matrices)," *Mat. Fiz. Lapok*, vol. 38, no. 1931, pp. 116–119, 1931.

[35] D. König, *Theorie der endlichen und unendlichen Graphen: Kombinatorische Topologie der Streckenkomplexe*, vol. 16. Akademische Verlagsgesellschaft mbh, 1936.

[36] F. Harary, *Graph Theory*. Reading, MA: Addison-Wesley, 1969.

[37] G. Sabidussi, "Graph multiplication," *Mathematische Zeitschrift*, vol. 72, no. 1, pp. 446–457, 1959.

[38] P. M. Weichsel, "The Kronecker product of graphs," *Proceedings of the American Mathematical Society*, vol. 13, no. 1, pp. 47–52, 1962.

[39] M. McAndrew, "On the product of directed graphs," *Proceedings of the American Mathematical Society*, vol. 14, no. 4, pp. 600–606, 1963.

[40] H. Teh and H. Yap, "Some construction problems of homogeneous graphs," *Bulletin of the Mathematical Society of Nanying University*, vol. 1964, pp. 164–196, 1964.

[41] A. Hoffman and M. McAndrew, "The polynomial of a directed graph," *Proceedings of the American Mathematical Society*, vol. 16, no. 2, pp. 303–309, 1965.

[42] F. Harary and C. A. Trauth, Jr, "Connectedness of products of two directed graphs," *SIAM Journal on Applied Mathematics*, vol. 14, no. 2, pp. 250–254, 1966.

[43] R. A. Brualdi, "Kronecker products of fully indecomposable matrices and of ultrastrong digraphs," *Journal of Combinatorial Theory*, vol. 2, no. 2, pp. 135–139, 1967.

[44] S. Parter, "The use of linear graphs in Gauss elimination," *SIAM Review*, vol. 3, no. 2, pp. 119–130, 1961.

[45] M. Fiedler, "Algebraic connectivity of graphs," *Czechoslovak Mathematical Journal*, vol. 23, no. 2, pp. 298–305, 1973.

[46] J. R. Gilbert, "Predicting structure in sparse matrix computations," *SIAM Journal on Matrix Analysis and Applications*, vol. 15, no. 1, pp. 62–79, 1994.

[47] T. Mattson, D. Bader, J. Berry, A. Buluc, J. Dongarra, C. Faloutsos, J. Feo, J. Gilbert, J. Gonzalez, B. Hendrickson, J. Kepner, C. Leiseron, A. Lumsdaine, D. Padua, S. Poole, S. Reinhardt, M. Stonebraker, S. Wallach, and A. Yoo, "Standards for graph algorithm primitives," in *High Performance Extreme Computing Conference (HPEC)*, IEEE, 2013.

[48] J. Kepner, "GraphBLAS special session," in *High Performance Extreme Computing Conference (HPEC)*, IEEE, 2013.

[49] T. Mattson, "Graph algorithms building blocks," in *International Parallel and Distributed Processing Symposium Workshops (IPDPSW)*, IEEE, 2014.

[50] T. Mattson, "GraphBLAS special session," in *High Performance Extreme Computing Conference (HPEC)*, IEEE, 2014.

[51] T. Mattson, "Graph algorithms building blocks," in *International Parallel and Distributed Processing Symposium Workshops (IPDPSW)*, IEEE, 2015.

[52] A. Buluç, "GraphBLAS special session," in *High Performance Extreme Computing Conference (HPEC)*, IEEE, 2015.

[53] T. Mattson, "Graph algorithms building blocks," in *International Parallel and Distributed Processing Symposium Workshops (IPDPSW)*, IEEE, 2016.

[54] A. Buluç and S. McMillan, "GraphBLAS special session," in *High Performance Extreme Computing Conference (HPEC)*, IEEE, 2016.

[55] A. Buluç and T. Mattson, "Graph algorithms building blocks," *International Parallel and Distributed Processing Symposium Workshops (IPDPSW)*, 2017.

[56] A. Buluç and J. R. Gilbert, "The combinatorial BLAS: Design, implementation, and applications," *The International Journal of High Performance Computing Applications*, vol. 25, no. 4, pp. 496–509, 2011.

[57] J. Kepner, W. Arcand, W. Bergeron, N. Bliss, R. Bond, C. Byun, G. Condon, K. Gregson, M. Hubbell, J. Kurz, A. McCabe, P. Michaleas, A. Prout, A. Reuther, A. Rosa, and C. Yee, "Dynamic Distributed Dimensional Data Model (D4M) database and computation system," in *2012 IEEE International Conference on Acoustics, Speech and Signal Processing (ICASSP)*, pp. 5349–5352, IEEE, 2012.

[58] K. Ekanadham, B. Horn, J. Jann, M. Kumar, J. Moreira, P. Pattnaik, M. Serrano, G. Tanase, and H. Yu, "Graph programming interface: Rationale and specification," tech. rep., IBM Research Report, RC25508 (WAT1411-052) November 19, 2014.

[59] D. Hutchison, J. Kepner, V. Gadepally, and A. Fuchs, "Graphulo implementation of server-side sparse matrix multiply in the Accumulo database," in *High Performance Extreme Computing Conference (HPEC)*, IEEE, 2015.

[60] M. J. Anderson, N. Sundaram, N. Satish, M. M. A. Patwary, T. L. Willke, and P. Dubey, "GraphPad: Optimized graph primitives for parallel and distributed platforms," in *International Parallel and Distributed Processing Symposium (IPDPSW)*, pp. 313–322, IEEE, 2016.

[61] Y. Wang, A. Davidson, Y. Pan, Y. Wu, A. Riffel, and J. D. Owens, "Gunrock: A high-performance graph processing library on the gpu," in *Proceedings of the 21st ACM SIGPLAN Symposium on Principles and Practice of Parallel Programming*, p. 11, ACM, 2016.

[62] P. Zhang, M. Zalewski, A. Lumsdaine, S. Misurda, and S. McMillan, "GBTL-CUDA: Graph algorithms and primitives for GPUs," in *International Parallel and Distributed Processing Symposium Workshops (IPDPSW)*, pp. 912–920, IEEE, 2016.

[63] D. Chakrabarti, Y. Zhan, and C. Faloutsos, "R-MAT: A recursive model for graph mining," in *Proceedings of the 2004 SIAM International Conference on Data Mining*, pp. 442–446, SIAM, 2004.

[64] J. Leskovec, D. Chakrabarti, J. Kleinberg, and C. Faloutsos, "Realistic, mathematically tractable graph generation and evolution, using Kronecker multiplication," in *European Conference on Principles of Data Mining and Knowledge Discovery*, pp. 133–145, Springer, 2005.

[65] C. F. Van Loan, "The ubiquitous Kronecker product," *Journal of Computational and Applied Mathematics*, vol. 123, no. 1, pp. 85–100, 2000.

7 Graph Analysis and Machine Learning Systems

Summary

Machine learning systems encompass the entire process of parsing, ingesting, querying, and analyzing data to make predictions. The ability of a system to make good predictions is bound by the quantity and quality of the data delivered by the data preparation steps to the modeling and analysis steps. In real-world machine learning systems, the data preparation typically consumes 90% of the effort and cost. Associative arrays provide a natural mathematical framework for representing data through all the steps of a machine learning system. Key steps in a machine learning system include graph construction, graph traversal, and identification of unusual vertices via eigenvalues, singular values, and other metrics (such as PageRank). The benefits of associative arrays for machine learning systems are not limited to data preparation. Associative arrays also provide a natural framework for modeling decision-making data, such as the network weights used to describe a deep neural network. This chapter describes in detail several examples of associative array-based algorithms for implementing these steps in machine learning systems.

7.1 Introduction

Machine learning describes the broad area of analysis and classification of data to create models for making predictions and has been a foundation of artificial intelligence since its inception [2–9]. Typical applications include categorizing document queries so that new queries provide more meaningful results, looking for correlations between patient outcomes in medical records, and detecting anomalies in computer networks. A complete machine learning system addresses all the tasks required to parse, ingest, query, and analyze data to make predictions (see Figure 7.1). In a machine learning system, machine learning algorithms are most commonly utilized in the data preparation that typically includes the analysis of the data and the resulting predictions. Data preparation processing in a machine learning system typically includes all steps from acquiring the data to getting the data ready for modeling and analysis.

The machine learning algorithms employed in the modeling and analysis steps in machine learning systems is well described in the academic literature. A wide range of algorithms are used to model the data and make predictions. These algorithms include decision trees [10], artificial neural networks [11], support vector machines [12], Bayesian networks [13], genetic algorithms [14], and clustering [15]. The ability of these analyses to make good predictions is limited by the quantity and quality of the data processed in the data preparation steps. Most machine learning algorithms rely on statistical techniques to estimate the values of model parameters. The ability to make statistically accurate model parameter estimates is directly related to the quantity of data. Furthermore, a machine learning model only can work if it accurately represents the data. Noise, errors, and clutter in the data that are outside the model statistical distributions need to be addressed in the data preparation. In addition, the more non-modeled data can be eliminated in the data preparation, the more the algorithm can use the data to focus on the problem of interest.

For real machine learning systems, the data preparation can dominate the processing and may consume 90% of the effort and cost. The general recognition that a significant fraction of raw data records have known flaws that need to be addressed by data preparation has led many professionals to spend the majority of their time cleaning up data [16]. The challenge of developing tools and technologies to address data cleaning is significantly aided by placing data into a common mathematical framework so that such tools can operate independent of the specific application domain [17]. The variety of data preparation approaches for enabling machine learning systems includes data representation, graph creation, graph traversal, and graph statistics. In many cases, well-designed data preparation can significantly reduce the complexity of the machine learning algorithm and allow a simpler algorithm to be used. The remainder of this chapter will be focused on how associative arrays can be used to implement these approaches.

7.2 Data Representation

An intuitive way to represent large unstructured data sets such as documents or social networks is through a graph representation. In such a representation, graph vertices can represent users or events, and edges can represent the relationship between vertices. Many recent efforts have looked at the mapping between graphs and linear algebra. In such a mapping, graphs are often represented as sparse arrays, such as associative arrays or sparse matrices using a graph schema.

This chapter looks at common classes of graph analytics that are used in machine learning applications and provides an initial set of graph algorithms recast as associative array operations. Further, these algorithms have been described so as to work on alternate semiring structures that replace traditional arithmetic addition and multiplication with more general binary functions denoted $\oplus$ and $\otimes$. This flexibility allows a wide variety of graph

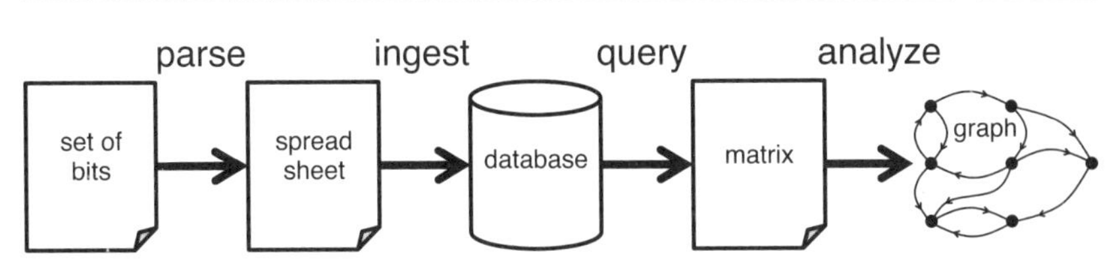

Figure 7.1
A machine learning system covers the whole process of parsing, ingesting, querying, and analyzing data to make predictions. Each part of the process uses data representations that can be encompassed by associative arrays. Data preparation processing typically covers all steps from acquiring the data to conditioning the data for analysis. Modeling usually includes all the analysis of the conditioned data.

algorithms to be represented by using the aforementioned building blocks of element-wise addition, element-wise multiplication, and array multiplication.

Associative arrays are used to describe the relationship between multidimensional entities using numeric/string keys and numeric/string values. Associative arrays provide a generalization of sparse matrices. Formally, an associative array $\mathbf{A}$ is a map from sets of keys $K_1 \times K_2$ to a value set V with a semiring structure

$$\mathbf{A} : K_1 \times K_2 \to V,$$

where $(V, \oplus, \otimes, 0, 1)$ is a semiring with addition operator $\oplus$, multiplication operator $\otimes$, additive-identity/multiplicative-annihilator 0, and multiplicative-identity 1. In the subsequent algorithms, V is the set of real numbers, $\oplus$ is standard arithmetic addition $+$ and $\otimes$ is standard arithmetic multiplication $\times$. The corresponding array operations are element-wise addition

$$\mathbf{C} = \mathbf{A} \oplus \mathbf{B} = \mathbf{A} + \mathbf{B}$$

or more explicitly

$$\mathbf{C}(k_1, k_2) = \mathbf{A}(k_1, k_2) + \mathbf{B}(k_1, k_2)$$

element-wise multiplication

$$\mathbf{C} = \mathbf{A} \otimes \mathbf{B}$$

or more explicitly

$$\mathbf{C}(k_1, k_2) = \mathbf{A}(k_1, k_2) \times \mathbf{B}(k_1, k_2)$$

scalar multiplication

$$\mathbf{C} = a \otimes \mathbf{B} = a\mathbf{B}$$

or more explicitly

$$\mathbf{C}(k_1, k_2) = a\mathbf{B}(k_1, k_2)$$

and array multiplication

$$\mathbf{C} = \mathbf{A} \oplus.\otimes \mathbf{B} = \mathbf{AB}$$

or more explicitly

$$\mathbf{C}(k_1,k_2) = \sum_k \mathbf{A}(k_1,k) \times \mathbf{B}(k,k_2)$$

As a data structure, associative arrays return a value, given a key tuple, and constitute a function between a set of key pairs and a value space. In practice, every associative array can be created from an empty associative array by adding values. With this definition, it is assumed that only a finite number of key tuples will have nonzero values and that all other tuples will have a default value of 0 (the additive-identity/multiplicative-annihilator). Further, the associative array mapping should support operations that resemble operations on ordinary vectors and matrices, such as matrix multiplication. In practice, associative arrays support a variety of linear algebraic operations, such as summation, union, intersection, and multiplication. Summation of two associative arrays, for example, that do not have any common row or column key performs a union of their underlying nonzero keys.

Database tables are exactly described using the mathematics of associative arrays [18]. In the D4M schema [19], a table in a database is an associative array. In this context, the primary difference between associative array data structures and sparse matrix data structures is that associative array entries always carry their global row and column labels while sparse matrices do not. Another difference between associative arrays and sparse matrices is that sparse matrices explicitly carry empty rows or columns while associative arrays do not. However, for practical purposes, associative arrays are often implemented using sparse matrices that have been augmented with additional metadata on the rows, columns, and values.

7.3 Graph Construction

In many machine learning applications, representing the data as a graph is the primary tool for categorizing data. For example, in social media it is desirable to know who a particular person's "friends" are. These friends are learned from the data by constructing the appropriate graph. Of course, there are often many ways to represent data as a graph, and determining the correct graph is a key challenge.

Constructing the right graph requires putting data into a common frame of reference so that similar entities can be compared. The associative arrays described in the previous subsection can be used to represent a variety of graphs using many different databases. A few commonly used graph schemas [18] are discussed below.

Adjacency Array

In this schema, data are organized as a graph adjacency array that can represent directed or undirected weighted graphs. Rows and columns of the adjacency array represent vertices, and values represent weighted edges between vertices. Adjacency arrays provide a great deal of functionality and are one of the more common ways to express graphs through matrices. For a graph with m edges, the adjacency array $\mathbf{A}$ is defined as

$$\mathbf{A}(k_1, k_2) = \begin{cases} v & \text{there is an edge from vertex } k_1 \text{ to } k_2 \\ 0 & \text{there is no edge from vertex } k_1 \text{ to } k_2 \end{cases}$$

where $v \in V$ is the weight of the edge from k_1 to k_2. The number of nonzero entries in $\mathbf{A}$ is m.

Incidence Array

The incidence array is capable of representing more complex graphs than an adjacency array. Specifically, the incidence array representation of a graph can represent both multi-graphs (graphs with multiple edges between the same set of vertices) and hyper-graphs (graphs with edges connecting more than two vertices). In the incidence array representation, array rows correspond to edges, and array columns represent vertices, with nonzero values in a row indicating vertices associated with the edge. The value at a particular row-column pair represents the edge weight. There are many representations for the incidence array, and a common format is described below.

Suppose a graph is a directed-weighted-multi-hyper-graph with m edges taken from an edge set K and vertex set $K_{\text{out}} \cup K_{\text{in}}$, where K_{out} is the set of vertices with outgoing edges and K_{in} is the set of vertices with incoming edges.

The *out-vertex* incidence array of the graph is denoted by

$$\mathbf{E}_{\text{out}} : K \times K_{\text{out}} \to V$$

and satisfies the condition that for all edges $k \in K$ and vertices $k_{\text{out}} \in K_{\text{out}}$

$$\mathbf{E}_{\text{out}}(k, k_{\text{out}}) \neq 0$$

if and only if edge k is directed outward from vertex k_{out}. The number of rows with nonzero entries in $\mathbf{E}_{\text{out}}$ is equal to the number of edges in the graph m. Likewise, the number of columns with nonzero entries in $\mathbf{E}_{\text{out}}$ is equal to the number of vertices with out-edges in the graph, n_{out}.

The *in-vertex* incidence array of the graph is denoted by

$$\mathbf{E}_{\text{in}} : K \times K_{\text{in}} \to V$$

and satisfies the condition that for all edges $k \in K$ and vertices $k_{\text{in}} \in K_{\text{in}}$

$$\mathbf{E}_{\text{in}}(k, k_{\text{in}}) \neq 0$$

if and only if edge k is directed inward to vertex k_{in}. The number of rows with nonzero entries in $\mathbf{E}_{\text{in}}$ is equal to the number of edges in the graph m. The number of columns with nonzero entries in $\mathbf{E}_{\text{in}}$ is equal to the number of vertices with in-edges in the graph n_{in}.

The incidence array and the adjacency array are often linked by the formula

$$\mathbf{A} = \mathbf{E}_{\text{out}}^{\mathsf{T}} \mathbf{E}_{\text{in}}$$

The above equations provide useful mechanisms for rewriting algorithms by using either the incidence array or the adjacency array.

D4M Schema

The D4M 2.0 Schema [19] described in Chapter 3.5 (see Figure 3.3) provides a four associative array solution $\{\mathbf{T}_{\text{edge}}, \mathbf{T}_{\text{edge}}^{\mathsf{T}}, \mathbf{T}_{\text{deg}}, \mathbf{T}_{\text{raw}}\}$ that can be used to represent diverse data as four database tables. The edge tables, $\mathbf{T}_{\text{edge}}$ and $\mathbf{T}_{\text{edge}}^{\mathsf{T}}$, contain the full semantic information of the data set using an adjacency array or incidence array representation. Storing both the array and its transpose allows rapid access to any row or any column for both row-oriented and column-oriented databases. From the schema described in [19], a dense database can be converted to a sparse representation by exploding each data entry into an associative array where each unique column-value pair is a column. The $\mathbf{T}_{\text{deg}}$ array maintains a count of the degrees of each of the columns of $\mathbf{T}_{\text{edge}}$. $\mathbf{T}_{\text{raw}}$ is used to store the raw data. A more thorough description of the schema is provided in [19]. Once the data are in sparse array form, the full machinery of linear algebraic graph processing and machine learning can be applied. Linear algebraic operations applied on associative arrays can produce many useful results. For example, addition of two associative arrays represents a union, and the multiplication of two associative arrays represents a correlation. In the D4M schema, either an adjacency array or incidence array representation of graph data can be operated on efficiently in real systems.

7.4 Adjacency Array Graph Traversal

A common machine learning use of graphs in social media applications is to suggest new friends to a user by analyzing their existing interests and friendships. Breadth-first search finds the nearest neighbors of vertices in a graph and is widely used for such applications. Array multiplication can be used to perform breadth-first search and many other graph traversal operations. This flexibility is one of the principal benefits of the array-based approach as it provides many different solutions to choose from to solve the same graph traversal problem. The ability to create many different algorithms to solve the same graph problem is a direct result of the linear system properties of associative arrays. Given an adjacency array of a directed graph $\mathbf{A}$ and a diagonal array of source vertices $\mathbf{V}$, the neighboring vertices of $\mathbf{V}$ can be found via many different combinations of $\mathbf{A}$ and $\mathbf{V}$.

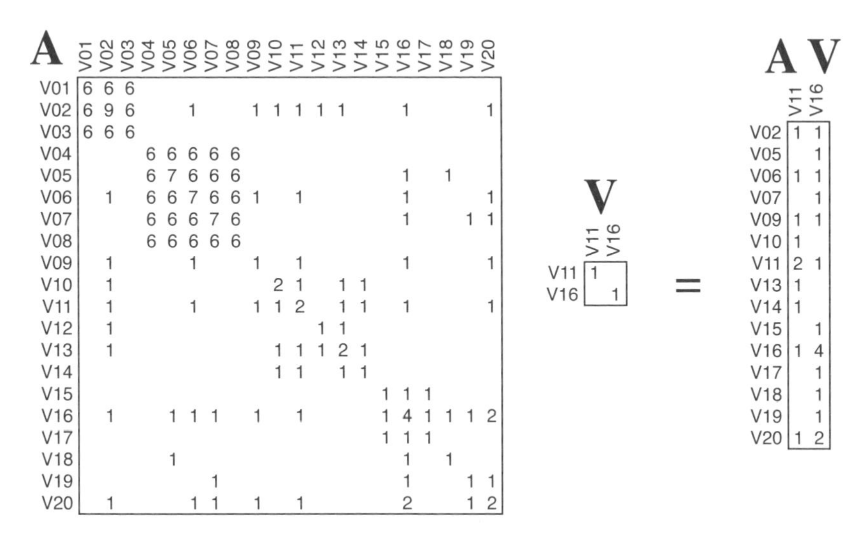

Figure 7.2
Breadth-first search of the graph represented by the adjacency array **A** of the graph in the painting shown in Figure 5.11. The values in **V** indicate the number of times a given pair of vertices share the same edge. The starting vertices of the search V11 and V16 are stored in the array **V**. The neighbors of the vertices in **V** are computed via the array multiplication **AV**.

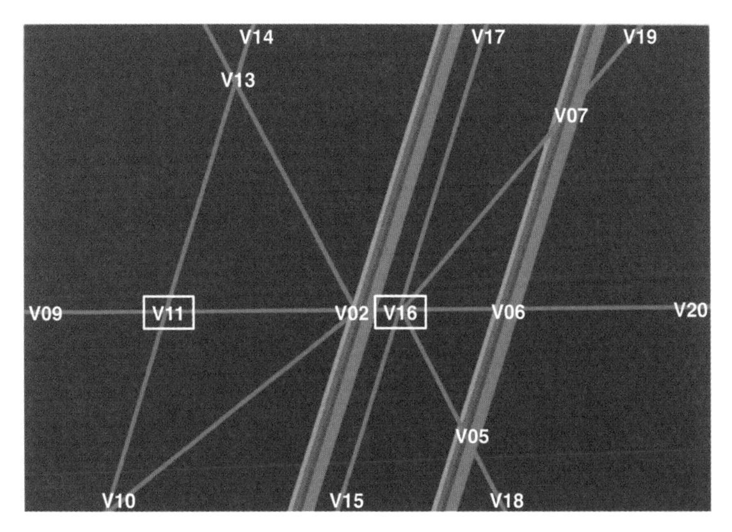

Figure 7.3
Result of breadth-first search starting at vertices V11 and V16.

The vertices of edges *ending* with *row* keys of $\mathbf{V}$ can by obtained via

$$\mathbf{AV}$$

The vertices of edges *starting* with *row* keys of $\mathbf{V}$ can by obtained via

$$\mathbf{A}^\mathsf{T}\mathbf{V} \quad \text{or} \quad \mathbf{V}^\mathsf{T}\mathbf{A}$$

The vertices of edges *ending* with the *column* keys of $\mathbf{V}$ can by obtained via

$$\mathbf{AV}^\mathsf{T} \quad \text{or} \quad \mathbf{VA}^\mathsf{T}$$

The vertices of edges *starting* with *column* keys of $\mathbf{V}$ can by obtained via

$$\mathbf{VA}$$

Furthermore, if the graph is undirected and the source array is symmetric so that

$$\mathbf{A} = \mathbf{A}^\mathsf{T} \quad \text{or} \quad \mathbf{V} = \mathbf{V}^\mathsf{T}$$

then all of the following graph traversal expressions can be used to obtain the neighbors of the vertices in $\mathbf{V}$

$$\mathbf{AV} \quad \text{or} \quad \mathbf{A}^\mathsf{T}\mathbf{V} \quad \text{or} \quad \mathbf{V}^\mathsf{T}\mathbf{A} \quad \text{or} \quad \mathbf{AV}^\mathsf{T} \quad \text{or} \quad \mathbf{VA}^\mathsf{T} \quad \text{or} \quad \mathbf{VA}$$

The above expressions allow different operations to be swapped in order to optimize a machine learning system.

Figures 7.2 and 7.3 illustrate breadth-first search of the undirected graph represented by the adjacency array $\mathbf{A}$ from Figure 5.11. The starting vertices for the search are stored in the array $\mathbf{V}$. The neighbors of the vertices in $\mathbf{V}$ are found by an array multiplication $\mathbf{AV}$. The values in $\mathbf{A}$ show how many times a given pair of vertices have the same edge. It is worth recalling that the adjacency array $\mathbf{A}$ represents a graph with *hyper-edges* that connect more than one vertex with the same edge. An undirected hyper-edge that connects 3 vertices looks the same in an adjacency array as 6 standard directed edges and 3 self-edges or 9 edges total. Likewise, an undirected hyper-edge that connects 5 vertices looks the same in an adjacency array as 20 standard directed edges and 5 self-edges or 25 edges total.

7.5 Incidence Array Graph Traversal

Graph traversal can also be performed with array multiplication by using incidence arrays $\mathbf{E}_{\text{out}}$ and $\mathbf{E}_{\text{in}}$. Using incidence arrays for graph traversal is a two-step process. The first step is obtaining the set of edges that contain a set of vertices. The second step is to use the set of edges to find the neighbor vertices associated with the first set of vertices.

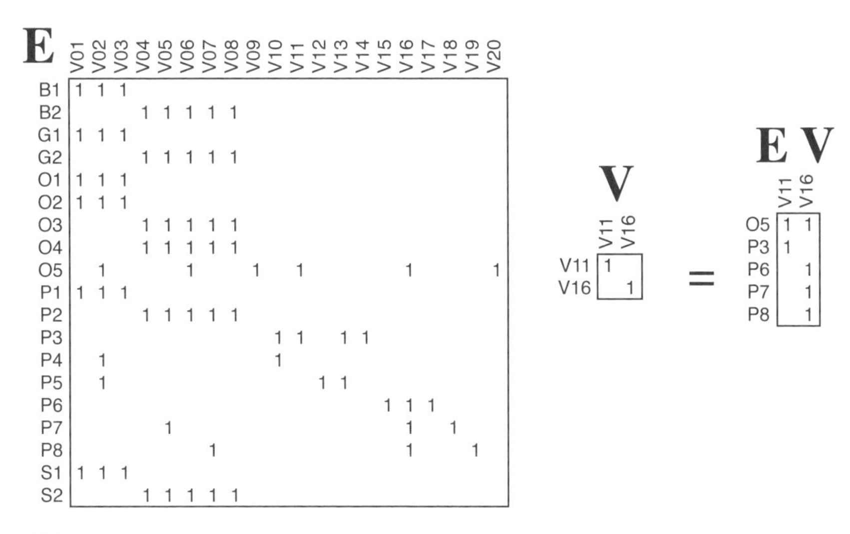

Figure 7.4
First half of a breadth-first search of a graph given by the incidence array **E**. The values in **E** indicate that the vertex key in the column shares the edge key labeled in the row. The starting vertices for the search V11 and V16 are stored in the array **V**. The edges of the vertices in **V** are computed using array multiplication **EV**.

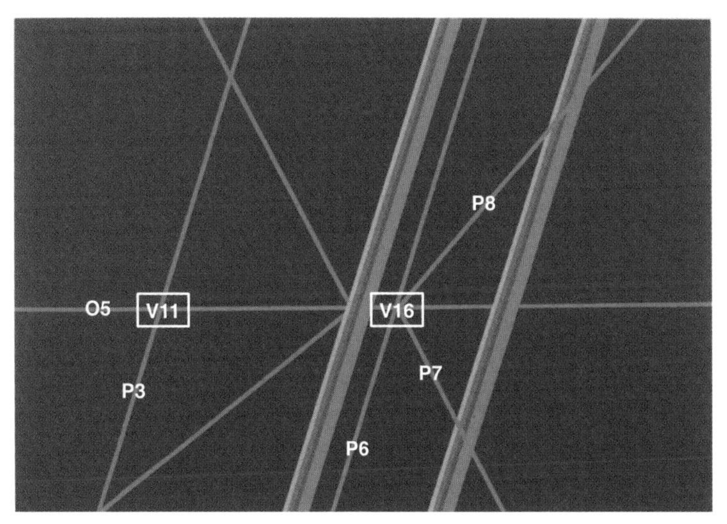

Figure 7.5
Result of the first half of a breadth-first search of an incidence array identifying the edges connected to vertices V11 and V16.

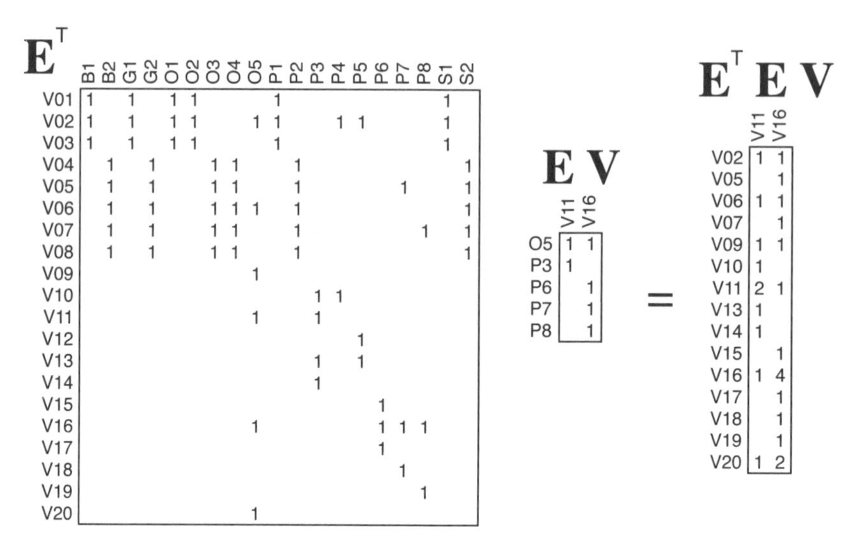

Figure 7.6
Second half of a breadth-first search of the graph shown by the incidence array **E**. The values in **EV** are the edges that contain the starting vertices V11 and V16. The vertex neighbors of the vertices in **V** are computed using the array multiplication $\mathbf{E}^\mathsf{T}\mathbf{E}\mathbf{V}$.

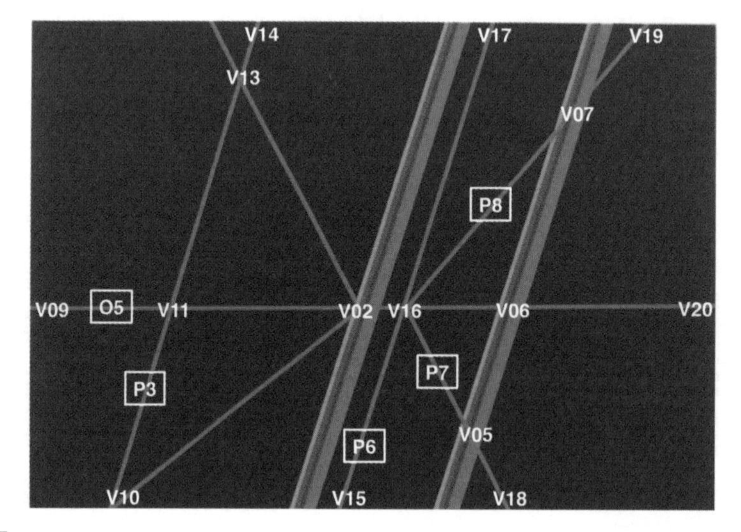

Figure 7.7
Result of the second half of the breadth-first search of the incidence array identifying the vertices connected to the set of edges {O5, P3, P6, P7, P8}.

Given a diagonal array of source vertices $\mathbf{V}$, the edges *ending* with *row* keys of $\mathbf{V}$ can by obtained via

$$\mathbf{E}_{\text{in}}\mathbf{V} \quad \text{or} \quad \mathbf{V}^{\mathsf{T}}\mathbf{E}_{\text{in}}^{\mathsf{T}}$$

The edges *starting* with *row* keys of $\mathbf{V}$ can by obtained via

$$\mathbf{E}_{\text{out}}\mathbf{V} \quad \text{or} \quad \mathbf{V}^{\mathsf{T}}\mathbf{E}_{\text{out}}^{\mathsf{T}}$$

The edges *ending* with the *column* keys of $\mathbf{V}$ can by obtained via

$$\mathbf{E}_{\text{in}}\mathbf{V}^{\mathsf{T}} \quad \text{or} \quad \mathbf{V}\mathbf{E}_{\text{in}}^{\mathsf{T}}$$

The edges *starting* with *column* keys of $\mathbf{V}$ can by obtained via

$$\mathbf{E}_{\text{out}}\mathbf{V}^{\mathsf{T}} \quad \text{or} \quad \mathbf{V}\mathbf{E}_{\text{out}}^{\mathsf{T}}$$

Furthermore, if the graph is undirected, then

$$\mathbf{E}_{\text{out}} = \mathbf{E}_{\text{in}} = \mathbf{E}$$

and the resulting source array is symmetric, which implies that

$$\mathbf{V} = \mathbf{V}^{\mathsf{T}}$$

then all of the following expressions are equivalent and can be used to find the neighbors of $\mathbf{V}$

$$\mathbf{E}^{\mathsf{T}}\mathbf{E}\mathbf{V} \quad \text{or} \quad (\mathbf{E}\mathbf{V})^{\mathsf{T}}\mathbf{E} \quad \text{or} \quad \mathbf{V}^{\mathsf{T}}\mathbf{E}^{\mathsf{T}}\mathbf{E}$$

The ability to exchange the above expressions can be very useful for optimizing the performance of a machine learning system.

Figures 7.4 and 7.5 show the initial part of a breadth-first search on an undirected graph described by the incidence array $\mathbf{E}$ from Figure 5.10. The values in $\mathbf{E}$ indicate whether or not a given vertex shares a given edge. The starting points of the search are held in the array $\mathbf{V}$. The edges of the vertices in $\mathbf{V}$ can be computed via the array multiplication $\mathbf{E}\mathbf{V}$.

After the set of edges containing a set of vertices has been obtained, these edges can be used to obtain the neighbor vertices, thus completing the graph traversal. The vertices of edges *ending* with *row* keys of $\mathbf{V}$ can by obtained via

$$\mathbf{E}_{\text{out}}^{\mathsf{T}}\mathbf{E}_{\text{in}}\mathbf{V} \quad \text{or} \quad \mathbf{V}^{\mathsf{T}}\mathbf{E}_{\text{in}}^{\mathsf{T}}\mathbf{E}_{\text{out}}$$

The vertices of edges *starting* with *row* keys of $\mathbf{V}$ can by obtained via

$$\mathbf{E}_{\text{in}}^{\mathsf{T}}\mathbf{E}_{\text{out}}\mathbf{V} \quad \text{or} \quad \mathbf{V}^{\mathsf{T}}\mathbf{E}_{\text{out}}^{\mathsf{T}}\mathbf{E}_{\text{in}}$$

The vertices of edges *ending* with the *column* keys of $\mathbf{V}$ can by obtained via

$$\mathbf{E}_{\text{out}}^{\mathsf{T}}\mathbf{E}_{\text{in}}\mathbf{V}^{\mathsf{T}} \quad \text{or} \quad \mathbf{V}\mathbf{E}_{\text{in}}^{\mathsf{T}}\mathbf{E}_{\text{out}}$$

The vertices of edges *starting* with *column* keys of $\mathbf{V}$ can by obtained via

$$\mathbf{E}_{\text{in}}^{\mathsf{T}}\mathbf{E}_{\text{out}}\mathbf{V}^{\mathsf{T}} \quad \text{or} \quad \mathbf{V}\mathbf{E}_{\text{out}}^{\mathsf{T}}\mathbf{E}_{\text{in}}$$

Additional expressions can be obtained by exploiting the transpose identity

$$\mathbf{A}^{\mathsf{T}}\mathbf{B}^{\mathsf{T}} = (\mathbf{BA})^{\mathsf{T}}$$

Likewise, because array multiplication is associative

$$\mathbf{A}(\mathbf{BC}) = (\mathbf{AB})\mathbf{C}$$

there are many other possible expressions for the above graph traversals. Furthermore, if the graph is undirected and the source array is symmetric, then the above graph traversal expressions are equivalent to following expressions

$$\mathbf{E}^{\mathsf{T}}\mathbf{E}\mathbf{V} \quad \text{or} \quad (\mathbf{E}\mathbf{V})^{\mathsf{T}}\mathbf{E} \quad \text{or} \quad \mathbf{V}^{\mathsf{T}}\mathbf{E}^{\mathsf{T}}\mathbf{E}$$

The above relations can be used to optimize a machine learning system by changing the order of operations.

Figures 7.6 and 7.7 illustrate the second half of the breadth-first-search of an undirected graph depicted by the incidence array $\mathbf{E}$ from Figure 5.10. The starting vertices for the search are stored in the array $\mathbf{V}$. The edges of the vertices in $\mathbf{V}$ are found by the array multiplication $\mathbf{EV}$. The neighbors of the vertices in $\mathbf{V}$ are computed $\mathbf{E}^{\mathsf{T}}\mathbf{EV}$.

7.6 Vertex Degree Centrality

A graph of relationships among vertices can be used to determine those vertices that are most important or most central to the graph. Algorithms that are used to determine the importance of different nodes are referred to as centrality metrics. In machine learning applications, centrality metrics can be used to categorize information and spot new trends. For example, if the most important vertices in a graph change over time, then this change can indicate an important new trend in the data or a new source of error in the data.

A simple way to categorize data in a graph is by identifying the most popular vertices (supernodes) in the graph and the least popular vertices (leaf nodes). Of the many centrality metrics, the simplest is degree centrality, which assumes the importance of a vertex is proportional to the number of edges going into or coming out of a vertex.

A	V01	V02	V03	V04	V05	V06	V07	V08	V09	V10	V11	V12	V13	V14	V15	V16	V17	V18	V19	V20
V01	6	6	6																	
V02	6	9	6			1			1	1	1	1	1			1				1
V03	6	6	6																	
V04				6	6	6	6	6												
V05				6	7	6	6	6								1		1		
V06		1		6	6	7	6	6	1		1					1				1
V07				6	6	6	7	6								1			1	1
V08				6	6	6	6	6												
V09		1				1			1		1					1				1
V10		1								2	1		1	1						
V11		1				1			1	1	2		1	1		1				1
V12		1										1	1							
V13		1								1	1	1	2	1						
V14										1	1		1	1						
V15															1	1	1			
V16		1			1	1	1		1		1				1	4	1	1	1	2
V17															1	1	1			
V18					1											1		1		
V19							1									1			1	1
V20		1				1	1		1		1					2			1	2

$\mathbb{1}$	Deg
V01	1
V02	1
V03	1
V04	1
V05	1
V06	1
V07	1
V08	1
V09	1
V10	1
V11	1
V12	1
V13	1
V14	1
V15	1
V16	1
V17	1
V18	1
V19	1
V20	1

=

$\mathbf{A}\mathbb{1}$	Deg
V01	18
V02	29
V03	18
V04	30
V05	33
V06	36
V07	34
V08	30
V09	6
V10	6
V11	10
V12	3
V13	7
V14	4
V15	3
V16	16
V17	3
V18	3
V19	4
V20	10

Figure 7.8
Vertex degree centrality of the graph depicted by the adjacency array **A** from the painting shown in Figure 5.11. The values in **A** indicate how many pairs of vertices share the same edge. The degree centrality of the vertices is computed by the array multiplication $\mathbf{A}\mathbb{1}$, where $\mathbb{1}$ is a vector of all 1's.

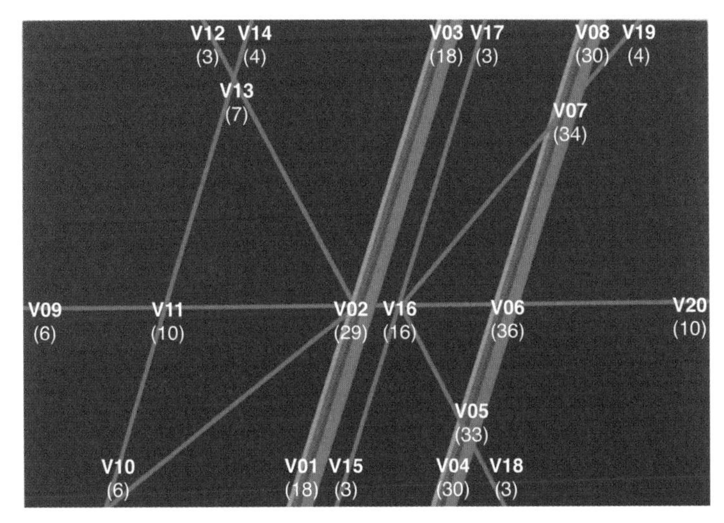

Figure 7.9
Output of the vertex degree centrality operation depicted in Figure 7.8. The degree centrality of each vertex is listed in parentheses below the vertex label. Vertex V06 has the highest degree centrality because it is linked to the most vertices via the most edges.

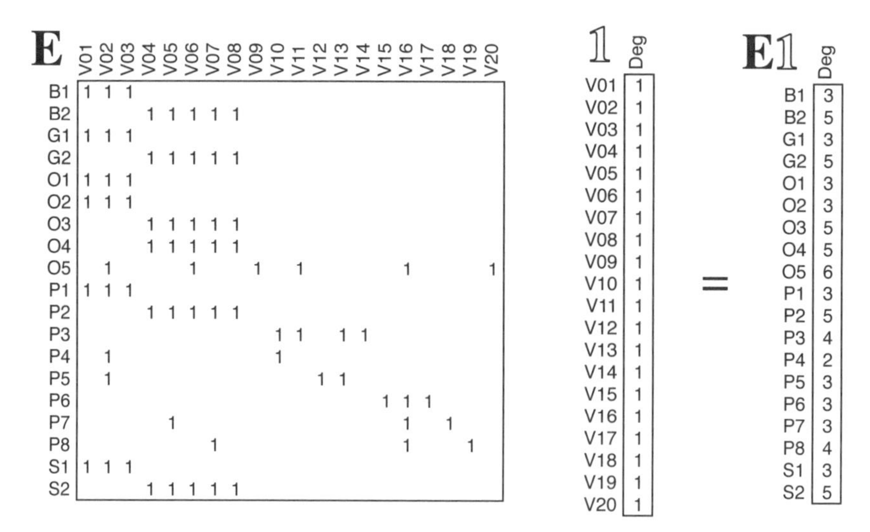

E	V01	V02	V03	V04	V05	V06	V07	V08	V09	V10	V11	V12	V13	V14	V15	V16	V17	V18	V19	V20
B1	1	1	1																	
B2				1	1	1	1	1												
G1	1	1	1																	
G2				1	1	1	1	1												
O1	1	1	1																	
O2	1	1	1																	
O3				1	1	1	1	1												
O4				1	1	1	1	1												
O5		1				1			1		1					1				1
P1	1	1	1																	
P2				1	1	1	1	1												
P3										1	1		1	1						
P4		1								1										
P5		1										1	1							
P6															1	1	1			
P7					1											1		1		
P8							1									1			1	
S1	1	1	1																	
S2				1	1	1	1	1												

$\mathbb{1}$	Deg
V01	1
V02	1
V03	1
V04	1
V05	1
V06	1
V07	1
V08	1
V09	1
V10	1
V11	1
V12	1
V13	1
V14	1
V15	1
V16	1
V17	1
V18	1
V19	1
V20	1

=

$\mathbf{E}\mathbb{1}$	Deg
B1	3
B2	5
G1	3
G2	5
O1	3
O2	3
O3	5
O4	5
O5	6
P1	3
P2	5
P3	4
P4	2
P5	3
P6	3
P7	3
P8	4
S1	3
S2	5

Figure 7.10
Edge degree centrality of the graph whose incidence array is **E**. The edge degree centrality of the vertices is calculated via the array multiplication $\mathbf{E}\mathbb{1}$, where $\mathbb{1}$ is a vector of all 1's.

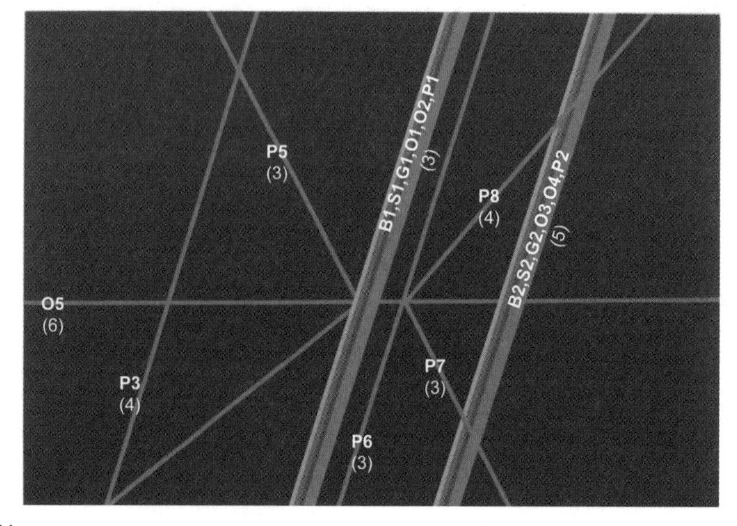

Figure 7.11
Output of the edge degree centrality operation depicted in Figure 7.10. The edge degree centrality of each edge is listed in parentheses below the edge label. Edge O5 has the largest edge degree centrality because this edge connects to the most vertices.

Given an adjacency array **A**, the degree centrality can be easily be computed by multiplying by a vector of all 1's (see Figures 7.8 and 7.9). Vertex V06 has the highest degree centrality because it is connected to the most vertices by the most edges. Vertex V06 is connected to vertices V04, V05, V06, V07, and V08 by six edges. Vertex V06 is connected to vertices V01, V06, V09, V11, V13, and V14 by one edge. Thus, the degree of V06 is

$$(5 \times 6) + (6 \times 1) = 36$$

7.7 Edge Degree Centrality

In addition to determining important vertices in a graph, it can also be of interest to determine important edges in a graph. In a hyper-graph where edges can connect more than one vertex, it is a simple matter to compute the number of vertices associated with each edge by multiplying the incidence array by a vector of 1's.

Figures 7.10 and 7.11 show the calculation of the edge centrality. Not surprisingly, edge O5 emerges as the edge with the highest edge centrality because it cuts across the length of the picture. As with degree centrality, edge centrality can be useful for identifying interesting edges, although it is most commonly used for identifying anomalies in the data. For example, an edge that connects to the most vertices may be the result of an error in how the data were collected.

7.8 Eigenvector Centrality

Other centrality metrics are explicitly linear algebraic in their formulation. For example, eigenvector centrality assumes that each vertex's centrality is proportional to the sum of its neighbors' centrality scores. Eigenvector centrality is equivalent to scoring each vertex based on its corresponding entry in the principal eigenvector. Recall that an eigenvector $\mathbf{v}$ of a square array $\mathbf{A}$ satisfies

$$\mathbf{A}\mathbf{v} = \lambda\mathbf{v}$$

The principal eigenvector is the vector that has the largest value of λ. In many cases, the principal eigenvector can be computed via the power method. Starting with an approximation of the principal eigenvector as a random positive vector $\mathbf{v}_0$ with entries between 0 and 1, this approximation can be iteratively improved via

$$\mathbf{v}_1 = \frac{\mathbf{A}\mathbf{v}_0}{||\mathbf{A}\mathbf{v}_0||}$$

or more generally

$$\mathbf{v}_{i+1} = \frac{\mathbf{A}\mathbf{v}_i}{||\mathbf{A}\mathbf{v}_i||}$$

where $|| \ ||$ denotes the Euclidean norm, given by the square root of the sum of squares. The above computation can be performed for a specific number of iterations or until $\mathbf{v}$ satisfies any of a number of convergence criteria (such as changes in $\mathbf{v}$ are below a specified threshold).

Figure 7.12 shows the first four and final iterations of such a calculation performed by using the adjacency array depicted in Figures 7.2 and 7.8. The final eigenvector highlights vertices V04, V05, V06, V07, and V08, which form a tightly coupled block in the adjacency array. This block of vertices is easily visible in the array because the vertices are listed sequentially in the array. If the vertices were listed randomly, the block would be difficult to see, but the eigenvector centrality would still highlight these vertices in the same way. Thus, the eigenvector centrality allows blocks of vertices to be pulled out independent of how they are ordered in the array. As with other centrality metrics, the eigenvector centrality can highlight vertices of interest as well as vertices to be ignored. It is not uncommon for the vertices identified by the largest principal eigenvector to be connected with a large background group of vertices that are not of great interest. It may be the case that the vertices associated with lower eigenvectors provide more useful information on what is happening in a graph.

It is not uncommon for different centrality metrics to produce results with some similarities. For example, in this particular graph, the first eigenvector places high value on one set of vertices with high degree centrality: {V04, V05, V06, V07, V08}. In contrast, the first eigenvector places a lower value on a number of vertices that have high degree centrality: {V01, V02, V03}. These vertices are highlighted in the *second* principal eigenvector (see Figure 7.13). The eigenvectors of the adjacency array can be thought of as the sets of vertices that will attract random travelers on the graph. The first principal eigenvector highlights the set of vertices a random traveler is most likely to visit. The second principal eigenvector highlights the set of vertices a random traveler is next most likely to visit.

Geometrically, the concept of how a graph affects a random traveler on a graph can be seen in how the adjacency array transforms a set of points that are on a sphere. The points on the sphere represent different starting vertex probabilities of a random traveler. In one dimension, such a sphere is just two points, -1 and +1. In two dimensions, such a sphere is a set of points lying on a circle of radius 1. In three dimensions, the unit sphere is a set of points lying on a sphere of radius 1. In higher dimensions, the unit sphere is the set of points that satisfy

$$||\mathbf{v}|| = 1$$

	$\mathbf{v}_0$	$\mathbf{v}_1$	$\mathbf{v}_2$	$\mathbf{v}_3$	...	$\mathbf{v}_\infty$
V01	0.56	0.29	0.24	0.17	...	0.01
V02	0.85	0.46	0.35	0.24	...	0.03
V03	0.35	0.29	0.24	0.17	...	0.01
V04	0.45	0.27	0.36	0.40	...	0.43
V05	0.05	0.30	0.38	0.42	...	0.45
V06	0.18	0.36	0.42	0.44	...	0.45
V07	0.66	0.32	0.39	0.42	...	0.45
V08	0.33	0.27	0.36	0.40	...	0.43
V09	0.90	0.09	0.06	0.04	...	0.02
V10	0.12	0.10	0.04	0.02	...	0.00
V11	0.99	0.17	0.08	0.04	...	0.02
V12	0.54	0.06	0.03	0.01	...	0.00
V13	0.71	0.13	0.04	0.02	...	0.00
V14	1.00	0.08	0.02	0.01	...	0.00
V15	0.29	0.03	0.01	0.00	...	0.00
V16	0.41	0.21	0.12	0.08	...	0.06
V17	0.46	0.03	0.01	0.00	...	0.00
V18	0.76	0.03	0.02	0.02	...	0.02
V19	0.82	0.05	0.03	0.02	...	0.02
V20	0.10	0.15	0.09	0.06	...	0.04

Figure 7.12
Iterative computation of the eigenvector centrality (first principal eigenvector) of the graph given by the adjacency array in Figure 5.11. The eigenvector centrality highlights the largest set of strongly connected vertices: {V04, V05, V06, V07, V08}.

It is difficult to draw the 20-dimensional sphere corresponding to the adjacency array in Figures 7.2 and 7.8. It is simpler to explain in two dimensions. Consider the 2×2 array

$$\mathbf{A} = \begin{array}{c} \\ 1 \\ 2 \end{array} \begin{array}{c} \begin{array}{cc} 1 & 2 \end{array} \\ \left[\begin{array}{cc} 0.5 & 1.1 \\ 1.1 & 0.5 \end{array}\right] \end{array}$$

Figure 7.14 shows how points on the unit circle are transformed by the adjacency array **A**. **A** is depicted as the blue parallelogram on which the uppermost point (1.6,1.6) corresponds to the sum of the rows (or columns) of **A**. The circle of radius 1 is depicted in green and is referred to as the unit circle.

Let **u** be any point lying on the green line. **Au** is the array multiplication of **A** with **u** and is shown as the green dashed ellipse. The green dashed ellipse represents how **A** distorts the unit circle. Note that the green dashed ellipse touches the corners of the blue parallelogram and is tangent to the sides of the blue parallelogram at that point. The long axis of the green dashed ellipse points directly at the upper rightmost point of the parallelogram.

	$\mathbf{v}^{(2)}$
V01	0.53
V02	0.65
V03	0.53
V04	-0.03
V05	-0.03
V06	0.02
V07	-0.02
V08	-0.03
V09	0.04
V10	0.04
V11	0.05
V12	0.04
V13	0.05
V14	0.01
V15	0.00
V16	0.05
V17	0.00
V18	0.00
V19	0.00
V20	0.05

Figure 7.13
Second principal eigenvector of the adjacency array **A** of the graph from Figure 5.11. The second principal eigenvector highlights the second largest set of strongly connected vertices: {V01, V02, V03}.

The points on the unit circle that align with the axis of the green dashed ellipse are shown by the black square. These points are given a special name called the *eigenvectors* of the matrix **A**. Multiplication of **A** by any point along these eigenvectors will only stretch the eigenvectors. All other points, when multiplied by **A**, will be stretched and twisted. The red box shows what happens when the eigenvectors of **A** are multiplied by **A**. The red box corresponds exactly to the axis of the green dashed ellipse. The lengths of the sides of the red box determine how much stretching occurs along each direction of the eigenvector and are called the eigenvalues.

The long axis means that points lying along this eigenvector will become longer. The short axis means that points lying along this eigenvector will become shorter. The ratio of the area of the red box to the are of the black box is equal to the ratio of the area of the green dashed ellipse to the area of the green solid circle. This ratio is equal to the area of the blue parallelogram and is called the *determinant*. While not all 2×2 matrices behave as shown in Figure 7.14, many do. In particular, many of the matrices that are relevant to the kinds of data represented with spreadsheets, databases, and graphs have the properties shown Figure 7.14. Furthermore, these properties hold as the array grows (they just get harder to draw).

The largest eigenvalue identifies the eigenvector that has the strongest impact on a random traveler on a graph. This eigenvector is referred to as the principal eigenvector. In

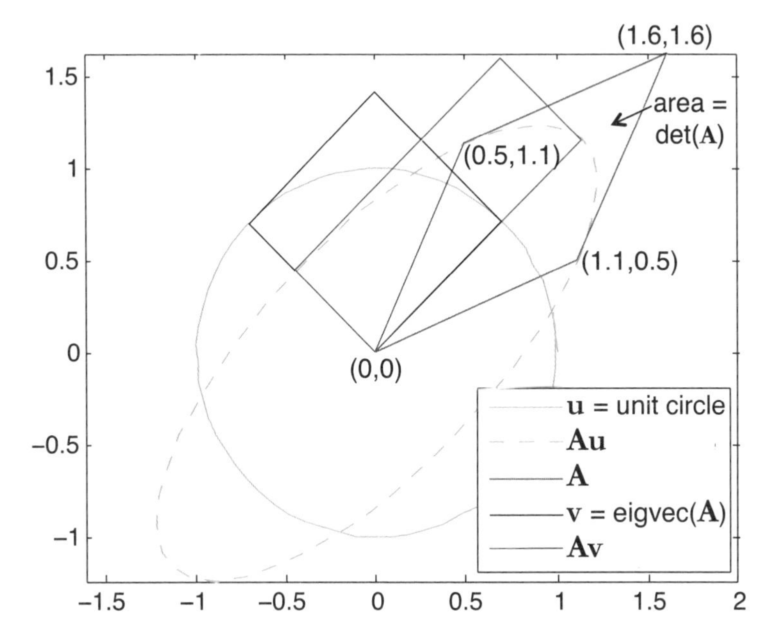

Figure 7.14

The properties of the 2×2 array $\mathbf{A} = \begin{bmatrix} 0.5 & 1.1 \\ 1.1 & 0.5 \end{bmatrix}$. Solid green line represents the points on the unit circle.

graph terms, the principal eigenvector can be thought of as the principal direction a random traveler will be pushed towards as they traverse the graph. The eigenvalue of the principal eigenvector shown in Figure 7.12 is 30.7. The eigenvalue of the second principal eigenvector shown in Figure 7.13 is 19.3. Thus, in the 20-dimensional space represented by the vertices of the graph, the longest axis will have a length of 30.7, and the second longest axis will have a length of 19.3.

Finally, it is worth mentioning that a 20-vertex graph can have no more than 20 eigenvectors and may have fewer than 20 eigenvectors. For example, the graphs in Figures 7.2 and 7.8 have approximately 9 eigenvectors. Any attempt to add additional eigenvectors would produce a vector that can be constructed by combining the eigenvectors that already exist.

7.9 Singular Value Decomposition

The eigenvector centrality can be readily computed on square adjacency arrays, but it is more difficult to apply to rectangular incidence arrays for which there is just as much need to identify vertices. Given an incidence array $\mathbf{E}$, there are two complementary ways to construct square arrays from $\mathbf{E}$

$$\mathbf{E}^\mathsf{T}\mathbf{E}$$

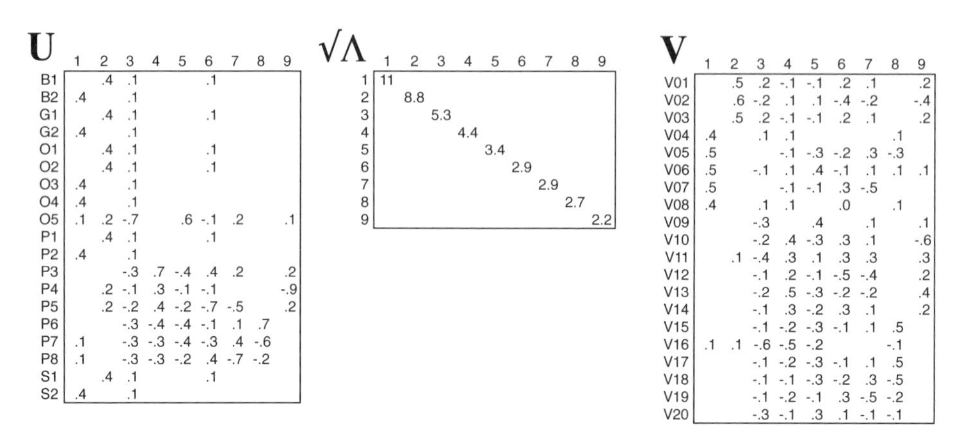

U

	1	2	3	4	5	6	7	8	9
B1		.4	.1			.1			
B2	.4		.1						
G1		.4	.1			.1			
G2	.4		.1						
O1		.4	.1			.1			
O2		.4	.1			.1			
O3	.4		.1						
O4	.4		.1						
O5	.1	.2	-.7		.6	-.1	.2		.1
P1		.4	.1			.1			
P2	.4		.1						
P3			-.3	.7	-.4	.4	.2		.2
P4		.2	-.1	.3	-.1	-.1			-.9
P5		.2	-.2	.4	-.2	-.7	-.5		.2
P6			-.3	-.4	-.4	-.1	.1	.7	
P7	.1		-.3	-.3	-.4	-.3	.4	-.6	
P8	.1		-.3	-.3	-.2	.4	-.7	-.2	
S1		.4	.1			.1			
S2	.4		.1						

√Λ

	1	2	3	4	5	6	7	8	9
1	11								
2		8.8							
3			5.3						
4				4.4					
5					3.4				
6						2.9			
7							2.9		
8								2.7	
9									2.2

V

	1	2	3	4	5	6	7	8	9
V01		.5	.2	-.1	-.1	.2	.1		.2
V02		.6	-.2	.1	.1	-.4	-.2		-.4
V03		.5	.2	-.1	-.1	.2	.1		.2
V04	.4		.1	.1				.1	
V05	.5			-.1	-.3	-.2	.3	-.3	
V06	.5		-.1	.1	.4	-.1	.1	.1	.1
V07	.5			-.1	-.1	.3	-.5		
V08	.4		.1	.1		.0		.1	
V09			-.3		.4		.1		.1
V10			-.2	.4	-.3	.3	.1		-.6
V11		.1	-.4	.3	.1	.3	.3		.3
V12			-.1	.2	-.1	-.5	-.4		.2
V13			-.2	.5	-.3	-.2	-.2		.4
V14			-.1	.3	-.2	.3	.1		.2
V15			-.1	-.2	-.3	-.1	.1	.5	
V16	.1	.1	-.6	-.5	-.2			-.1	
V17			-.1	-.2	-.3	-.1	.1	.5	
V18			-.1	-.1	-.3	-.2	.3	-.5	
V19			-.1	-.2	-.1	.3	-.5	-.2	
V20			-.3	-.1	.3	.1	-.1	-.1	

Figure 7.15

Singular value decomposition (SVD) of the incidence array **E** shown in Figure 7.4. The columns of **U** are the left singular vectors of **E**, the columns of **V** are the right singular vectors of **E**, and the diagonal values of the diagonal array $\sqrt{\Lambda}$ are referred to as the singular values of **E**. The arrays are dense. All the values in the arrays are rounded to the nearest tenth to highlight the structure of the arrays. The quantities are linked by the formula $\mathbf{E} = \mathbf{U}\sqrt{\Lambda}\mathbf{V}^{\mathsf{T}}$.

and

$$\mathbf{E}\mathbf{E}^{\mathsf{T}}$$

Specific examples of the computation of these arrays are shown in Figures 5.11 and 5.12.

The eigenvalues and eigenvectors of the square array $\mathbf{E}^{\mathsf{T}}\mathbf{E}$ satisfy

$$(\mathbf{E}^{\mathsf{T}}\mathbf{E})\mathbf{u} = \lambda\mathbf{u}$$

or

$$(\mathbf{E}^{\mathsf{T}}\mathbf{E})\mathbf{U} = \Lambda\mathbf{U}$$

where the diagonal elements of the diagonal array Λ are the eigenvalues λ and the columns of array **U** are the eigenvectors **u**. Likewise, the eigenvalues and eigenvectors of the square array $\mathbf{E}\mathbf{E}^{\mathsf{T}}$ satisfy

$$(\mathbf{E}\mathbf{E}^{\mathsf{T}})\mathbf{v} = \lambda\mathbf{v}$$

or

$$(\mathbf{E}^{\mathsf{T}}\mathbf{E})\mathbf{V} = \Lambda\mathbf{V}$$

where the diagonal elements of the diagonal array Λ are the eigenvalues λ and the columns of array **V** are the eigenvectors **v**.

The eigenvalues and eigenvectors of the square arrays $\mathbf{E}^{\mathsf{T}}\mathbf{E}$ are $\mathbf{E}\mathbf{E}^{\mathsf{T}}$ are connected by the formula

$$\mathbf{E} = \mathbf{U}\sqrt{\Lambda}\,\mathbf{V}^{\mathsf{T}}$$

$$\mathbf{U}(:,1)\,\sqrt{\Lambda(1,1)}\,\mathbf{V}(:,1)^{\mathsf{T}} =$$

	V01	V02	V03	V04	V05	V06	V07	V08	V09	V10	V11	V12	V13	V14	V15	V16	V17	V18	V19	V20
B1																				
B2				2	2	2	2	2												
G1																				
G2				2	2	2	2	2												
O1																				
O2																				
O3				2	2	2	2	2												
O4				2	2	2	2	2												
O5				1	1	1	1	1												
P1																				
P2				2	2	2	2	2												
P3																				
P4																				
P5																				
P6																				
P7																				
P8																				
S1																				
S2				2	2	2	2	2												

Figure 7.16
Approximation of incidence array **E** shown in Figure 7.4 via the array multiplication of the largest left singular vector $\mathbf{U}(:,1)$ and the largest right singular vector $\mathbf{V}(:,1)$. The array is dense. All values have been rounded to the nearest integer to more clearly show the structure of the array.

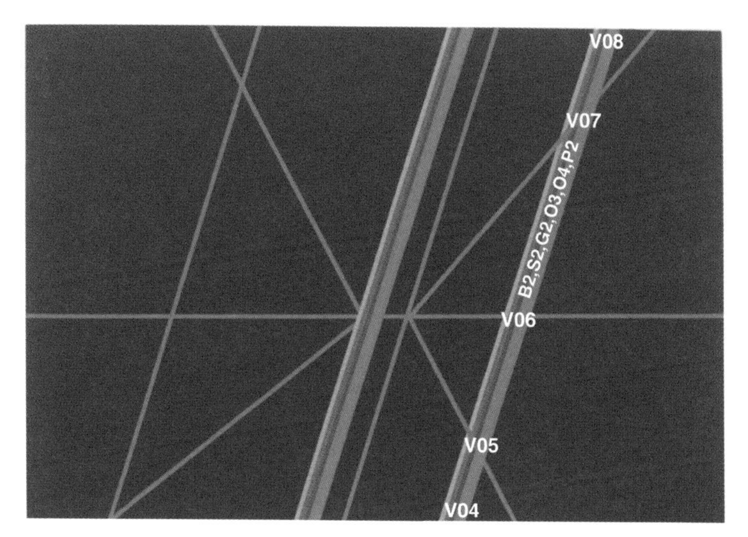

Figure 7.17
The specific sets of vertices {V04,V05, V06, V07, V08} and edges {B2,S2, G2, O3, O4, P2} that are highlighted by the computation shown in Figure 7.16.

In the above formula, the columns of $\mathbf{U}$ are referred to as the left singular vectors of $\mathbf{E}$, the columns of $\mathbf{V}$ are referred to as the right singular vectors of $\mathbf{E}$, and the diagonal values of the diagonal array $\sqrt{\Lambda}$ are referred to as the singular values of $\mathbf{E}$. The entire formula is referred to as the singular value decomposition, or SVD, of $\mathbf{E}$ [20–22]. The SVD provides information on the structure of non-square arrays in a manner similar to how the eigenvalues and eigenvectors can be used to describe the structure of a square array. Figure 7.15 provides an example of the SVD for the array $\mathbf{E}$ shown in Figure 7.4. As with the eigenvalues, there are only 9 singular vectors.

The SVD can be a useful tool for creating an approximation of an overall graph by multiplying the largest left and right singular values via

$$\mathbf{U}(:,1)\ \sqrt{\Lambda}(1,1)\ \mathbf{V}(:,1)^{\mathsf{T}}$$

Figure 7.16 shows this calculation for the incidence array $\mathbf{E}$ in Figure 7.4. The specific sets of vertices {V04,V05, V06, V07, V08} and edges {B2,S2, G2, O3, O4, P2} that are highlighted by this computation are shown in Figure 7.17. Thus, from the perspective of the SVD, this combination of vertices and edges is the strongest single component in the graph.

7.10 PageRank

Perhaps the most well-known centrality metric is PageRank, which was developed by Google founders Sergey Brin and Larry Page to rank webpages [23]. The algorithm was originally applied to rank webpages for keyword searches. The algorithm measures each webpage's relative importance by assigning a numerical rating from the most important to the least important page within the set of identified webpages. The PageRank algorithm analyzes the topology of a graph representing links among webpages and outputs a probability distribution used to represent the likelihood that a person randomly clicking on links will arrive at any particular page.

This algorithm was originally applied specifically to rank webpages within a Google search. However, the mathematics can be applied to any graph or network [24]. The algorithm is applicable to social network analysis [25], recommender systems [26], biology [27], chemistry [28], and neuroscience [29]. In chemistry, this algorithm is used in conjunction with molecular dynamics simulations that provides geometric locations for a solute in water. The graph contains edges between the water molecules and can be used to calculate whether the hydrogen bond potential can act as a solvent. In neuroscience, the brain represents a highly complex vertex/edge graph. PageRank has recently been applied to evaluate the importance of brain regions from observed correlations of brain activity. In network analysis, PageRank can analyze complex networks and sub-networks that can reveal behavior that could not be discerned by traditional methods.

	PageRank
V01	0.056
V02	**0.098**
V03	0.056
V04	0.062
V05	0.073
V06	**0.082**
V07	0.076
V08	0.062
V09	0.030
V10	0.038
V11	0.050
V12	0.024
V13	0.044
V14	0.030
V15	0.029
V16	0.074
V17	0.029
V18	0.021
V19	0.024
V20	0.043

Figure 7.18
PageRank of the adjacency array $\mathbf{A}$ of the graph derived from the painting from Figure 5.11. PageRank highlights two vertices that are the centers of their respective blocks of vertices: V02 and V06.

The adjacency matrix for the original PageRank algorithm used webpages for the rows and the webpages to which they linked for the columns. Furthermore, each value in the adjacency matrix is normalized by its out-degree centrality, which prevents webpages with lots of links from having more influence than those with just a few links. PageRank simulates a random walk on a graph with the added possibility of jumping to an arbitrary vertex. Each vertex is then ranked according to the probability of landing on it at an arbitrary point in an infinite random walk. If the probability of jumping to any vertex is 0, then this rank is simply the principal eigenvector of

$$\tilde{\mathbf{A}}(k_1,k_2) = \frac{\mathbf{A}(k_1,k_2)}{\mathbf{d}_{\text{out}}(k_1)}$$

where $\mathbf{d}_{\text{out}}$ is the array of out-degree centralities

$$\mathbf{d}_{\text{out}} = \mathbf{A}\mathbb{1}$$

The PageRank algorithm assumes the probability of randomly jumping to a different vertex is $1-c \sim 0.15$. In this case, the principal eigenvector can be calculated using the power

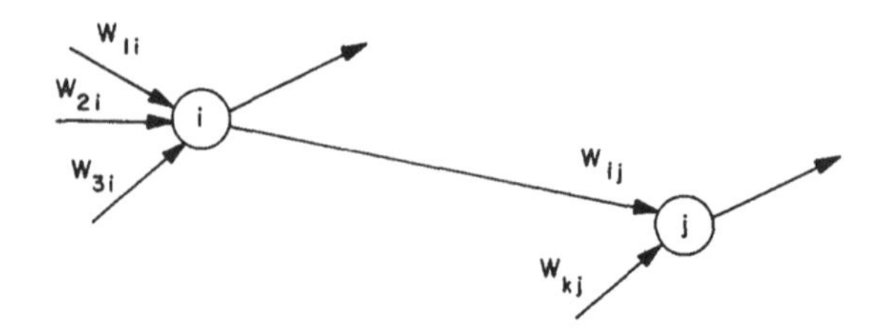

Figure 7.19
Typical network elements i and j showing connection weights w (reproduced from [3]).

method. Rearranging terms, the power method computation for PageRank becomes

$$\mathbf{v}_{i+1} = c\mathbf{v}_i^{\mathsf{T}}\tilde{\mathbf{A}} + \frac{1-c}{n}\sum_{k_1}\mathbf{v}_i(k_1)$$

PageRank integrates the effects of multiple blocks of vertices that are represented by different eigenvectors. Figure 7.18 shows the PageRank of the adjacency array $\mathbf{A}$ of the graph in the painting depicted in Figure 5.11. PageRank highlights vertices V02 and V06 that are at the center of the blocks of vertices selected by the principal and second eigenvectors.

PageRank is closely related to the eigenvector centrality. In fact, let $\mathbf{v}^{(1)}$ be the first eigenvector of the array

$$c\mathbf{A}^{\mathsf{T}} + \frac{1-c}{m}$$

where m is the number of rows (or columns) in the array. The eigenvector $\mathbf{v}^{(1)}$ will agree with PageRank $\mathbf{v}$ when each is normalized by their absolute value norm, which is the sum of the absolute value of all of their elements.

$$\frac{\mathbf{v}^{(1)}}{\sum_k |\mathbf{v}^{(1)}(k)|} = \frac{\mathbf{v}}{\sum_k |\mathbf{v}(k)|}$$

7.11 Deep Neural Networks

The final stage of a machine learning system involves using all the processed inputs to make a decision or provide a label on the inputs. Deep neural networks (DNNs) are one of the most common approaches used to implement this final decision stage in a machine learning system.

Using biologically inspired neural networks to implement machine learning was the topic of the first paper presented at the first machine learning conference in 1955 [2, 3] (see Figure 7.19). At this time, it was recognized that direct computational training of neural networks was computationally infeasible [8]. The subsequent manyfold improvement in neural network computation and theory has made it possible to train neural networks

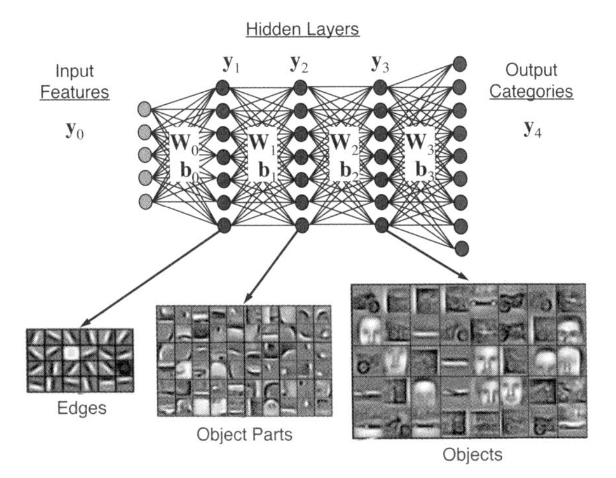

Figure 7.20
Four layer ($L = 4$) deep neural network architecture for categorizing images. The input features $\mathbf{y}_0$ of an image are passed through a series of network layers $\mathbf{W}_{k=0,1,2,3}$, with bias terms $\mathbf{b}_{k=0,1,2,3}$, that produce scores for categories $\mathbf{y}_{L=4}$. (Figure adapted from [43]).

that are capable of better-than-human performance in a variety of important artificial intelligence problems [11, 30–32]. Specifically, the availability of large corpora of validated data sets [33–35] and the increases in computation spurred by games [36–39] have allowed the effective training of large deep neural networks with 100,000s of input features, N, and hundreds of layers, L, that are capable of choosing from among 100,000s of categories, M (see Figure 7.20).

The connection between graphs and DNNs lies in the standard matrix representation of graphs. In theory, the flexible semiring computations provided by associative arrays may provide a convenient way for representing massive DNNs on both conventional and custom computing hardware [40–42].

The primary mathematical operation performed by a DNN is the inference, or forward propagation, step. During inference, an input feature vector $\mathbf{Y}_0$ is turned into a vector $\mathbf{Y}_L$ containing the score on each of the potential categories. Inference is executed repeatedly during the training phase to determine both the weight matrices $\mathbf{W}_k$ and the bias vectors $\mathbf{b}_k$ of the DNN. The inference computation shown in Figure 7.20 is given by

$$\mathbf{y}_{k+1} = h(\mathbf{W}_k \mathbf{y}_k + \mathbf{b}_k)$$

where $h()$ is a non-linear function applied to each element of the vector. A commonly used function is the rectified linear unit (ReLU) given by

$$h(\mathbf{y}) = \max(\mathbf{y}, 0)$$

which sets values less than 0 to 0 and leaves other values unchanged. When training a DNN, it is common to compute multiple $\mathbf{y}_k$ vectors at once in a batch that can be denoted as the matrix $\mathbf{Y}_k$. In matrix form, the inference step becomes

$$\mathbf{Y}_{k+1} = h(\mathbf{W}_k \mathbf{Y}_k + \mathbf{B}_k)$$

where $\mathbf{B}_k$ is a replication of $\mathbf{b}_k$ along columns.

If $h()$ were a linear function, then the above equation could be solved directly and the computation could be greatly simplified. However, current evidence suggests that the non-linearity of $h()$ is required for a DNN to be effective. Interestingly, the inference computation can be rewritten as a linear function over two different semirings

$$\mathbf{y}_{k+1} = \mathbf{W}_k \mathbf{y}_k \otimes \mathbf{b}_k \oplus 0$$

or in matrix form

$$\mathbf{Y}_{k+1} = \mathbf{W}_k \mathbf{Y}_k \otimes \mathbf{B}_k \oplus 0$$

where the $\oplus = \max$ and $\otimes = +$. The computations of $\mathbf{W}_k \mathbf{y}_k$ and $\mathbf{W}_k \mathbf{Y}_k$ are computed over the standard arithmetic $+.\times$ semiring

$$S_1 = (\mathbb{R}, +, \times, 0, 1)$$

while the $\oplus$ and $\otimes$ operations are performed over the max.+ semiring

$$S_2 = (\mathbb{R} \cup \{-\infty\}, \max, +, -\infty, 0)$$

Thus, the ReLU DNN can be written as a linear system that oscillates over two semirings S_1 and S_2. S_1 is the most widely used semiring and performs standard correlation between vectors. S_2 is also a commonly used semiring for selecting optimal paths in graphs. Thus, the inference step of a ReLU DNN can be viewed as combining correlations of inputs to select optimal paths through the neural network.

7.12 Conclusions, Exercises, and References

The flexibility of associative arrays to represent data makes them ideal for capturing the flow of data through a machine learning system. A common first step in a machine learning system is the construction of a graph. Associative arrays readily support both adjacency array and incidence array representations of graphs. Traversing a graph in either representation can be performed via array multiplication. Likewise, computing the statistics of a graph can also be accomplished via array multiplication. The analysis of the graph to find and categorize vertices of interest (or vertices to be eliminated) can be performed using a variety of algorithms, such as eigenvalues, singular values, PageRank, and deep neural networks. All of these algorithms are readily implemented with associative array algebra.

Exercises

Exercise 7.1 — Refer to Figure 7.1.
(a) Describe some of the typical steps in a machine learning system.
(b) Compare data preparation vs. modeling.
(c) Comment on where most of the effort goes in building a machine learning system.

Exercise 7.2 — What is the formal mathematical definition of an associative array?

Exercise 7.3 — Compare the kinds of graphs that can be represented with adjacency arrays and incidence arrays.

Exercise 7.4 — Write six equivalent array multiplication expressions for doing graph traversal with an adjacency array **A** of an undirected graph and an array of starting vertices **V**.

Exercise 7.5 — Refer to Figure 7.2. Describe how the results of the array multiplication **VA** differ from those in the figure.

Exercise 7.6 — Take a graph from your own experience and write down its adjacency matrix.

Exercise 7.7 — Using the adjacency matrix from the previous exercise, find the nearest neighbors of two vertices by using array multiplication.

Exercise 7.8 — Using the adjacency matrix from the previous exercise, compute the out-degree centrality and the in-degree centrality by using array multiplication. Comment on the significance of the high out-degree and in-degree vertices.

Exercise 7.9 — Using the adjacency matrix from the previous exercise, employ the power method to compute the first eigenvector. Comment on the significance of the vertices with highest first eigenvector values.

Exercise 7.10 — Using the adjacency matrix from the previous exercise, employ the power method to compute the PageRank. Comment on the significance of the vertices with highest PageRank values.

References

[1] J. Kepner, M. Kumar, J. Moreira, P. Pattnaik, M. Serrano, and H. Tufo, "Enabling massive deep neural networks with the GraphBLAS," in *High Performance Extreme Computing Conference (HPEC)*, IEEE, 2017.

[2] W. H. Ware, "Introduction to session on learning machines," in *Proceedings of the March 1-3, 1955, Western Joint Computer Conference*, pp. 85–85, ACM, 1955.

[3] W. A. Clark and B. G. Farley, "Generalization of pattern recognition in a self-organizing system," in *Proceedings of the March 1-3, 1955, Western Joint Computer Conference*, pp. 86–91, ACM, 1955.

[4] O. G. Selfridge, "Pattern recognition and modern computers," in *Proceedings of the March 1-3, 1955, Western Joint Computer Conference*, pp. 91–93, ACM, 1955.

[5] G. Dinneen, "Programming pattern recognition," in *Proceedings of the March 1-3, 1955, Western Joint Computer Conference*, pp. 94–100, ACM, 1955.

[6] A. Newell, "The chess machine: an example of dealing with a complex task by adaptation," in *Proceedings of the March 1-3, 1955, Western Joint Computer Conference*, pp. 101–108, ACM, 1955.

[7] J. McCarthy, M. L. Minsky, N. Rochester, and C. E. Shannon, "A proposal for the dartmouth summer research project on artificial intelligence, august 31, 1955," *AI Magazine*, vol. 27, no. 4, p. 12, 2006.

[8] M. Minsky and O. G. Selfridge, "Learning in random nets," in *Information Theory: Papers Read at a Symposium on Information Theory Held at the Royal Institution, London, August 29th to September 2nd*, pp. 335–347, London: Butterworths, 1960.

[9] M. Minsky, "Steps toward artificial intelligence," *Proceedings of the IRE*, vol. 49, no. 1, pp. 8–30, 1961.

[10] J. R. Quinlan, "Induction of decision trees," *Machine Learning*, vol. 1, no. 1, pp. 81–106, 1986.

[11] R. Lippmann, "An introduction to computing with neural nets," *IEEE ASSP Magazine*, vol. 4, no. 2, pp. 4–22, 1987.

[12] J. A. Suykens and J. Vandewalle, "Least squares support vector machine classifiers," *Neural Processing L etters*, vol. 9, no. 3, pp. 293–300, 1999.

[13] N. Friedman, D. Geiger, and M. Goldszmidt, "Bayesian network classifiers," *Machine Learning*, vol. 29, no. 2-3, pp. 131–163, 1997.

[14] K. Deb, A. Pratap, S. Agarwal, and T. Meyarivan, "A fast and elitist multiobjective genetic algorithm: NSGA-II," *IEEE Transactions on Evolutionary Computation*, vol. 6, no. 2, pp. 182–197, 2002.

[15] B. Karrer and M. E. Newman, "Stochastic blockmodels and community structure in networks," *Physical Review E*, vol. 83, no. 1, p. 016107, 2011.

[16] M. Soderholm, "Big data's dirty little secret," *Datanami*, July 2 2015.

[17] M. Soderholm, "Five steps to fix the data feedback loop and rescue analysis from 'bad' data," *Datanami*, Aug. 17 2015.

[18] J. Kepner and V. Gadepally, "Adjacency matrices, incidence matrices, database schemas, and associative arrays," in *International Parallel and Distributed Processing Symposium Workshops (IPDPSW)*, IEEE, 2014.

[19] J. Kepner, C. Anderson, W. Arcand, D. Bestor, B. Bergeron, C. Byun, M. Hubbell, P. Michaleas, J. Mullen, D. O'Gwynn, A. Prout, A. Reuther, A. Rosa, and C. Yee, "D4M 2.0 schema: A general purpose high performance schema for the Accumulo database," in *High Performance Extreme Computing Conference (HPEC)*, IEEE, 2013.

[20] E. Beltrami, "Sulle funzioni bilineari," *Giornale di Matematiche ad Uso degli Studenti Delle Universita*, vol. 11, no. 2, pp. 98–106, 1873.

[21] C. Jordan, "Mémoire sur les formes bilinéaires," *Journal de Mathématiques Pures et Appliquées*, vol. 19, pp. 35–54, 1874.

[22] G. W. Stewart, "On the early history of the singular value decomposition," *SIAM Review*, vol. 35, no. 4, pp. 551–566, 1993.

[23] S. Brin and L. Page, "The anatomy of a large-scale hypertextual web search engine," *Computer Networks and ISDN Systems*, vol. 30, no. 1, pp. 107–117, 1998.

[24] D. F. Gleich, "PageRank beyond the web," *SIAM Review*, vol. 57, no. 3, pp. 321–363, 2015.

[25] H. Kwak, C. Lee, H. Park, and S. Moon, "What is Twitter, a social network or a news media?," in *Proceedings of the 19th International Conference on World Wide Web*, pp. 591–600, ACM, 2010.

[26] Y. Song, D. Zhou, and L.-w. He, "Query suggestion by constructing term-transition graphs," in *Proceedings of the Fifth ACM International Conference on Web Search and Data Mining*, pp. 353–362, ACM, 2012.

[27] J. L. Morrison, R. Breitling, D. J. Higham, and D. R. Gilbert, "GeneRank: using search engine technology for the analysis of microarray experiments," *BMC Bioinformatics*, vol. 6, no. 1, p. 233, 2005.

[28] B. L. Mooney, L. R. Corrales, and A. E. Clark, "MoleculaRnetworks: An integrated graph theoretic and data mining tool to explore solvent organization in molecular simulation," *Journal of Computational Chemistry*, vol. 33, no. 8, pp. 853–860, 2012.

[29] X.-N. Zuo, R. Ehmke, M. Mennes, D. Imperati, F. X. Castellanos, O. Sporns, and M. P. Milham, "Network centrality in the human functional connectome," *Cerebral Cortex*, vol. 22, no. 8, pp. 1862–1875, 2012.

[30] D. A. Reynolds, T. F. Quatieri, and R. B. Dunn, "Speaker verification using adapted Gaussian mixture models," *Digital Signal Processing*, vol. 10, no. 1-3, pp. 19–41, 2000.

[31] A. Krizhevsky, I. Sutskever, and G. E. Hinton, "ImageNet classification with deep convolutional neural networks," in *Advances in Neural Information Processing Systems*, pp. 1097–1105, 2012.

[32] Y. LeCun, Y. Bengio, and G. Hinton, "Deep learning," *Nature*, vol. 521, no. 7553, pp. 436–444, 2015.

[33] J. P. Campbell, "Testing with the YOHO CD-ROM voice verification corpus," in *International Conference on Acoustics, Speech, and Signal Processing (ICASSP)*, vol. 1, pp. 341–344, IEEE, 1995.

[34] Y. LeCun, C. Cortes, and C. J. Burges, "The MNIST database of handwritten digits," 1998.

[35] J. Deng, W. Dong, R. Socher, L.-J. Li, K. Li, and L. Fei-Fei, "ImageNet: A large-scale hierarchical image database," in *IEEE Conference on Computer Vision and Pattern Recognition (CVPR)*, pp. 248–255, IEEE, 2009.

[36] M. Campbell, A. J. Hoane, and F.-h. Hsu, "Deep Blue," *Artificial Intelligence*, vol. 134, no. 1-2, pp. 57–83, 2002.

[37] M. P. McGraw-Herdeg, D. P. Enright, and B. S. Michel, "Benchmarking the NVIDIA 8800GTX with the CUDA development platform," *High Performance Embedded Computing Workshop (HPEC)*, 2007.

[38] A. Kerr, D. Campbell, and M. Richards, "GPU performance assessment with the HPEC challenge," in *High Performance Embedded Computing Workshop (HPEC)*, 2008.

[39] E. A. Epstein, M. I. Schor, B. Iyer, A. Lally, E. W. Brown, and J. Cwiklik, "Making Watson fast," *IBM Journal of Research and Development*, vol. 56, no. 3.4, pp. 15–1, 2012.

[40] W. S. Song, J. Kepner, H. T. Nguyen, J. I. Kramer, V. Gleyzer, J. R. Mann, A. H. Horst, L. L. Retherford, R. A. Bond, N. T. Bliss, E. Robinson, S. Mohindra, and J. Mullen, "3-D graph processor," in *High Performance Embedded Computing Workshop (HPEC)*, MIT Lincoln Laboratory, 2010.

[41] W. S. Song, J. Kepner, V. Gleyzer, H. T. Nguyen, and J. I. Kramer, "Novel graph processor architecture," *Lincoln Laboratory Journal*, vol. 20, no. 1, pp. 92–104, 2013.

[42] W. S. Song, V. Gleyzer, A. Lomakin, and J. Kepner, "Novel graph processor architecture, prototype system, and results," in *High Performance Extreme Computing Conference (HPEC)*, IEEE, 2016.

[43] H. Lee, R. Grosse, R. Ranganath, and A. Y. Ng, "Convolutional deep belief networks for scalable unsupervised learning of hierarchical representations," in *Proceedings of the 26th Annual International Conference on Machine Learning*, pp. 609–616, ACM, 2009.

II MATHEMATICAL FOUNDATIONS

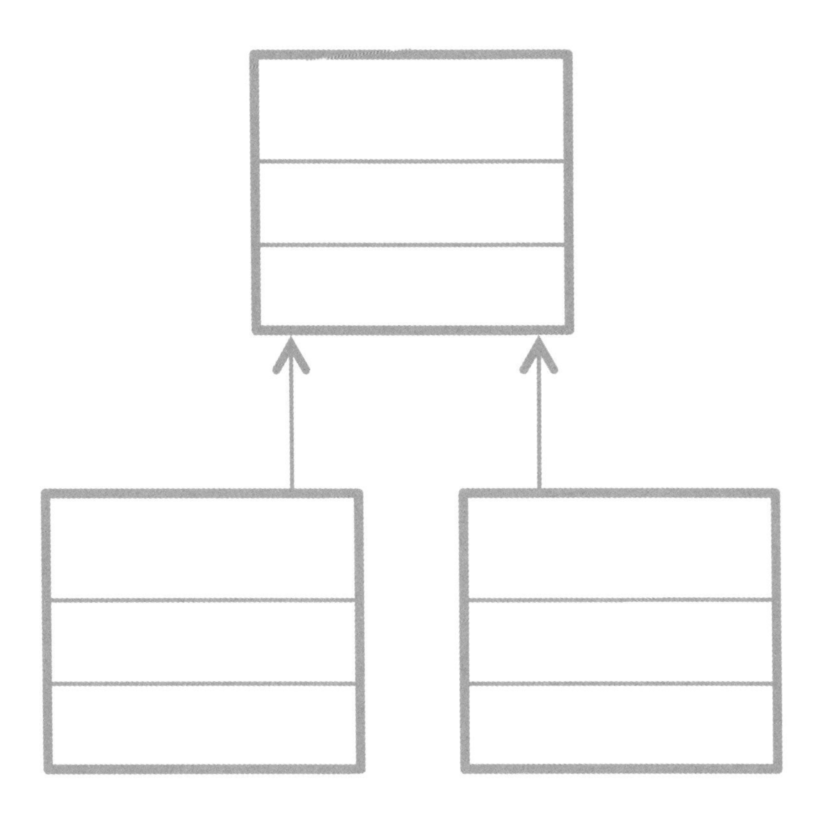

8 Visualizing the Algebra of Associative Arrays

Summary

Associative arrays have analogs of most matrix operations. Demonstrating this analogy formally requires a variety of abstract algebra objects, such as sets, semirings, linear algebra, ordered sets, and Boolean algebra. Building on these abstract algebra objects leads to the formal concept of an associative array algebra. The relationships between and among these mathematical objects can be visually depicted using class diagrams, which are used to depict similar relationships in software. This chapter graphically describes the interrelationships among the mathematical objects necessary for defining associative algebra and provides the intuition necessary to formally understand these objects.

8.1 Associative Array Analogs of Matrix Operations

Chapter 6 provided an overview of the primary matrix operations in terms of graph manipulations. Given non-negative integer indices

$$i \in I = \{1, \ldots, m\}$$

and

$$j \in J = \{1, \ldots, n\}$$

and a set of values V (such as the real numbers $\mathbb{R}$), then matrices can be viewed as mappings between these sets of indices and their corresponding scalars. Using this terminology, the $m \times n$ matrices $\mathbf{A}$, $\mathbf{B}$, and $\mathbf{C}$, can be written as

$$\mathbf{A} : I \times J \to V$$
$$\mathbf{B} : I \times J \to V$$
$$\mathbf{C} : I \times J \to V$$

The product of a scalar $v \in V$ with a matrix can be written

$$\mathbf{C} = v\mathbf{A} = v \otimes \mathbf{A} \quad \text{where} \quad \mathbf{C}(i, j) = v\mathbf{A}(i, j) = v \otimes \mathbf{A}(i, j)$$

The corresponding element-wise addition of the above matrices is written

$$\mathbf{C} = \mathbf{A} \oplus \mathbf{B} \quad \text{where} \quad \mathbf{C}(i, j) = \mathbf{A}(i, j) \oplus \mathbf{B}(i, j)$$

Likewise, the corresponding element-wise multiplication (or Hadamard product) of the above matrices is written

$$\mathbf{C} = \mathbf{A} \otimes \mathbf{B} \quad \text{where} \quad \mathbf{C}(i, j) = \mathbf{A}(i, j) \otimes \mathbf{B}(i, j)$$

Finally, matrix multiplication, which combines element-wise addition and element-wise multiplication, is written

$$\mathbf{C} = \mathbf{A}\mathbf{B} = \mathbf{A} \oplus.\otimes \mathbf{B} \quad \text{where} \quad \mathbf{C}(i, j) = \bigoplus_{k \in K} \mathbf{A}(i,k) \otimes \mathbf{B}(k, j)$$

where $\mathbf{A}$, $\mathbf{B}$, and $\mathbf{C}$ are matrices (or mappings)

$$\mathbf{A} : I \times K \to V$$
$$\mathbf{B} : K \times J \to V$$
$$\mathbf{C} : I \times J \to V$$

When the value set V of an associative array is equipped with two binary operations $\oplus$ and $\otimes$, the associative array equivalents of the matrix operations can be written with the identical notation.

Associative arrays extend matrices in many ways. A few of the more notable extensions are as follows. First, the value set V is not limited to a *field*, the numbers for which the ordinary rules of arithmetic apply, and can likewise be much broader to include words or strings. Finally, associative arrays have an implicit sparseness such that only the nonzero entries in an associative array are stored. For an associative array, the dimensions of row keys (K_1) or column keys (K_2) are often large, but the actual sizes of the rows or columns that have nonzero entries are often small compared to the dimensions of the row or column keys.

Second, associative array row and column indices (or keys) are not limited to integers (such as with I, J, and K above) and can be any finite set with ordered values, meaning that the values are ordered with some understood relation. The transition from integer sets I and J to more general key sets $k_1 \in K_1$ and $k_2 \in K_2$ results in associative arrays $\mathbf{A}$, $\mathbf{B}$, and $\mathbf{C}$, that are written as

$$\mathbf{A} : K_1 \times K_2 \to V$$
$$\mathbf{B} : K_1 \times K_2 \to V$$
$$\mathbf{C} : K_1 \times K_2 \to V$$

The product of a scalar $v \in V$ with an associative array can be written

$$\mathbf{C} = v\mathbf{A} = v \otimes \mathbf{A} \quad \text{where} \quad \mathbf{C}(k_1, k_2) = v\mathbf{A}(k_1, k_2) = v \otimes \mathbf{A}(k_1, k_2)$$

The equivalent element-wise addition of the above array is written

$$\mathbf{C} = \mathbf{A} \oplus \mathbf{B} \quad \text{where} \quad \mathbf{C}(k_1,k_2) = \mathbf{A}(k_1,k_2) \oplus \mathbf{B}(k_1,k_2)$$

Similarly, the corresponding element-wise multiplication of the above arrays is written

$$\mathbf{C} = \mathbf{A} \otimes \mathbf{B} \quad \text{where} \quad \mathbf{C}(k_1,k_2) = \mathbf{A}(k_1,k_2) \otimes \mathbf{B}(k_1,k_2)$$

Finally, array multiplication, which combines element-wise addition and element-wise multiplication, is written

$$\mathbf{C} = \mathbf{AB} = \mathbf{A} \oplus.\otimes \mathbf{B} \quad \text{where} \quad \mathbf{C}(k_1,k_2) = \bigoplus_{k_3 \in K_3} \mathbf{A}(k_1,k_3) \otimes \mathbf{B}(k_3,k_2)$$

where **A**, **B**, and **C** are associative arrays

$$\mathbf{A} : K_1 \times K_3 \to V$$
$$\mathbf{B} : K_3 \times K_2 \to V$$
$$\mathbf{C} : K_1 \times K_2 \to V$$

Example 8.1

In Figure 4.3, the top three operations +, |, − correspond to element-wise addition $\oplus$ while the bottom operation & corresponds to the element-wise multiplication $\otimes$. Likewise, the *, CatKeyMul, and CatKeyVal operations depicted in Figures 4.5, 4.7, and 4.8 all correspond to array multiplication $\oplus.\otimes$.

Users of associative arrays are particularly interested in operations where the underlying operations $\oplus$ and $\otimes$ are "well-behaved," and where the array multiplication is "linear" relative to scalar multiplication and element-wise addition. Operations are deemed "well-behaved" when they satisfy the associative identities

$$\mathbf{A} \oplus (\mathbf{B} \oplus \mathbf{C}) = (\mathbf{A} \oplus \mathbf{B}) \oplus \mathbf{C}$$
$$\mathbf{A} \otimes (\mathbf{B} \otimes \mathbf{C}) = (\mathbf{A} \otimes \mathbf{B}) \otimes \mathbf{C}$$
$$\mathbf{A}(\mathbf{BC}) = (\mathbf{AB})\mathbf{C}$$

and the commutative identities

$$\mathbf{A} \oplus \mathbf{B} = \mathbf{B} \oplus \mathbf{A}$$
$$\mathbf{A} \otimes \mathbf{B} = \mathbf{B} \otimes \mathbf{A}$$

Element-wise multiplcation and array multiplication are considered to be linear if they also satisfy the distributive identities

$$\mathbf{A} \otimes (\mathbf{B} \oplus \mathbf{C}) = (\mathbf{A} \otimes \mathbf{B}) \oplus (\mathbf{A} \otimes \mathbf{C})$$
$$\mathbf{A}(\mathbf{B} \oplus \mathbf{C}) = \mathbf{AB} \oplus \mathbf{AC}$$

When defined using two operations $\oplus$ and $\otimes$ on the value set V, the array operations only satisfy the above conditions if associativity, commutativity, and distributivity of $\otimes$ over $\oplus$ hold for V as well. These properties are desirable because they allow one to carry a large amount of algebraic intuition from linear algebra into the realm of associative arrays. They also imply the following properties of scalar multiplication

$$u(v\mathbf{A}) = (uv)\mathbf{A}$$
$$u(v\mathbf{A}) = v(u\mathbf{A})$$
$$(u \oplus v)\mathbf{A} = (u\mathbf{A}) \oplus (v\mathbf{A})$$
$$u(\mathbf{A} \oplus \mathbf{B}) = (u\mathbf{A}) \oplus (u\mathbf{B})$$

where $u, v \in V$.

Determining when two operations $\oplus$ and $\otimes$ on the value set V result in well-behaved associative arrays is the subject of much of the remainder of this text. Delving into this question requires understanding the properties of V and then proving the appropriate abstract algebra concepts. Ultimately, these proofs lead to a well-defined associative array algebra.

8.2 Abstract Algebra for Computer Scientists and Engineers

The foundational properties of an associative array value set V are those of mathematical sets. Sets are based on specific mathematical principles that are deeply ingrained into human thinking [1]. The first is the concept of distinct or different objects or elements. If there are two kitchens or two bedrooms in a building, it is intuitively understood that they are distinct [2]. People do not confuse one kitchen with two kitchens or one bedroom with two bedrooms. Likewise, when there are two kitchens and another kitchen is added, it is clear that there are three kitchens and not three of something else. Adding another kitchen does not change the fact that these are kitchens. Furthermore, it doesn't matter what order the kitchens are added together. If there are two kitchens and one kitchen is added that is the same as having one kitchen and adding another two kitchens. Finally, if two kitchens are added to one bedroom, it is clear that there are not three kitchens or three bedrooms, but three rooms. Thus "threeness" is something that is somehow independent of whether something is a kitchen or a bedroom.

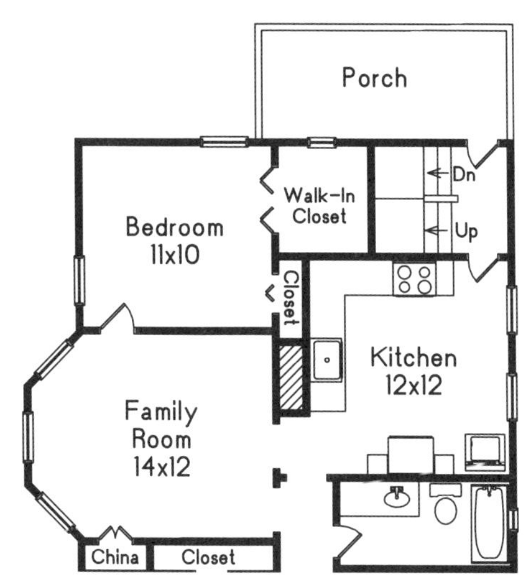

Figure 8.1
A floor plan of a house using standard architectural symbols for walls, doors, and windows. A floor plan provides a simple depiction of how the rooms are laid out in a house.

These mathematical principles are so intuitive that they require little explanation. In fact, other species are believed to share these mathematical concepts with humans [3]. However, for more difficult questions, further explanations are required. Such questions include what does it mean to have zero bricks? Can zero bricks be added to two bricks? What happens if bricks are taken away? What if more bricks are taken away than are already there? What does it mean to split a brick into two halves? When two bricks are added to one hammer, how does the result become three objects?

Mathematicians explore and answer these questions through a process of creating *definitions* of mathematical concepts, making *conjectures* about the properties that can be inferred from these concepts, and creating logical arguments that *prove* the conjectures, at which point the conjectures become *theorems*. A theorem can then be used as the basis of other conjectures or proofs. This process has been likened to building a house: the builder starts with a foundation and builds each floor (theorem) one on top of the other. The house analogy is only somewhat accurate because in building a real house there is generally a good idea of what the final house will look like and what its end purpose is. In mathematics, the house is being built, but no one really knows what it will end up looking like.

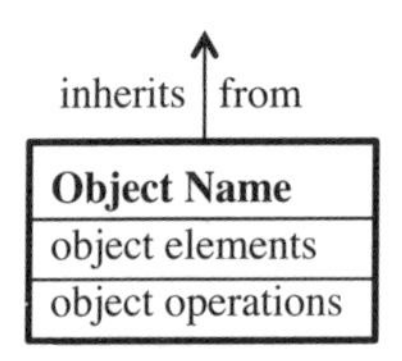

Figure 8.2
A class diagram represents a class of objects as a rectangle with three specific areas. The top area lists the name of the object in bold. The middle area lists the elements attributed to the object. The bottom area lists the operations that can be performed on the elements. Arrows indicate that one object inherits elements and operations from another object.

8.3 Depicting Mathematics

Computer programmers construct computer programs using a process with some similarities to how mathematicians construct mathematics. Classical computers only do binary operations on zeros and ones. Computer programs are merely complex combinations of these much simpler binary operations. If computer programmers could only write programs directly in terms of binary operations, then it would only be possible for humans to write the very simplest of programs. Computer programmers have become masters of creating classes of programming objects that build on top of each other [4–6]. These classes allow programmers to create more complex programs by creating and extending a few objects. In particular, all modern computer programming extensively utilizes a mechanism by which two different classes of objects can be viewed as extensions of a single more general class of object. The process by which two rocks and one bird can be viewed as three objects is at the very foundation of modern computer software.

Building a house to suit a purpose requires architectural drawings such as floor, site, elevation, and cross-section drawings (see Figure 8.1). Architectural drawings are used by architects for a number of purposes: to develop a design idea, to communicate concepts, to establish the merits of a design, to enable construction, to create a record of a completed building, and to make a record of a building that already exists. To effectively communicate these ideas, architectural drawings use agreed-upon formats and symbols [7]. A floor plan of house is not a house, nor does the floor plan tell a builder how to build a house, but a floor plan does provide a way to visually describe the rooms that will be in the house.

Building a computer program to suit a purpose requires diagrams such as class, activity, component, deployment, and use-case diagrams. Diagrams of computer programs are used by programmers for several purposes: to develop a design idea, to communicate concepts, to establish the merits of a design, to enable construction, to create a record of a completed program, and to make a record of a program that already exists. To effectively communicate these ideas, computer programmers use agreed-upon formats and symbols, such as the Universal Markup Language (UML) [8]. One of the most commonly used UML diagrams is a class diagram (see Figure 8.2). A UML class diagram depicts a class of objects as a

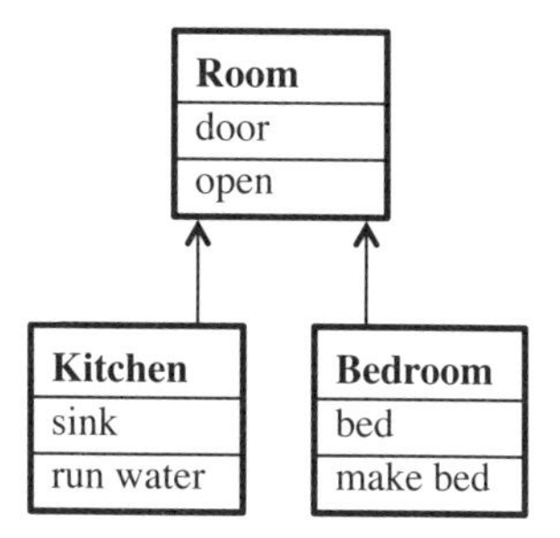

Figure 8.3
A class diagram showing that kitchens and bedrooms are both rooms, but are also distinct objects. All rooms have doors that can also be opened. Both kitchens and bedrooms inherit from rooms, so they also have doors that can be opened. In addition, kitchens have sinks that run water, and bedrooms have beds than can be made.

rectangle with three areas. The top area lists the name of the object in bold. The middle area lists the elements attributed to the object. The bottom area lists the operations that can be performed on the elements. Arrows indicate that one object inherits elements and operations from another object. Class diagrams of computer programs are not programs, nor do they tell the programmer how to write a program, but class diagrams provide a simple visual description of the objects that will be in the program.

UML class diagrams can be effective tools for representing the relationships among objects in other disciplines. Figure 8.3 depicts that kitchens and bedrooms are both two types of rooms. Rooms have doors that can be opened. Because kitchens and bedrooms inherit characteristics from rooms, they also have doors that can be opened. Kitchens also have sinks that run water, and bedrooms also have beds than can be made.

8.4 Associative Array Class Diagrams

Mathematics does not have a standardized diagram for depicting the elements of a mathematical object or proof. However, class diagrams can be used to describe mathematical objects in a simplified manner. Class diagrams of mathematics are not proofs, nor do they tell the mathematician how to write the proof, but class diagrams do provide a way to visualize all of the objects that will be in the proof. Using class diagrams, it is possible to summarize the interrelationships among the 32 mathematical objects that are used to define the algebra of associative arrays. For those who want to understand the algebra of associative arrays at a practical level, the visual depiction of these objects enables the understanding of the properties and eliminates the need to memorize the detailed definitions of each object.

A rapid overview of the important mathematical objects for defining associative arrays is presented subsequently. Wherever possible, the individuals most associated with these mathematical objects are mentioned to provide additional context. The development of these mathematical objects spans much of the mathematics that was developed in the 1800s and 1900s. The principal centers of work in these areas were found in Germany, France, and England, and later in the United States.

8.5 Set

The foundation of modern mathematics is modern set theory, which traces its beginning to the seminal paper by Russian-German mathematician George Cantor [9]. Some of Cantor's methods were different from the approaches advocated by Polish-German mathematician Leopold Kronecker, who was a leading mathematician at the University of Berlin (the center of mathematics research at the time). These disagreements were a significant challenge for Cantor and exacerbated his mental illness, but ultimately Cantor's approach to set theory became widely accepted. It would not be possible to rigorously encompass all the aspects of set theory that are relevant to this work. Thus, only a cursory description can be provided here.

The early development of set theory led to the observation of a number of paradoxes. The most well-known paradox is attributed to the British earl, humanitarian, Nobel laureate, and mathematician Bertrand Russell, who proposed the paradox "the set of all sets that are not members of themselves" [10]. The initial reaction to Russell's paradox was skepticism [11], but ultimately his paradox was accepted and led to a better definition of set theory. The foundations, or axioms, of set theory used as the basis of associative arrays were developed by German mathematician Ernst Zermelo [12] and German-Israeli mathematician Abraham Fraenkel [13]. Zermelo was a lecturer and assistant of Nobel physicist Max Planck at the University of Gottingen before he switched his interests to set theory. Fraenkel went on to became the first dean of mathematics at Hebrew University.

Sets are mathematical objects that contain elements and support several operations. Examples of set operations include equality, subsets, intersection, union, set difference, cartesian product (the creation of pairs of elements), the power set (the set of all subsets), and cardinality (the number of elements). For a standard set V with distinct element $v \in V$, there are a number of standard operations

- $U = V$ means that U and V have the same elements
- $U \subset V$ denotes that U is a subset of V
- $U \cap V$ denotes the intersection of sets U and V
- $U \cup V$ denotes the union of sets U and V
- $U \setminus V$ denotes the difference of sets U and V
- $U \times V$ denotes the cartesian product of sets U and V
- $\mathcal{P}(V)$ is the power set of V
- $|V|$ is the cardinality of V

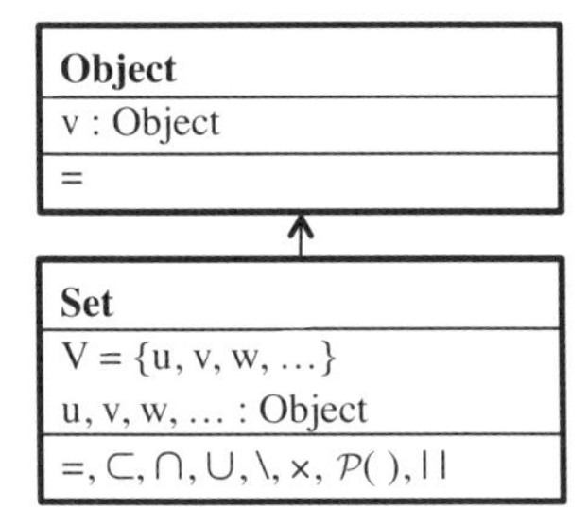

Figure 8.4
Class diagram depicting the relationship between an object and a set.

In a standard set, order is not important, nor are duplicates counted, and there is no underlying structure. The fact that sets have no underlying structure provides an empty slate from which to build upon and is why sets are the basis of modern mathematics.

Sets are built from elements, which are themselves mathematical objects that admit an equality operation so that an element can know itself. The relationship of sets and objects can be summarized in the class diagram shown in Figure 8.4. The definition is recursive. The element of a set is a mathematical object that has one element, which is also an object.

8.6 Semiring

Semirings are the primary mathematical object of interest for associative arrays. It is unclear if the term ring was meant to evoke an association among or a looping back of the elements of a set. *Semi* in this context means inverse elements are not required, such as fractions or negative numbers, but they may exist. A zero-sum-free semiring implies that there are no inverses. Thus, a semiring implies a collection of elements and corresponding operations without inverse elements. The term ring is attributed to German mathematician David Hilbert [14]. Hilbert was the leading professor of mathematics at the University of Gottingen during its zenith as the world's leading center of mathematical research. Much of the mathematics that will be subsequently described is a result of the work done at the University of Gottingen. Tragically, Hilbert had to witness the dismantling of his life's work during the 1930s.

The journey from sets to semirings is done in a well-defined series of steps. Each step adds properties that result in the definition of a new mathematical object. The class diagram shown in Figure 8.5 summarizes the relationships among these mathematical objects.

The journey to a semiring begins by defining a closed binary operation $\otimes$ such that

$$u \otimes v \in V$$

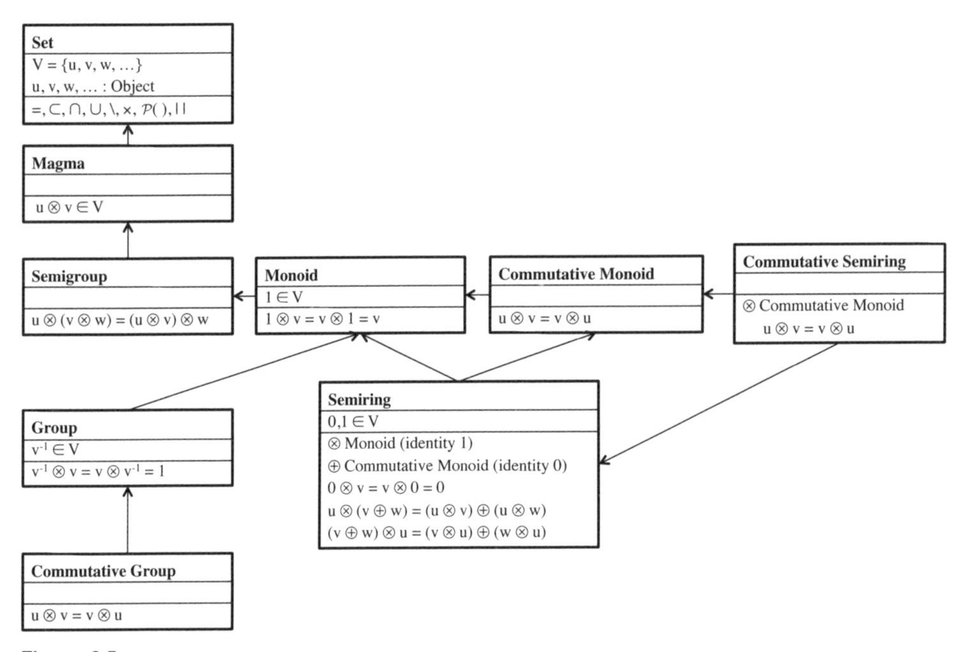

Figure 8.5
Class diagram depicting the construction of a semiring mathematical object.

A set with this property is referred to as a *magma*. Regrettably, the origin of the term magma is unclear. Magma is mentioned by the pseudonymous Nicolas Bourbaki collective of French mathematicians [15] and may come from the French word for a jumble or unstructured collection, which is consistent with its mathematical usage. If the binary operation in a magma has the additional property of being associative so that

$$u \otimes (v \otimes w) = (u \otimes v) \otimes w$$

then the mathematical object is a *semigroup*. In this case, a semigroup can be viewed as a group but with fewer properties. In this context, the prefix *semi* (as in semigroup, semiring, semimodule, or semialgebra) generally indicates that the object does not require inverse elements v^{-1} where

$$v \otimes v^{-1} = 1$$

The first documented use of the term *semi-groupe* is attributed to the French mathematician and independent scholar Jean Armand Marie Joseph de Seguier [16].

Including in a set the identity element 1 implies that

$$v \otimes 1 = v$$

A semigroup with an identity element is referred to as a *monoid*. Mono is a prefix meaning 1 and so it is natural that a monoid is distinguished by having an identity element (often denoted 1). The first recorded use of the term is associated with the French-American mathematician Claude Chevalley [17]. Extending a monoid to include the commutative property so that

$$u \otimes v = v \otimes u$$

results in a *commutative monoid*.

The aforementioned mathematical objects have so far been defined with respect to a single operation ⊗. Matrix mathematics and associative arrays require two operations ⊗ and ⊕. A *semiring* has two operations with two identity elements 1 and 0, where 1 is the identity element with respect to ⊗ and 0 is the identity element with respect to ⊕. More formally, a semiring contains a monoid ⊗ and a commutative monoid ⊕. To complete the definition requires specifying how ⊗ and ⊕ interact. For a semiring, 0 has the annihilator property

$$v \otimes 0 = 0$$

and ⊗ distributes over ⊕ such that

$$u \otimes (v \oplus w) = (u \otimes v) \oplus (u \otimes w)$$

Traditional arithmetic with $\oplus \equiv +$, $\otimes \equiv \times$ includes the properties of a semiring (along with other properties that make it a different mathematical object). One of the more interesting semirings is the tropical semiring ($\oplus \equiv \max$, $\otimes \equiv +$), which derives its name from the tropical location of one of its early pioneers, Brazilian mathematician Imre Simon [18]. Finally, a semiring where ⊗ is commutative is referred to as a *commutative semiring*.

A *group* is one of the foundational objects of abstract mathematics. A group extends a monoid by including inverse elements v^{-1} (or $-v$) where

$$v \otimes v^{-1} = 1$$

or

$$v \oplus -v = 0$$

The French mathematican Evariste Galois is credited as the first to use the modern term group (groupe in French). Galois' most formal definition of the group is spelled out in a letter written to a colleague on the eve of the duel that at age 29 would take his life [19]. A *commutative group* extends a group by requiring the binary operation to be commutative. Commutative groups are also called *Abelian* groups, an honor bestowed by French mathematician Camille Jordan on Norwegian mathematician Niels Henrik Abel for his early contributions to group theory. Tragically, like Galois, Abel also died at a young age (26) on the eve of obtaining his first stable job as a professor at the University of Berlin.

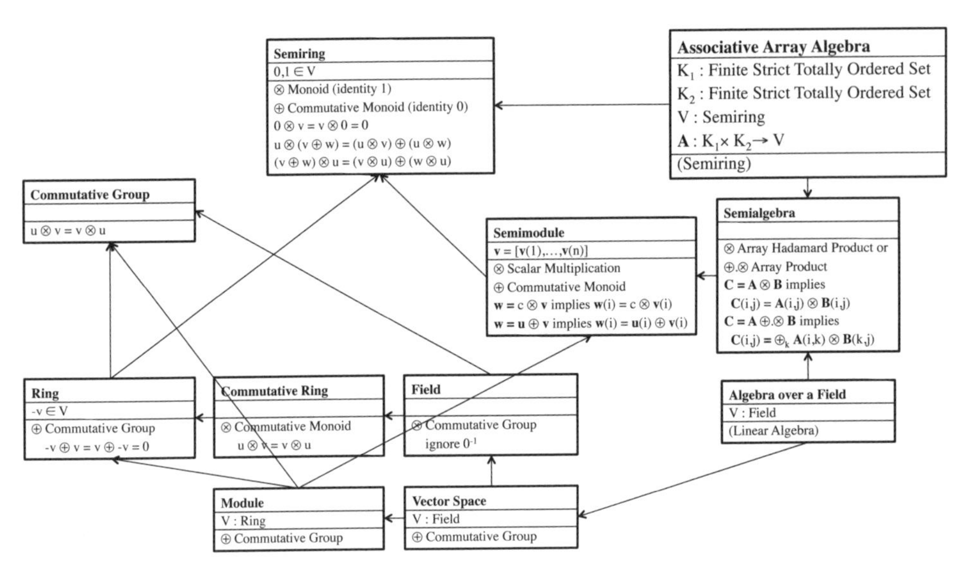

Figure 8.6
Class diagram of an algebra over a field (linear algebra).

8.7 Linear Algebra

Associative arrays have many properties that are similar to linear algebra. The journey from semirings to associative arrays is informed by linear algebra, which is more formally referred to as an algebra over a field. Figure 8.6 shows the class diagram depicting the relationships among the mathematical objects that start with a semiring and end up with linear algebra.

So far, the descriptions of the mathematical objects have been limited to operations on elements of sets. Organizing these elements into vectors **u**, **v**, and **w** with appropriate scalar multiplication

$$\mathbf{w} = c \otimes \mathbf{v} \quad \text{where} \quad \mathbf{w}(i) = c \otimes \mathbf{v}(i)$$

and element-wise addition

$$\mathbf{w} = \mathbf{u} \oplus \mathbf{v} \quad \text{where} \quad \mathbf{w}(i) = \mathbf{u}(i) \oplus \mathbf{v}(i)$$

results in a new mathematical object called a *semimodule*.

Extending a semimodule to matrices by including an element-wise (Hadamard) product

$$\mathbf{C} = \mathbf{A} \otimes \mathbf{B} \quad \text{where} \quad \mathbf{C}(i,j) = \mathbf{A}(i,j) \otimes \mathbf{B}(i,j)$$

or a matrix product

$$\mathbf{C} = \mathbf{AB} = \mathbf{A} \ \oplus.\otimes \ \mathbf{B} \quad \text{where} \quad \mathbf{C}(i,j) = \bigoplus_k \mathbf{A}(i,k) \otimes \mathbf{B}(k,j)$$

results in a *semialgebra*.

The term matrix was first coined by English mathematician James Joseph Sylvester in 1848 while he was working as an actuary with fellow English mathematician and lawyer Arthur Cayley [20]. Sylvester would later be the tutor of Florence Nightingale and go on to become the first mathematician at Johns Hopkins University. The term matrix was taken from the Latin word for womb, perhaps to evoke the concept of a womb of numbers.

A *ring* extends a semiring by replacing its commutative monoid operation $\oplus$ with a commutative group. The commutative group $\oplus$ in a ring also requires that additive inverses satisfying

$$v \oplus -v = 0$$

be included in the set V. A *commutative ring* extends a ring by replacing the monoid operation $\otimes$ with a commutative monoid that satisfies

$$u \otimes v = v \otimes u$$

A commutative ring is nearly equivalent to the normal arithmetic that is used in everyday transactions. The mathematics of adding, subtracting, multiplying, and dividing decimal numbers is done with a mathematical object referred to as a *field*. A field extends a commutative ring by replacing the commutative monoid operation $\otimes$ with the commutative group operation. The commutative group $\otimes$ in a field also requires that multiplicative inverses satisfying $v \otimes v^{-1} = 1$ be included in the set V. Furthermore, the zero inverse 0^{-1} is excluded from the set V. The term field is attributed to American mathematician and founder of the University of Chicago mathematics department Eliakim Hastings Moore [21]. Moore chose the English word field as a substitute for the German word zahlenkorper (body of numbers) coined by German mathematician Richard Dedekind (Gauss' last student) in 1858 [22].

Extending a semimodule by replacing the commutative monoid operation $\oplus$ with the commutative group operation satisfying

$$\mathbf{u} \oplus \mathbf{v} = \mathbf{v} \oplus \mathbf{u}$$

where the value set V also includes inverse elements

$$\mathbf{v} \oplus -\mathbf{v} = 0$$

results in a *module*. The term module is also attributed to Dedekind, who translated his own German term modul into the French term module [23].

The most commonly used form of vector mathematics is associated with the *vector space* mathematical object that extends a module by replacing the ring in the module with a field. The term vector space was used in the 1880s by English physicist Oliver Heaviside [24], who first wrote Maxwell's equations of electromagnetism in terms of vector terminology. The mathematical definition of a vector space (linear space) was first described by Italian

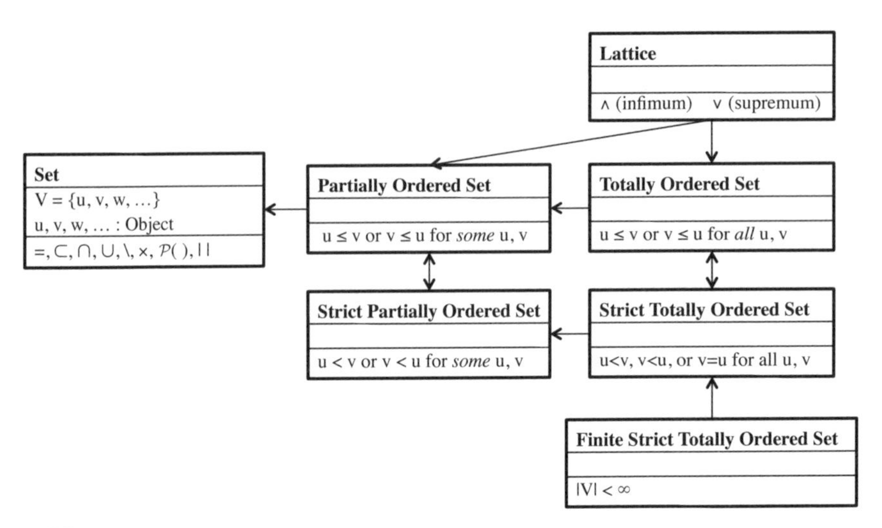

Figure 8.7
Class diagram depicting the relationships between various ordered sets.

mathematician Giuseppe Peano [25], who made many contributions to the axiomatization of mathematics.

Linear algebra (algebra over a field) is a short step from a semialgebra or a vector space. Linear algebra extends a semialgebra by replacing the semiring with a field. Likewise, linear algebra extends a vector space by including either an element-wise product or matrix product. The first book titled *Linear Algebra* was written in 1882 by Ottoman mathematician and general Hussein Tevfik Pacha [26]. The formalization of linear algebra was developed by many others, including Giuseppe Peano and David Hilbert.

8.8 Ordered Sets

Many of the interesting semirings for associative arrays depend strongly on the ordering of the elements in the value set. Likewise, for practical reasons, the row and column keys of an associative array are also ordered sets. The journey from a set to a finite strict totally ordered set passes through a number of mathematical objects. Figure 8.7 shows the class diagram of the relationships among the mathematical objects starting with a set and ending up with a strict totally ordered set.

The key concept for ordering the elements of set is extending an ordinary set with the $\leq$ relation. A set is said to be *partially ordered* if

$$v \leq v$$

for every $v \in V$ (reflexitivity), if

$$u \leq v \quad \text{and} \quad v \leq u$$

implies

$$v = u$$

for every $u, v \in V$ (anti-symmetry), and if

$$u \leq v \quad \text{and} \quad v \leq w$$

implies

$$u \leq w$$

for every $u, v, w \in V$ (transitivity). While some elements in the set V can be compared, not all pairs u, v must be comparable, and it is not required that

$$u \leq v \quad \text{or} \quad v \leq u$$

for every $u, v \in V$.

The term poset [27] is an abbreviation for partially ordered set and is credited to American mathematician and Harvard professor Garrett Birkhoff (who was the son of American mathematician and Harvard professor George Birkhoff). If any two $u, v \in V$ are comparable, so $u \leq v$ or $v \leq u$, then the set is said to be *totally ordered* [27, p2]. A totally ordered set extends a partially ordered set by stipulating that all elements can be compared.

The $\leq$ operator does not distinguish between $=$ and $<$. A *strict partially ordered set* extends (in the UML sense) a partially ordered set by specifying that there are some elements $u, v \in V$ where

$$u \leq v \quad \text{but not} \quad v \leq u$$

so that $u < v$. The above condition implies that some elements in the set V are less than other elements. If for every pair $u, v \in V$ there is exactly one of

$$u = v \quad \text{or} \quad u < v \quad \text{or} \quad v < u$$

then the set is defined to be a *strict totally ordered set*. Including the condition that the set V has a finite number of elements results in a *finite strict totally ordered set* of the type that is required to make the row keys and column keys of an associative array work in practice. Specifically, the fact that all elements are comparable means that the elements can be sorted and enables the efficient retrieval of the rows or columns of an associative array.

The $\leq$ relation naturally leads to the question of whether in a partially or totally ordered set V two elements have a least upper bound $\vee$ (supremum) and a greatest lower bound $\wedge$ (infimum). The existence of these bounds extends partially and totally ordered sets to become mathematical objects referred to as *lattices*. Lattice theory began when Dedekind

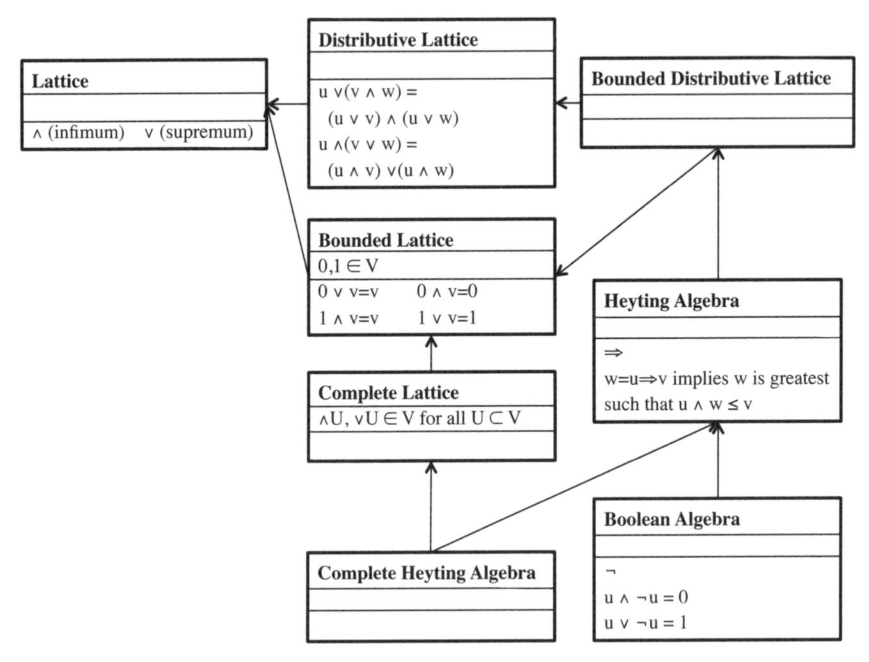

Figure 8.8
Class diagram depiction of Boolean algebra.

published the two fundamental papers that brought the theory to life well over 100 years ago [28, 29]. Lattices are better behaved than partially ordered sets lacking these bounds [30].

8.9 Boolean Algebra

Boolean algebra is one of the most important mathematical objects of the digital era. The term Boolean algebra is named for English-Irish mathematician and University of Cork professor George Boole, who laid the foundations of the field [31, 32]. In the preface of his 1847 book, Boole states

> What may be the final estimate of the value of the system, I have neither the wish nor the right to anticipate.

Little did Boole know that all of the functionality of digital computers would be based on the mathematics he pioneered.

Associative arrays work effectively with a wide range of value sets with various properties, including Boolean algebra. Boolean algebras have two values 0 and 1; two binary operations $\wedge$ (called *infimum*) and $\vee$ (called *supremum*); and a negation operation $\neg$. Figure 8.8 shows the class diagram that starts by extending lattices and leads to Boolean algebra.

A lattice has a pair of operations, and how these operations interact defines much of a lattice's behavior. Given elements $u, v, w \in V$ in a lattice, the lattice is a *distributive lattice* if the $\vee$ operation distributes over the $\wedge$ operation

$$u \vee (v \wedge w) = (u \vee v) \wedge (u \vee w)$$

and if the $\wedge$ operation distributes over the $\vee$ operation

$$u \wedge (v \vee w) = (u \wedge v) \vee (u \wedge w)$$

This behavior is beyond that of what is typically found in a semiring, whereby $\otimes$ distributes over $\oplus$

$$u \otimes (v \oplus w) = (u \otimes v) \oplus (u \otimes w)$$

but $\oplus$ does not distributes over $\otimes$

$$u \oplus (v \otimes w) \neq (u \oplus v) \otimes (u \oplus w)$$

If the elements 0 and 1 are the bounds of the lattice such that for $v \in V$

$$v \vee 0 = v$$

and

$$1 \wedge v = v$$

then the lattice is said to be a *bounded lattice*. Combining the properties of a distributive lattice and a bounded lattice produces a *bounded distributive lattice*.

A lattice is said to be a *complete lattice* if the infimum $\bigwedge U$ and supremum $\bigvee U$ exist for every subset $U \subset V$.

Given a pair of elements $u, v \in V$ that are part of a bounded distributive lattice, it is reasonable to ask what is the greatest element $w \in V$ such that

$$u \wedge w \leq v$$

This element is denoted

$$w = u \Rightarrow v$$

and the inclusion of the $\Rightarrow$ operator defined above results in a *Heyting algebra* (pronounced hating algebra) named after Dutch mathematician and University of Amsterdam professor Arend Heyting, who pioneered the concept [33]. If the lattice in a Heyting algebra is complete, then it is a *complete Heyting algebra*.

A Boolean algebra $(V, \wedge, \vee, \neg, 0, 1)$ is formed from a Heyting algebra by adding the $\neg$ operation with the complementary properties

$$v \vee \neg v = 1 \quad , \quad v \wedge \neg v = 0$$

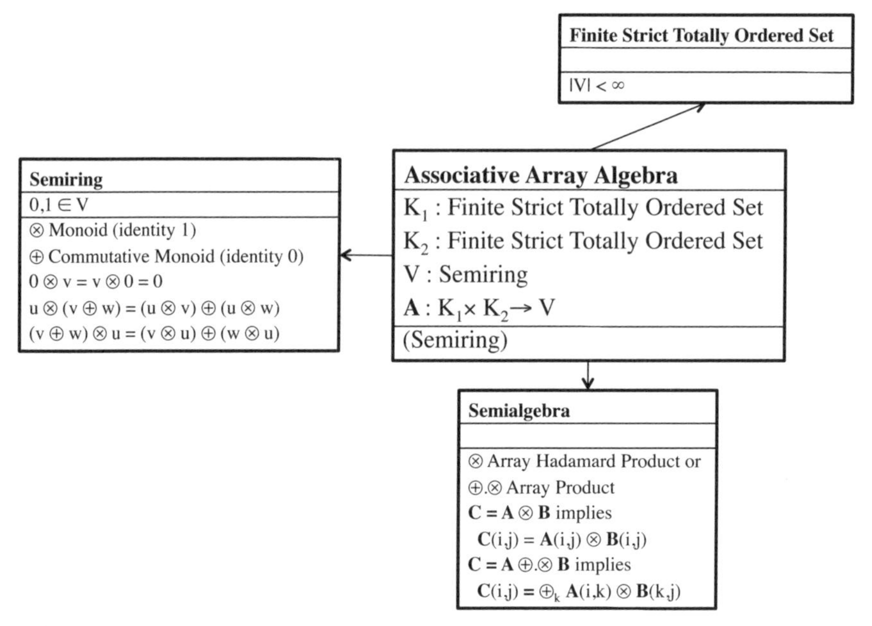

Figure 8.9
Class diagram depiction of an associative array algebra.

8.10 Associative Array Algebra

The previously described mathematical objects lay the foundation for associative arrays, which combines three mathematical objects: semiring, semialgebra, and finite strict totally ordered sets (see Figure 8.9). Associative arrays extend a semialgebra by allowing the row and column sets to be any strict totally ordered sets. Likewise, the value set of an associative array is a semiring. Furthermore, associative arrays themselves are also semirings.

The class diagram of all the mathematical objects needed to build an associative array algebra is shown in Figure 8.10. Class diagrams of mathematics are not proofs, nor do they show the mathematician how to create the proof, but class diagrams do provide a visual guide to the objects that will be in the proof. The class diagrams can serve as helpful reminders of the larger context of these objects and their relationship to associative arrays.

8.11 Conclusions, Exercises, and References

The properties of matrix mathematics are grounded in the abstract algebra objects whose formalization was the focus of much of mathematics in the past century. Demonstrating that these properties carry over into associative arrays requires defining associative array algebra in a similar manner. Associative array operations and matrix operations have many

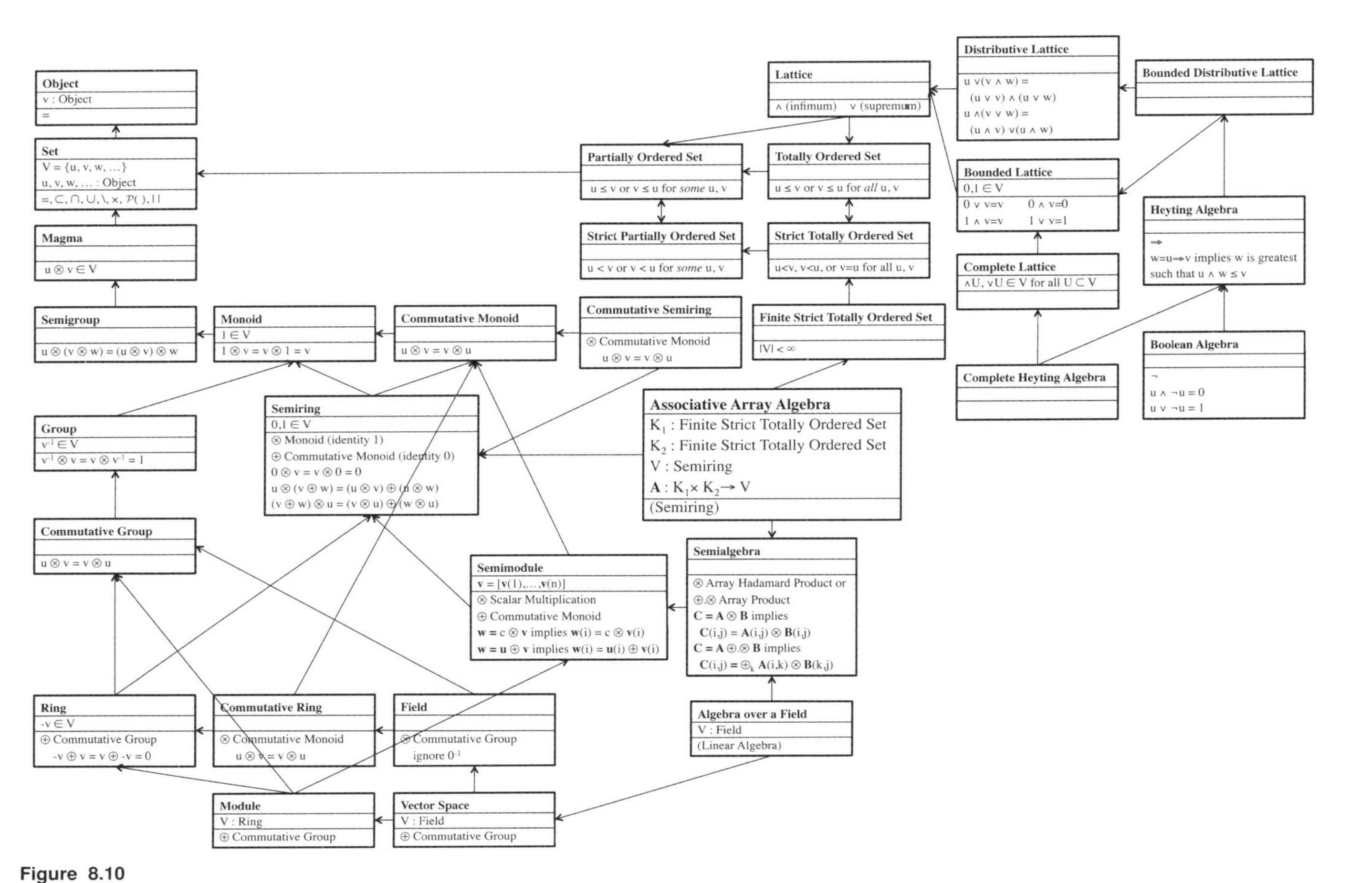

Figure 8.10
Complete UML depiction of mathematical objects underpinning associative array algebra. This diagram integrates the UML depiction of sets (Figure 8.4), semirings (Figure 8.5), semimodules (Figure 8.6), ordered sets (Figure 8.7), Boolean algebra (Figure 8.8), and associative array algebra (Figure 8.9).

similarities. Sets, semirings, linear algebra, ordered sets, and Boolean algebra all have properties that build upon each other. The relationships between and among these mathematical objects can be shown visually with class diagrams, which are used to depict similar relationships in software.

Exercises

Exercise 8.1 — Assume bathrooms have doors and sinks, but also have toilets that flush. Draw a picture showing how a bathroom could be added to the class diagram in Figure 8.3.

Exercise 8.2 — List the standard properties of a set.

Exercise 8.3 — Based on Figure 8.5, describe the properties that are added to the following objects when they go from

(a) semigroup to monoid

(b) monoid to commutative monoid

(c) monoid to group

(d) group to commutative group

Exercise 8.4 — Based on Figure 8.6, describe the objects that each of the following objects inherits from

(a) ring

(b) commutative ring

(c) field

(d) module

(e) vector space

Exercise 8.5 — Based on Figure 8.7, describe the properties that are added to the following objects when they go from

(a) set to partially ordered set

(b) partially ordered set to totally ordered set

(c) partially ordered set to strict partially ordered set

(d) strict partially ordered set to strict totally ordered set

Exercise 8.6 — Can Figure 8.10 be drawn on sheet of paper such that no lines cross?

Exercise 8.7 — Given the formulation for associative array multiplication where $\oplus$ is standard arithmetic addition and $\otimes$ is standard arithmetic multiplication

$$\mathbf{C}(k_1,k_2) = \sum_{k_3} \mathbf{A}(k_1,k_3) \times \mathbf{B}(k_3,k_2)$$

use the associative array **A** in Figure 5.2 to compute the following via array multiplication

(a) **AA**

(b) $(\mathbf{AA})^{\mathsf{T}}$

Exercise 8.8 — Using small associative arrays and $\oplus \equiv$ max and $\otimes \equiv$ min, write down examples that illustrate the associativity property of element-wise addition, element-wise multiplication, and array multiplication.

Exercise 8.9 — Using small associative arrays and $\oplus \equiv$ max and $\otimes \equiv$ min, write down examples that illustrate the commutativity property of element-wise addition and element-wise multiplication.

Exercise 8.10 — Using small associative arrays and $\oplus \equiv$ max and $\otimes \equiv$ min, write down examples that illustrate the distributivity property of element-wise multiplication with element-wise addition and array multiplication with element-wise addition.

Exercise 8.11 — Drawing from your own experience, select a topic you know well and create a class diagram illustrating relationships in that topic.

References

[1] C. R. Gallistel and R. Gelman, "Non-verbal numerical cognition: From reals to integers," *Trends in Cognitive Sciences*, vol. 4, no. 2, pp. 59–65, 2000.

[2] C. B. Boyer and U. C. Merzbach, *A History of Mathematics*. Hoboken, NJ: John Wiley & Sons, 2011.

[3] K. Macpherson and W. A. Roberts, "Can dogs count?," *Learning and Motivation*, vol. 44, no. 4, pp. 241–251, 2013.

[4] A. Goldberg and D. Robson, *Smalltalk-80: the Language and its Implementation*. Boston: Addison-Wesley Longman, 1983.

[5] B. Stroustrup, *The C++ Programming Language*. Addison-Wesley, 1986.

[6] K. Arnold and J. Gosling, *The Java Programming Language*. Boston: Addison Wesley, 1996.

[7] "United States National CAD Standard." http://www.nationalcadstandard.org. Accessed: 2017-04-08.

[8] G. Booch, J. Rumbaugh, and I. Jacobson, *The Unified Modeling Language User Guide*. Boston: Addison Wesley, 1998.

[9] G. Cantor, "Ueber eine eigenschaft des inbegriffs aller reellen algebraischen zahlen," *Journal für die Reine und Angewandte Mathematik*, vol. 77, pp. 258–262, 1874.

[10] B. Russell, "Letter to Gottlob Frege," June 16 1902.

[11] G. Frege, "Letter to Bertand Russell," June 22 1902.

[12] E. Zermelo, "Untersuchungen über die grundlagen der mengenlehre. i," *Mathematische Annalen*, vol. 65, no. 2, pp. 261–281, 1908.

[13] A. Fraenkel, "Zu den grundlagen der cantor-zermeloschen mengenlehre," *Mathematische Annalen*, vol. 86, no. 3, pp. 230–237, 1922.

[14] D. Hilbert, *Zahlbericht (Report on the theory of numbers)*. Berlin, Heidelberg: Springer, 1897.

[15] N. Bourbaki, P. M. Cohn, and J. Howie, *Algebra*. Paris: Hermann, 1974.

[16] J.-A. de Séguier, *Éléments de la théorie des groupes abstraits*. Paris: Gauthier-Villars, 1904.

[17] C. Chevalley, *Fundamental Concepts of Algebra*, vol. 7. Academic Press, 1957.

[18] I. Simon, "Recognizable sets with multiplicities in the tropical semiring," in *Mathematical Foundations of Computer Science 1988*, pp. 107–120, Berlin, Heidelberg: Springer, 1988.

[19] E. Galois, "Lettre à auguste chevalier," May 29 1832.

[20] A. Cayley, "A memoir on the theory of matrices," *Philosophical Transactions of the Royal Society of London*, vol. 148, pp. 17–37, 1858.

[21] E. H. Moore, "A doubly-infinite system of simple groups," *Bulletin of the American Mathematical Society*, vol. 3, no. 3, pp. 73–78, 1893.

[22] W. Scharlau, "Unpublished algebraic works of Dedekind, Richard from his Gottingen era 1855-1858," *Archive for History of Exact Sciences*, vol. 27, no. 4, pp. 335–367, 1982.

[23] R. Dedekind, "Sur la théorie des nombres entiers algébriques," *Bulletin des Sciences Mathématiques et Astronomiques*, vol. 1, no. 1, pp. 207–248, 1877.

[24] O. Heaviside, "Lxii. on resistance and conductance operators, and their derivatives, inductance and permittance, especially in connexion with electric and magnetic energy," *The London, Edinburgh, and Dublin Philosophical Magazine and Journal of Science*, vol. 24, no. 151, pp. 479–502, 1887.

[25] G. Peano, *Calcolo geometrico secondo l'Ausdehnungslehre di H. Grassmann: preceduto dalla operazioni della logica deduttiva*, vol. 3. fratelli Bocca, 1888.

[26] H. T. Pacha, *Linear Algebra*. Constantinople: A.H. Boyajian, 1882.

[27] G. Birkhoff, *Lattice Theory*, vol. 25. New York: American Mathematical Society, 1948.

[28] R. Dedekind, "Über zerlegungen von zahlen durch ihre grössten gemeinsamen theiler," in *Fest-Schrift der Herzoglichen Technischen Hochschule Carolo-Wilhelmina*, pp. 1–40, Berlin, Heidelberg: Springer, 1897.

[29] R. Dedekind, "Über die von drei moduln erzeugte dualgruppe," *Mathematische Annalen*, vol. 53, no. 3, pp. 371–403, 1900.

[30] G.-C. Rota, "The many lives of lattice theory," *Notices of the AMS*, vol. 44, no. 11, pp. 1440–1445, 1997.

[31] G. Boole, *The Mathematical Analysis of Logic: Being an Essay Towards a Calculus of Deductive Reasoning*. Cambridge: MacMillan, Barclay, & MacMillan, 1847.

[32] G. Boole, *An Investigation of the Laws of Thought: On which are Founded the Mathematical Theories of Logic and Probabilities*. New York: Dover Publications, 1854.

[33] A. Heyting, *Die formalen Regeln der intuitionistischen Mathematik*. Verlag der Akademie der Wissenschaften, 1930.

9 Defining the Algebra of Associative Arrays

Summary

The overarching properties of associative arrays are determined by the underlying properties of the values of the entries in an associative array. The properties of the values are determined by the formal definitions of the underlying mathematical objects. Understanding these mathematical objects requires some of the terminology of modern abstract algebra, with particular emphasis on semirings. Many of the interesting semirings for associative arrays depend strongly on the ordering of the elements in the value set. For this reason, the general terminology and theory of lattices are explained. This chapter rigorously defines the mathematical objects necessary to construct associative arrays and culminates with definitions of the semirings of greatest interest for practical data analysis applications.

9.1 Operations on Sets

The previous chapter provided an overview of the mathematical objects necessary for associative arrays. This chapter begins the process of defining these objects more rigorously.

The foundational object of abstract algebra is a set that is defined by axioms. The set axioms that are most relevant to associative arrays are the Zermelo-Fraenkel-Choice (ZFC) axioms (see [1–3] for definitions and examples, therein).

Extensionality — Two sets U and V are equal if and only if they have the same elements.

Regularity — Every set V contains an element v disjoint from V so that $v \cap V = \emptyset$.

Schema of Specification — Any definable subclass of a set is a set. If f is a definable rule on the elements of a set V, then $V_f = \{v \in V \mid f(v)\}$ is also a set.

Pairing — If U and V are sets, then the set of this pair of sets $\{U, V\}$ is also a set.

Union — The union of all of the elements of a set $\bigcup_{v \in V} v$ is a set.

Schema of Replacement — If f is a definable function on the elements of set V, then $f(V)$ is also a set.

Infinity — The set of natural numbers $\mathbb{N} = \{0, 1, 2, \ldots\}$ exists.

Power Set — The power set $\mathcal{P}(V)$ of a set V exists.

Choice — For every set V, there is binary relation that well orders the elements of V.

A binary operation $\otimes$ on a set V is a map $V \times V \to V$. An arbitrary binary operation is of limited utility, and it is helpful to put conditions on these maps. For any elements $u,v,w \in V$, these properties typically include associativity

$$(u \otimes v) \otimes w = u \otimes (v \otimes w)$$

commutativity

$$u \otimes v = v \otimes u$$

the existence of an identity

$$v \otimes 1 = 1 \otimes v = v$$

and the existence of inverses

$$v \otimes v^{-1} = v^{-1} \otimes v = 1$$

These properties are particularly nice because they make computation much simpler and provide for algebraic manipulation of expressions and equations. There exists specific terminology for sets equipped with binary operations that satisfy some or all of the above properties.

Associativity is particularly useful because it allows expressions like

$$u \otimes v \otimes w$$

to be regarded unambiguously, allowing the definition of a repeated $\otimes$ operation and making exponents possible, such as

$$v \otimes v \otimes v = v^3$$

For this reason, associativity will be one of the first assumptions applied to any binary operation being considered. The existence of an identity is useful because it is a precursor to the existence of inverse elements and shows up regularly even when there are not inverses. The utility of associativity and an identity motivates the following definitions [4].

Definition 9.1

Monoid

A set V with a binary operator $\otimes$ is a *monoid*, denoted $(V, \otimes, 1)$, if

1. $(u \otimes v) \otimes w = u \otimes (v \otimes w)$ for every $u,v,w \in V$ (*associativity*) and
2. there exists an element $1 \in V$ such that $v \otimes 1 = 1 \otimes v = v$ (*existence of an identity*).

V is typically called the *underlying set* of the monoid $(V, \otimes, 1)$.

If $\otimes$ also satisfies the identity

$$u \otimes v = v \otimes u$$

for every $u,v \in V$ (*commutativity*), then $(V, \otimes, 1)$ is said to be a *commutative monoid*.

Definition 9.2

Group

If $(V, \otimes, 1)$ is a monoid, and v is an element of V, then an element v^{-1} of V is said to be an *inverse* of v if

$$v \otimes v^{-1} = v^{-1} \otimes v = 1$$

When there exists an inverse for every element of V, the monoid $(V, \otimes, 1)$ is said to be a *group*.
If $\otimes$ also satisfies the identity

$$u \otimes v = v \otimes u$$

for every $u, v \in V$ (*commutativity*), then group $(V, \otimes, 1)$ is said to be a *commutative group*.

Example 9.1

The set of natural numbers $\mathbb{N} = \{0, 1, 2, \ldots\}$ combined with the binary addition operator $+$ and identity element 0 forms a commutative monoid $(\mathbb{N}, +, 0)$.

Example 9.2

The set of natural numbers $\mathbb{N} = \{0, 1, 2, \ldots\}$ combined with the binary multiplication operator $\times$ and identity element 1 forms a commutative monoid $(\mathbb{N}, \times, 1)$.

The commutative monoids $(\mathbb{N}, +, 0)$ and $(\mathbb{N}, \times, 1)$ are not groups, as no element has an inverse except the identity element. No natural number can be added to another natural number and equal 0 except $0 + 0 = 0$. Likewise, no natural number can be multiplied by another natural number and equal 1 except $1 \times 1 = 1$.

Having looked at a single binary operation on a set, it is now useful to look at sets with two binary operations. A common property used to connect a pair of binary operations $\otimes$ and $\oplus$ is the distributive property. A pair of binary operations is said to be left distributive when

$$u \otimes (v \oplus w) = (u \otimes v) \oplus (u \otimes w)$$

and right distributive when

$$(u \oplus v) \otimes w = (u \otimes w) \oplus (v \otimes w)$$

where

$$u, v, w \in V$$

A pair of operations is called *distributive* if it is both left and right distributive.

It is also typical to require that any identity element 0 of $\oplus$ satisfies the property that

$$0 \otimes v = v \otimes 0 = 0$$

The above properties motivate the formal definition of a semiring.

Definition 9.3

Semiring

[5] A *semiring* denoted

$$(V, \oplus, \otimes, 0, 1)$$

is a set V equipped with two binary operations, addition

$$\oplus : V \times V \to V$$

and multiplication

$$\otimes : V \times V \to V$$

which satisfy the following axioms for any $u, v, w \in V$

1. $(V, \oplus, 0)$ is a commutative monoid and $(V, \otimes, 1)$ is a monoid.
2. $u \otimes (v \oplus w) = (u \otimes v) \oplus (u \otimes w)$ and $(v \oplus w) \otimes u = (v \otimes u) \oplus (w \otimes u)$ (*distributivity*).
3. 0 annihilates V: $v \otimes 0 = 0 \otimes v = 0$.

When $(V, \otimes, 1)$ is a commutative monoid, the semiring $(V, \oplus, \otimes, 0, 1)$ is said to be a *commutative semiring*.

Definition 9.4

Ring

When $(V, \oplus, 0)$ is a commutative group, the semiring $(V, \oplus, \otimes, 0, 1)$ is said to be a ring.
When the ring $(V, \oplus, \otimes, 0, 1)$ is also a commutative semiring, then it is called a *commutative ring*.

Definition 9.5

Field

If both $(V, \oplus, 0)$ and $(V \setminus \{0\}, \otimes, 1)$ are commutative groups, then $(V, \oplus, \otimes, 0, 1)$ is said to be a *field*.

Example 9.3

The standard arithmetic that is first taught to children is a semiring, and this is the "standard" semiring structure on $\mathbb{N}$. More specifically, the set of natural numbers $\mathbb{N} = \{0,1,2,\ldots\}$ combined with standard addition + and multiplication $\times$ has additive identity 0 and multiplicative identity 1. Combined together, the set, operations, and properties form the commutative semiring

$$(\mathbb{N},+,\times,0,1)$$

Example 9.4

The set integers $\mathbb{Z} = \{\ldots,-1,0,1,\ldots\}$ with standard addition + and multiplication $\times$ is a ring

$$(\mathbb{Z},+,\times,0,1)$$

$\mathbb{Z}$ supports a ring because all $\mathbb{Z}$ includes all the additive inverses, such as $1+-1=0$.

Example 9.5

The set of matrices $\mathbf{A} \in \mathbb{S}^{m\times m}$ with entries in the ring $(\mathbb{S},\oplus,\otimes,0,1)$ is also a ring. Specifically, these matrices with element-wise matrix addition $\otimes$, element-wise matrix multiplication $\otimes$, additive identity 0, and ones matrix $\mathbb{1}$ form a ring

$$(\mathbb{S}^{n\times m},\otimes,\oplus,0,\mathbb{1})$$

Example 9.6

The set of matrices $\mathbf{A} \in \mathbb{S}^{m\times m}$ with entries in the ring $(\mathbb{S},\oplus,\otimes,0,1)$ is also a ring. Specifically, the $m \times m$ matrices with element-wise matrix addition $\otimes$, matrix multiplication $\mathbf{C} = \mathbf{A}{\oplus}.{\otimes}\mathbf{B}$, additive identity 0, and identity matrix $\mathbb{I}$ form a ring

$$(\mathbb{S}^{n\times n},\oplus,\oplus.\otimes,0,\mathbb{I})$$

Example 9.7

The set $\mathbb{R}_+ = [0,+\infty)$ of non-negative real numbers with standard addition +, multiplication $\times$, additive identity 0, and multiplicative identity 1 forms a commutative semiring

$$(\mathbb{R}_+,+,\times,0,1)$$

but not a ring. This is the "standard" semiring structure on $\mathbb{R}_+$.

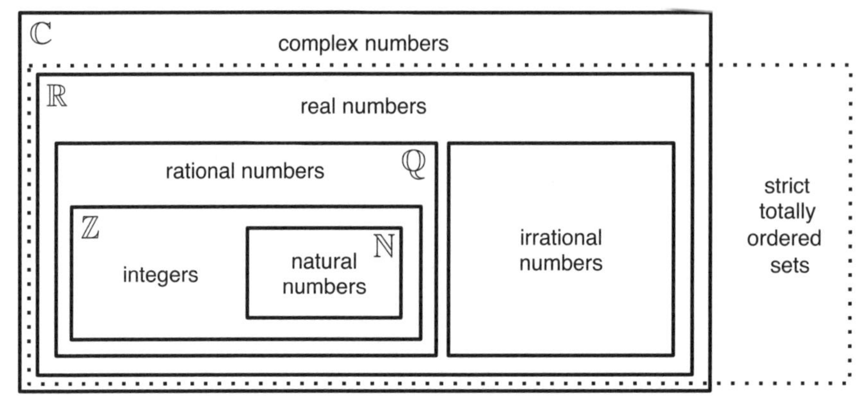

Figure 9.1
Venn diagram depiction of the relationship between various sets of numbers: $\mathbb{N} \subset \mathbb{Z} \subset \mathbb{Q} \subset \mathbb{R} \subset \mathbb{C}$. The sets that have a unique ordering are shown inside the dotted region.

Example 9.8

The set of rational numbers $\mathbb{Q} = \{m/n \mid n \neq 0, m, n \in \mathbb{N}\}$, with standard addition $+$ and multiplication $\times$ is a field

$$(\mathbb{Q}, +, \times, 0, 1)$$

Example 9.9

The set of real numbers $\mathbb{R} = (-\infty, +\infty)$, with standard addition $+$ and multiplication $\times$ is a field

$$(\mathbb{R}, +, \times, 0, 1)$$

Example 9.10

The set of complex numbers $\mathbb{C} = \{x + y\sqrt{-1} \mid x, y \in \mathbb{R}\}$, with standard complex addition $+$ and complex multiplication $\times$ is a field

$$(\mathbb{C}, +, \times, 0, 1)$$

The above examples deal with a variety of sets that will be important in the subsequent chapters. The sets include the natural numbers

$$\mathbb{N} = \{0, 1, 2, \ldots\}$$

the integers

$$\mathbb{Z} = \{\ldots, -1, 0, 1, \ldots\}$$

the rational numbers

$$\mathbb{Q} = \{m/n \mid n \neq 0, m, n \in \mathbb{N}\}$$

(where equivalent fractions are identified), the real numbers

$$\mathbb{R} = (-\infty, +\infty)$$

and the complex numbers

$$\mathbb{C} = \{x + y\sqrt{-1} \mid x, y \in \mathbb{R}\}$$

The above sets have the following subset relationships

$$\mathbb{N} \subset \mathbb{Z} \subset \mathbb{Q} \subset \mathbb{R} \subset \mathbb{C}$$

These subset relationships are shown in Figure 9.1. These sets are all common choices for the value set V in an associative array. An additional important characteristic on sets is the ordering of the elements of the set with respect to the $<$ operation. A set with a unique order to the elements is referred to as a strict totally ordered set. These sets can include non-numeric elements, such as words or strings, and are a common choice for the value set V in an associative array. Furthermore, the row key and column key sets K_1 and K_2 must be strict totally ordered sets in order to guarantee efficient implementations. The properties of different orderings on sets are discussed in greater detail in the next section.

9.2 Ordered Sets

The examples given above of semirings are all extremely interconnected, being based on subsets of the complex numbers $\mathbb{C}$. Other examples make use of the ordering properties of the underlying sets [6]. As with binary operators on a set, the possible orders on a set are restricted by additional properties [7].

Definition 9.6

Partially Ordered Set

A partially ordered set V, also denoted $(V, \leq)$, is a set V with a relation $\leq$, called the *partial order*, on V satisfying

Reflexivity — $v \leq v$ for every $v \in V$,

Antisymmetry — $u \leq v$ and $v \leq u$ implies $u = v$ for every comparable pair $u, v \in V$, and

Transitivity — $u \leq v$ and $v \leq w$ implies $u \leq w$ for every comparable triple $u, v, w \in V$.

Definition 9.7

Totally Ordered Set

A totally ordered set $(V, \leq)$ is a partially ordered set $(V, \leq)$ such that for every pair $u, v \in V$, either $u \leq v$ or $v \leq u$, a property called *totality*. If neither $u \leq v$ nor $v \leq u$ holds, then u and v are *incomparable*. When the partial order $\leq$ is total, $\leq$ is called a *total order*.

Example 9.11

The natural numbers with standard less-than-or-equal-to are a totally ordered set $(\mathbb{N}, \leq)$.

Example 9.12

The integers with standard less-than-or-equal-to are a totally ordered set $(\mathbb{Z}, \leq)$.

Example 9.13

The rational numbers with standard less-than-or-equal-to are a totally ordered set $(\mathbb{Q}, \leq)$.

Example 9.14

The real numbers with standard less-than-or-equal-to are a totally ordered set $(\mathbb{R}, \leq)$.

Example 9.15

The complex numbers with the dictionary order $x + y\sqrt{-1} \leq z + w\sqrt{-1}$ (so either $x \leq z$ or $x = z$ and $y \leq w$) are a partially ordered set $(\mathbb{C}, \leq)$.

Example 9.16

The words {blue, green, orange, pink, red, silver, white} sorted lexicographically are a totally ordered set.

Figure 9.2 depicts some of the properties of ordered sets from Figure 5.10. Recall that Figure 5.10 depicted a series of colored lines whose layering implies constraints on the order that the colors were painted. These colors form a partially ordered set. From the painting, certain orderings are imposed as a requirement. For example, the order must

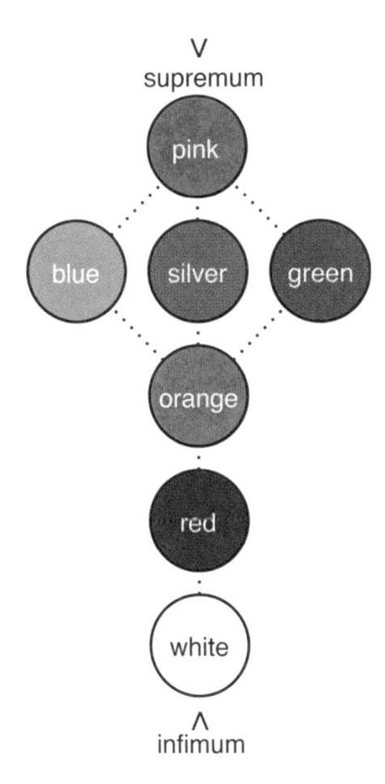

Figure 9.2
Hasse diagram of partially ordered sets from the color orderings listed in Figure 5.10. The sets form a partially ordered set. Comparable pairs are marked with dotted lines. The top of the diagram is denoted the supremum and the bottom is denoted the infimum.

begin with the white of the canvas, followed by the red of the background, followed by orange, then some ordering of blue, silver, green, and finally pink. This constraint means that a compatible ordered list of colors can be generated by first starting with

$$\text{white} < \text{red} < \text{orange} < \text{pink}$$

and then inserting any ordering of blue, silver, and green between orange and pink. There are six possible total orderings constructible from the partial ordering. In Figure 9.3, the colors are arranged to form all the totally ordered sets in which all of the colors are now comparable. The proposed orders are compatible with that in Figure 9.2.

9.3 Supremum and Infimum

Ultimately, it is intended that the ordering on a partially ordered set give rise to two binary operations that make the partially ordered set into a semiring. While at first it may not be clear what these operations might be in the general case, in the specific case of totally ordered sets, it is simpler. Given two elements $u, v \in V$ in a totally ordered set with total order operation $\leq$, the totality condition states that either $u \leq v$ or $v \leq u$. In particular, it is guaranteed that one element is at least as large as the other and that one is at most as small

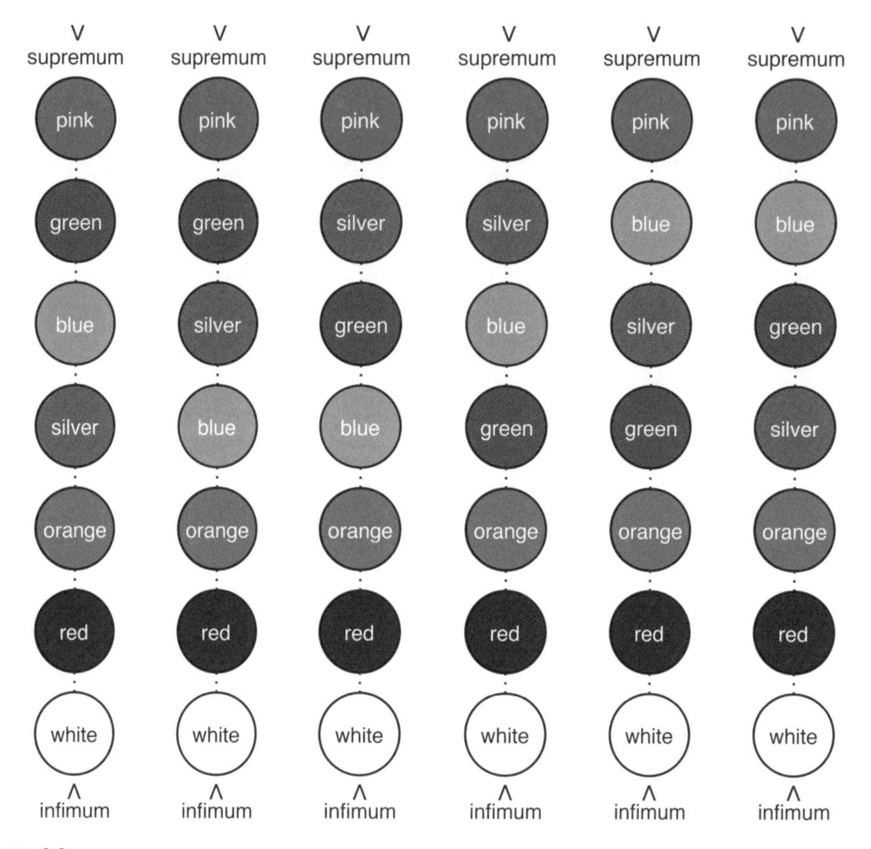

Figure 9.3
Hasse diagrams of totally ordered sets from the color ordering listed in Figure 5.10. The orders are compatible with those in Figure 9.2. The top of each Hasse diagram is the supremum and the bottom is the infimum.

as the other. In effect, there exists a *maximum* and *minimum* of the set $\{u, v\}$. The maximum of $\{u, v\}$ is denoted $\max(u, v)$ while the minimum is denoted $\min(u, v)$.

The above description of maximum and minimum leads naturally to the question: When do such a maximum and minimum exist? More specifically, given two elements $u, v \in V$ in a partially ordered set, does there exist a maximum or minimum? If there were such a maximum, then either $u \leq v$ or $v \leq u$, depending on whether v or u were the maximum, respectively. Such a maximum implies that $(V, \leq)$ is totally ordered. However, there are partial orders that are not total orders, such as is shown in Figure 9.2. A simple mathematical example is the power set of a simple set like $\{0, 1\}$ with respect to the subset operation $\subset$

$$\mathcal{P}(\{0, 1\}) = \{\emptyset, \{0\}, \{1\}, \{0, 1\}\}$$

In the above set,

$$\{0\} \subset \{0,1\} \quad \text{and} \quad \{0,1\} \not\subset \{0\}$$

which satisfies the requirements of a total order. Likewise,

$$\{1\} \subset \{0,1\} \quad \text{and} \quad \{0,1\} \not\subset \{1\}$$

also satisfies the requirements of a total order. However,

$$\{0\} \not\subset \{1\} \quad \text{and} \quad \{1\} \not\subset \{0\}$$

shows that $\subset$ does not satisfy the requirements of a total order. Thus $(\mathcal{P}(\{0,1\}),\subset)$ is partially ordered but not totally ordered. In this case, it is still possible to ask whether there exists a smallest element greater than u and v, and likewise a greatest element smaller than u and v.

Definition 9.8

Supremum

Let $(V,\leq)$ be a partially ordered set and U a subset of V. The notion of the least element in V greater than all of U is as follows: v is an *upper bound* of U if $u \leq v$ for all $u \in U$ and is the *supremum* or *least upper bound* if it is the smallest among all upper bounds. It is denoted by

$$\bigvee U \quad \text{or} \quad \bigvee_{u \in U} u \quad \text{or} \quad \sup U$$

Definition 9.9

Infimum

Let $(V,\leq)$ be a partially ordered set and U a subset of V. The notion of the greatest element in V less than all of U is as follows: v is a *lower bound* of U if $v \leq u$ for all $u \in U$ and is the *infimum* or *greatest lower bound* if it is the largest among all lower bounds. It is denoted by

$$\bigwedge U \quad \text{or} \quad \bigwedge_{u \in U} u \quad \text{or} \quad \inf U$$

The concepts of supremum and infimum extend to functions. Figure 9.4 depicts a monotonically increasing function f that preserves supremum and infimum. In particular, given inputs x_i and outputs $f(x_i)$

$$x_1 = \bigwedge_{n=1}^{4} x_n \quad \text{and} \quad x_4 = \bigvee_{n=1}^{4} x_n$$

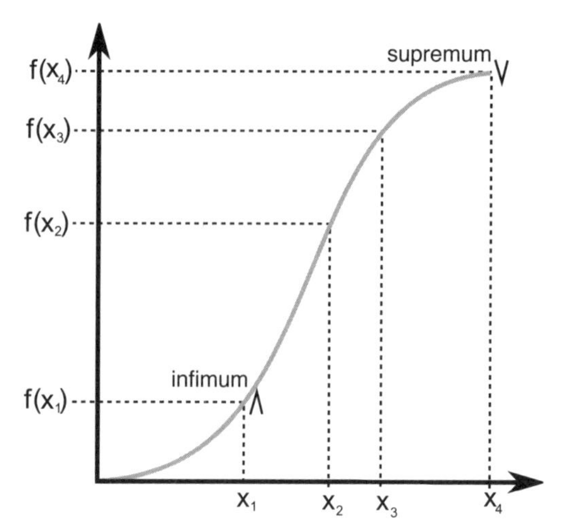

Figure 9.4
Monotonically increasing function that preserves supremum and infimum.

and

$$f(x_1) = \bigwedge_{n=1}^{4} f(x_n) \quad \text{and} \quad f(x_4) = \bigvee_{n=1}^{4} f(x_n)$$

so

$$f\left(\bigwedge_{n=1}^{4} x_n\right) = \bigwedge_{n=1}^{4} f(x_n) \quad \text{and} \quad f\left(\bigvee_{n=1}^{4} x_n\right) = \bigvee_{n=1}^{4} f(x_n).$$

Therefore, f preserves suprema and infima.

Figure 9.5 depicts a function that is not monotonically increasing and does not preserve suprema and infima . In particular, given inputs x_i and outputs $f(x_i)$

$$x_1 = \bigwedge_{n=1}^{4} x_n \quad \text{and} \quad x_4 = \bigvee_{n=1}^{4} x_n$$

but

$$f(x_3) = \bigwedge_{n=1}^{4} f(x_n) \quad \text{and} \quad f(x_2) = \bigvee_{n=1}^{4} f(x_n)$$

so

$$f\left(\bigwedge_{n=1}^{4} x_n\right) \neq \bigwedge_{n=1}^{4} f(x_n) \quad \text{and} \quad f\left(\bigvee_{n=1}^{4} x_n\right) \neq \bigvee_{n=1}^{4} f(x_n).$$

Therefore, f does not preserve suprema or infima.

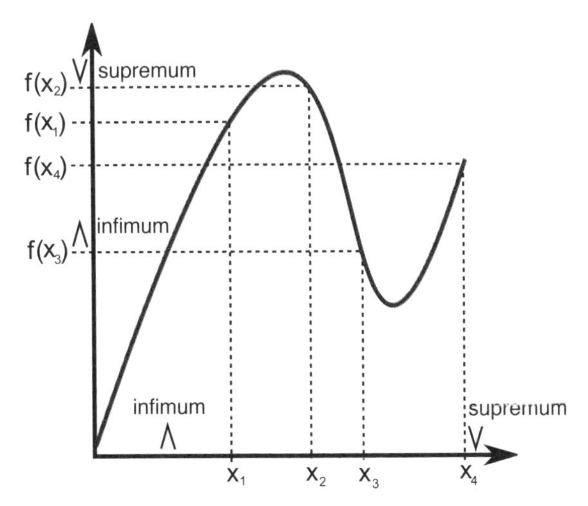

Figure 9.5
Example of a function that is not monotonically increasing and does not preserve supremum and infimum.

9.4 Lattice

Not every partially ordered set has a supremum or an infimum for all subsets of the underlying set. For example, the set

$$V = \{\emptyset, \{0\}, \{1\}\}$$

with the operation $\subset$ is a partially ordered set $(V, \subset)$, but $\{\{0\}, \{1\}\}$ is a subset of V which has no supremum.

For this reason, it is necessary to restrict the partially ordered sets being examined in order to ensure the existence of such infima and suprema.

Definition 9.10

Supremum Semilattice

A *supremum semilattice* $(U, \vee)$ is a partially ordered set $(U, \leq)$ in which for every two elements $u, v \in U$ the least upper bound $u \vee v$ exists. $\vee$ is referred to as the *supremum* operation.

Definition 9.11

Infimum Semilattice

An *infimum semilattice* $(U, \wedge)$ is a partially ordered set $(U, \leq)$ in which for every two elements $u, v \in U$ the greatest lower bound $u \wedge v$ exists. $\wedge$ is referred to as the *infimum* operation.

Definition 9.12

Lattice

A *lattice* [8]

$$(U, \wedge, \vee)$$

is a partially ordered set $(U, \leq)$ that is both an infimum semilattice and a supremum semilattice.

Note the following elementary properties of $\vee$ and $\wedge$.

Lemma 9.1

Monotonicity of Supremum and Infimum with Fixed Element

Suppose u, v, w are elements in a lattice U and $u \leq v$. Then

$$w \wedge u \leq w \wedge v \quad \text{and} \quad w \vee u \leq w \vee v$$

That is, taking $\wedge$ or $\vee$ with a fixed element is a monotonic function.

Proof. See Exercise 9.7. □

There is an equivalent way to describe lattices via an algebraic definition rather than the above order-based definition. The algebraic properties of the infimum and supremum operations may determine the partial order $\leq$. These properties are

1. Associativity for all $u, v, w \in U$

$$u \vee (v \vee w) = (u \vee v) \vee w \quad \text{and} \quad u \wedge (v \wedge w) = (u \wedge v) \wedge w$$

2. Commutativity for all $u, v \in U$

$$u \vee v = v \vee u \quad \text{and} \quad u \wedge v = v \wedge u$$

3. Idempotence for all $u \in U$

$$u \wedge u = u \vee u = u$$

4. Absorbency for all $u, v \in U$

$$u \wedge (u \vee v) = u = u \vee (u \wedge v)$$

Given any set U and operations $\vee, \wedge$ that satisfy the above properties, it is possible to define $\leq$ by declaring $u \leq v$ if and only if

$$u \wedge v = u$$

or equivalently

$$u \vee v = v$$

The above properties imply that $\leq$ is a partial order, as shown in [9].

Example 9.17

Every totally ordered set is a lattice with supremum max and infimum min.

Example 9.18

Every power set $\mathcal{P}(V)$ (the set of subsets of V) is a lattice with supremum operation $\cup$, infimum operation $\cap$, and partial order operation $\subset$.

Example 9.19

$\mathbb{N}$ equipped with the divisibility relation is a lattice with supremum the least common multiple and infimum the greatest common divisor.

The infimum and supremum operations within a lattice are well-behaved, but these are insufficient to ensure that a lattice is a semiring. In particular, the existence of identities and distributivity shrinks the class of partially ordered sets that fulfill the necessary properties. To get an idea of the order properties a lattice must have in order to have identities, note that if for every $u \in U$

$$u \vee 0 = u$$

then it follows that

$$0 \leq u$$

meaning that 0 is the minimum of the lattice when considered as a partially ordered set. Similary, if for every $u \in U$

$$u \wedge 1 = u$$

then

$$u \leq 1$$

and 1 is the maximum.

Definition 9.13

Bounded Partial Order

A partially ordered set $(U, \leq)$ is said to be *bounded* if there exist elements 0 and 1 in U such that $0 \leq u \leq 1$ for every element $u \in U$.

Definition 9.14

Bounded Lattice

A *bounded lattice* U, denoted

$$(U, \vee, \wedge, 0, 1)$$

contains an element 0 and an element 1 such that $u \wedge 0 = u \vee 1 = u$ for any $u \in U$. Notice here that it follows from the above axioms that $u \vee 1 = 1$ and $u \wedge 0 = 0$. Equivalently, a bounded lattice is a bounded partially ordered set that is also a lattice.

Example 9.20

The power set $\mathcal{P}(U)$ is an example of a bounded lattice with maximum U and minimum $\emptyset$.

Example 9.21

$\mathbb{N}$ ordered by divisibility is bounded with maximum 0 and minimum 1.

Example 9.22

Every bounded totally ordered set is a bounded lattice, such as $\mathbb{R} \cup \{-\infty, \infty\}$.

Distributivity, unlike the existence of identity elements, does not have an enlightening order characterization, prompting the following straightforward definition.

Definition 9.15

Distributive Lattice

A lattice U is *distributive* if it obeys the distributive property, whereby

$$u \wedge (v \vee w) = (u \wedge v) \vee (u \wedge w) \quad \text{and} \quad u \vee (v \wedge w) = (u \vee v) \wedge (u \vee w)$$

A bounded distributive lattice $(U, \vee, \wedge, 0, 1)$ forms two commutative monoids $(U, \vee, 0)$ and $(U, \wedge, 1)$ in which $\vee$ and $\wedge$ distribute over each other. The only remaining property that $(U, \vee, \wedge, 0, 1)$ would need to fulfill to be a semiring is for it to satisfy the identities

$$u \wedge 0 = 0 \quad \text{and} \quad u \vee 1 = 1$$

which follow immediately from the fact that 0 is the minimum and 1 the maximum. Thus, $(U, \vee, \wedge, 0, 1)$ and $(U, \wedge, \vee, 1, 0)$ are both semirings.

A supremum semilattice $(U, \vee)$ with a minimum 0 or an infimum semilattice $(L, \wedge)$ with a maximum 1 are nearly semirings since, by idempotence, associativity, and commutativity, it follows that

$$\begin{aligned} u \vee (v \vee w) &= (u \vee u) \vee (w \vee v) \\ &= u \vee (u \vee (w \vee v)) \\ &= u \vee ((u \vee w) \vee v) \\ &= u \vee (v \vee (u \vee w)) \\ &= (u \vee v) \vee (u \vee w) \end{aligned}$$

and similarly for $\wedge$, showing that $\vee$ and $\wedge$ naturally distribute over themselves. However, it is not the case that $0 \vee u = 0$ or $1 \wedge u = 1$ in general, so $(L, \vee, \vee, 0, 0)$ and $(L, \wedge, \wedge, 1, 1)$ are not semirings, only lacking the property that the additive identity be a multiplicative annihilator.

Topping off this hierarchy of lattice structures is the Boolean algebra [10].

Definition 9.16

Boolean Algebra

A *Boolean algebra U*, also denoted

$$(U, \vee, \wedge, \neg, 0, 1)$$

is a bounded distributive lattice $(U, \vee, \wedge, 0, 1)$ with a unary operation

$$\neg : U \to U$$

satisfying

De Morgan's Law — [11]

$$\neg(u \wedge v) = \neg u \vee \neg v$$

Double Negation Elimination —

$$\neg\neg v = v$$

Law of the Excluded Middle —

$$\neg v \vee v = 1$$

Law of Noncontradiction —

$$\neg v \wedge v = 0$$

Example 9.23

$\mathcal{P}(U)$ ordered by $\subset$ is an example of a Boolean algebra, where $\neg u$ is given by u^c, the complement of u relative to U.

Pushing the existence of binary suprema and infima to arbitrary suprema and infima produces a slightly different kind of lattice structure.

Definition 9.17

Complete Lattice

A *complete lattice* U is a lattice in which for every subset $U' \subset U$ the least upper bound $\bigvee U'$ and the greatest lower bound $\bigwedge U'$ exist.

Example 9.24

$\mathcal{P}(V)$ with the subset ordering and $\mathbb{R} \cup \{-\infty, \infty\}$ are complete lattices.

9.5 The Semirings of Interest

Currently, the D4M [12, 13] software instantiation of associative arrays provides the following well-behaved binary operations on the entries of associative arrays: $+$, $\times$, max, and min on the real numbers; max and min on bounded totally ordered (BTO) sets; $\cup$ and $\cap$ on power sets; logical & and logical | on the set $\{0, 1\}$ and more generally set suprema $\vee$ and set infimum $\wedge$ on a distributive lattice. Combining certain pairs of these operations implies that V forms a commutative semiring under the $\oplus$ and $\otimes$ operations, denoted by $(V, \oplus, \otimes, 0, 1)$ where 0 is the additive identity and 1 is the multiplicative identity, which ensures that the resulting algebra will be well-behaved. "Semi" in the context of rings, modules, and algebras implies all the properties of these objects except that the existence of additive inverses is not required. Which of these pairs of operations (insofar as such pairs make sense) distribute over each other has largely been dealt with in Sections 9.1 and 9.4 or involve simple counterexamples. Table 9.1 shows the relevant sets, their corresponding operations (and their D4M equivalents), and the combinations that form semirings.

This table can be checked easily, and since the proofs confirming them are elementary or by definition, the proofs are only outlined here. The *no* entries can be checked via simple counterexamples, and the *yes* entries are mostly straightforward to see. Since $\mathbb{F}$ is a field, $\times$ distributes over $+$. Likewise, it is a well-known and elementary fact that max and min are mutually distributive, as are $\cup$ and $\cap$, and $\vee$ and $\wedge$. $+$ distributes over max and min because, by the definition of an ordered field, for any elements u, v, w such that $u \leq v$, then

$$u + w \leq v + w$$

$\times$ does not distribute over min and max when negative numbers are involved since multiplication by a negative number reverses the ordering of two numbers. However, when dealing with a set of non-negative real numbers, it holds that for non-negative reals u, v, w that $u \geq v$ implies

$$uw \geq vw$$

Thus, over the set of non-negative real numbers, multiplication distributes over min and max.

A main objective is to study the properties of well-behaved associative arrays that are relevant to practical implementation (such as D4M), and it is reasonable to focus on proving statements about the algebraic properties of the semirings induced by pairs of operations exhibiting linear behavior. Table 9.1 shows these semirings of interest; every "yes" is an instance of a semiring structure.

Example 9.25

The max-plus algebras

$$(\mathbb{F} \cup \{\text{-}\infty, \infty\}, \max, +, \text{-}\infty, 0)$$

and

$$(\mathbb{F} \cup \{\text{-}\infty, \infty\}, \min, +, \infty, 0)$$

where $\mathbb{F}$ is an ordered field such as $\mathbb{Q}$ or $\mathbb{R}$.
$+$ is defined on $\mathbb{F} \cup \{\text{-}\infty, \infty\}$ by setting

$$\text{-}\infty + u = u + \text{-}\infty = \text{-}\infty$$

for every

$$u \in \mathbb{F} \cup \{\text{-}\infty, \infty\}$$

and

$$u + \infty = \infty + u = \infty$$

for all

$$u \in \mathbb{F} \cup \{\infty\}$$

Then $-\infty$ is a multiplicative annihilator, and it can be verified by case-by-case computation that addition remains associative, commutative, and distributes over max.

Table 9.1

The semirings of interest. Distributivity of $\otimes$ over $\oplus$ over fields, bounded totally ordered sets (BTO set), power sets ($\mathcal{P}$ set), and bounded distributive lattices (BD Lattices), and relevant intersections of these domains. The intended intersection of the Field and BTO Set domains are the extensions of ordered fields where a formal minimum $-\infty$ and formal maximum ∞ have been added. The algebraic properties of $-\infty$ and ∞ are explained in Example 9.25. A blank cell denotes that there is no relevant domain in which these operations exist, "no" implies that $\otimes$ does not distribute over $\oplus$, and "yes" implies that $\otimes$ does distribute over $\oplus$.
$*$: multiplication in $\mathbb{R}$ distributes over max and min when the domain is restricted to the non-negative reals.
$\dagger$: although distributivity holds, the additive and multiplicative identities are the same, so the additive identity cannot be a multiplicative annihilator, so this pair of operations fails to give a semiring.

Domain			Field	Field	BTO Set	BTO Set	$\mathcal{P}$ Set	$\mathcal{P}$ Set	BD Lattice	BD Lattice
	MATLAB		+	.*	max	min	union	intersect	$\vert$	&
		$\oplus \backslash \otimes$	+	$\times$	max	min	$\cup$	$\cap$	$\vee$	$\wedge$
Field	+	+	no	yes	no	no				
Field	.*	$\times$	no	no	no	no				
BTO Set	max	max	yes	yes*	yes†	yes				
BTO Set	min	min	yes	yes*	yes	yes†				
$\mathcal{P}$ Set	union	$\cup$					yes†	yes		
$\mathcal{P}$ Set	intersect	$\cap$					yes	yes†		
BD Lattice	$\vert$	$\vee$							yes†	yes
BD Lattice	&	$\wedge$							yes	yes†

Example 9.26

The max-min and min-max algebras

$$(V, \max, \min, -\infty, \infty)$$

and

$$(V, \min, \max, \infty, -\infty)$$

where V is a totally ordered set with order $\leq$, minimum value $-\infty$, and maximum value ∞. In Figure 4.1, the set V is defined as the set of all entries in the array A, and it is given the max-min algebra by equipping V with the lexicographical ordering.

Example 9.27

The power set algebras

$$(\mathcal{P}(V), \cup, \cap, \emptyset, V)$$

and

$$(\mathcal{P}(V), \cap, \cup, V, \emptyset)$$

where V is an arbitrary set and $\mathcal{P}(V)$ is the power set of V. With V as the set of all entries in the array A of Figure 4.1, the power set algebra can be defined on $\mathcal{P}(V)$, as is done in Figure 4.3.

In later chapters, the semirings in Table 9.1 will be referred to collectively as the *semirings of interest*. Understanding the behavior of the array operations defined using these semirings will largely be a matter of investigating questions similar to those found in linear algebra.

9.6 Conclusions, Exercises, and References

Proving the properties of associative arrays requires that the formal definitions of the underlying mathematical objects be specified. These definitions include operations on sets, ordered sets, supremum and infimum, and lattices. From these properties, the various semirings of interest can be selected for further study.

Exercises

Exercise 9.1 — For the constant $\alpha = 3$ and associative arrays **A**, **B**, and **C** given by

$$\mathbf{A} = \begin{array}{c} \\ 1 \\ 2 \end{array}\begin{array}{c} \begin{array}{ccc} \mathsf{a} & \mathsf{b} & \mathsf{c} \end{array} \\ \begin{bmatrix} 7 & 2 & 1 \\ 0 & 3 & 3 \end{bmatrix} \end{array}, \quad \mathbf{B} = \begin{array}{c} \\ 1 \\ 2 \end{array}\begin{array}{c} \begin{array}{ccc} \mathsf{a} & \mathsf{b} & \mathsf{c} \end{array} \\ \begin{bmatrix} 4 & 2 & 5 \\ 1 & 0 & 1 \end{bmatrix} \end{array}, \quad \mathbf{C} = \begin{array}{c} \\ \mathsf{a} \\ \mathsf{b} \\ \mathsf{c} \end{array}\begin{array}{c} \begin{array}{cccc} 1 & 2 & 3 & 4 \end{array} \\ \begin{bmatrix} 3 & 2 & 1 & 0 \\ 0 & 1 & 2 & 3 \\ 0 & 1 & 0 & 1 \end{bmatrix} \end{array}$$

perform the following computations over the value set $V = \mathbb{N}$ with operations $\oplus \equiv +$ and $\otimes \equiv \times$

(a) $\alpha \otimes \mathbf{A}$

(b) $\mathbf{A} \oplus \mathbf{B}$

(c) $\mathbf{A} \otimes \mathbf{B}$

(d) $\mathbf{A} \oplus.\otimes \mathbf{C}$

Exercise 9.2 — For the constant $\alpha = 3$ and associative arrays **A**, **B**, and **C** given by

$$\mathbf{A} = \begin{array}{c} \\ 1 \\ 2 \end{array}\begin{array}{c} \begin{array}{ccc} \mathsf{a} & \mathsf{b} & \mathsf{c} \end{array} \\ \begin{bmatrix} 7 & 2 & 1 \\ -\infty & 3 & 3 \end{bmatrix} \end{array}, \quad \mathbf{B} = \begin{array}{c} \\ 1 \\ 2 \end{array}\begin{array}{c} \begin{array}{ccc} \mathsf{a} & \mathsf{b} & \mathsf{c} \end{array} \\ \begin{bmatrix} 4 & 2 & 5 \\ 1 & -\infty & 1 \end{bmatrix} \end{array}, \quad \mathbf{C} = \begin{array}{c} \\ \mathsf{a} \\ \mathsf{b} \\ \mathsf{c} \end{array}\begin{array}{c} \begin{array}{cccc} 1 & 2 & 3 & 4 \end{array} \\ \begin{bmatrix} 3 & 2 & 1 & -\infty \\ -\infty & 1 & 2 & 3 \\ -\infty & 1 & -\infty & 1 \end{bmatrix} \end{array}$$

perform the following computations over the value set $V = \mathbb{R} \cup \{-\infty, \infty\}$ with operations $\oplus \equiv \max$ and $\otimes \equiv \min$

(a) $\alpha \otimes \mathbf{A}$

(b) $\mathbf{A} \oplus \mathbf{B}$

(c) $\mathbf{A} \otimes \mathbf{B}$

(d) $\mathbf{A} \oplus.\otimes \mathbf{C}$

Exercise 9.3 — For the constant $\alpha = 3$ and associative arrays **A**, **B**, and **C** given by

$$\mathbf{A} = \begin{array}{c} \\ 1 \\ 2 \end{array}\begin{array}{c} \begin{array}{ccc} \mathsf{a} & \mathsf{b} & \mathsf{c} \end{array} \\ \begin{bmatrix} 7 & 2 & 1 \\ \infty & 3 & 3 \end{bmatrix} \end{array}, \quad \mathbf{B} = \begin{array}{c} \\ 1 \\ 2 \end{array}\begin{array}{c} \begin{array}{ccc} \mathsf{a} & \mathsf{b} & \mathsf{c} \end{array} \\ \begin{bmatrix} 4 & 2 & 5 \\ 1 & \infty & 1 \end{bmatrix} \end{array}, \quad \mathbf{C} = \begin{array}{c} \\ \mathsf{a} \\ \mathsf{b} \\ \mathsf{c} \end{array}\begin{array}{c} \begin{array}{cccc} 1 & 2 & 3 & 4 \end{array} \\ \begin{bmatrix} 3 & 2 & 1 & \infty \\ \infty & 1 & 2 & 3 \\ \infty & 1 & \infty & 1 \end{bmatrix} \end{array}$$

perform the following computations over the value set $V = \mathbb{R} \cup \{-\infty, \infty\}$ with operations $\oplus \equiv \min$ and $\otimes \equiv \max$

(a) $\alpha \otimes \mathbf{A}$

(b) $\mathbf{A} \oplus \mathbf{B}$

(c) $\mathbf{A} \otimes \mathbf{B}$

(d) $\mathbf{A} \oplus.\otimes \mathbf{C}$

Exercise 9.4 — For the constant $\alpha = 3$ and associative arrays **A**, **B**, and **C** given by

$$\mathbf{A} = \begin{array}{c} \\ 1 \\ 2 \end{array} \begin{array}{c} \begin{array}{ccc} \text{a} & \text{b} & \text{c} \end{array} \\ \begin{bmatrix} 7 & 2 & 1 \\ \text{-}\infty & 3 & 3 \end{bmatrix} \end{array}, \quad \mathbf{B} = \begin{array}{c} \\ 1 \\ 2 \end{array} \begin{array}{c} \begin{array}{ccc} \text{a} & \text{b} & \text{c} \end{array} \\ \begin{bmatrix} 4 & 2 & 5 \\ 1 & \text{-}\infty & 1 \end{bmatrix} \end{array}, \quad \mathbf{C} = \begin{array}{c} \\ \text{a} \\ \text{b} \\ \text{c} \end{array} \begin{array}{c} \begin{array}{cccc} 1 & 2 & 3 & 4 \end{array} \\ \begin{bmatrix} 3 & 2 & 1 & \text{-}\infty \\ \text{-}\infty & 1 & 2 & 3 \\ \text{-}\infty & 1 & \text{-}\infty & 1 \end{bmatrix} \end{array}$$

perform the following computations over the value set $V = \mathbb{R} \cup \{\text{-}\infty, \infty\}$ with operations $\oplus \equiv \max$ and $\otimes \equiv +$

(a) $\alpha \otimes \mathbf{A}$

(b) $\mathbf{A} \oplus \mathbf{B}$

(c) $\mathbf{A} \otimes \mathbf{B}$

(d) $\mathbf{A} \oplus.\otimes \mathbf{C}$

Exercise 9.5 — For the constant $\alpha = \{0\}$ and associative arrays **A**, **B**, and **C** given by

$$\mathbf{A} = \begin{array}{c} \\ \text{a} \\ \text{b} \end{array} \begin{array}{c} \begin{array}{cc} 1 & 2 \end{array} \\ \begin{bmatrix} \emptyset & \{0\} \\ \{1\} & \emptyset \end{bmatrix} \end{array}, \quad \mathbf{B} = \begin{array}{c} \\ \text{a} \\ \text{b} \end{array} \begin{array}{c} \begin{array}{cc} 1 & 2 \end{array} \\ \begin{bmatrix} \{0,1\} & \{2\} \\ \emptyset & \{0\} \end{bmatrix} \end{array}, \quad \mathbf{C} = \begin{array}{c} \\ 1 \\ 2 \end{array} \begin{array}{c} \text{a} \\ \begin{bmatrix} \{0\} \\ \{0,2\} \end{bmatrix} \end{array}$$

perform the following computations over the value set $V = \mathcal{P}(\{0,1,2\})$ with operations $\oplus \equiv \cup$ and $\otimes \equiv \cap$

(a) $\alpha \otimes \mathbf{A}$

(b) $\mathbf{A} \oplus \mathbf{B}$

(c) $\mathbf{A} \otimes \mathbf{B}$

(d) $\mathbf{A} \oplus.\otimes \mathbf{C}$

Exercise 9.6 — Show that $\oplus.\otimes$ distributes over $\oplus$ (element-wise addition) if and only if $\otimes$ distributes over $\oplus$ in V.

Exercise 9.7 — Prove Lemma 9.1.

Exercise 9.8 — It is possible to combine several semirings

$$(V_1, \oplus_1, \otimes_1, 0_1, 1_1), \ldots, (V_n, \oplus_n, \otimes_n, 0_n, 1_n)$$

by taking the product $\prod_{i=1}^{n} V_i$ and defining $\oplus$ and $\otimes$ component-wise

$$(v_1,\ldots,v_n)\oplus(w_1,\ldots,w_n) = (v_1 \oplus_1 w_1,\ldots,v_n \oplus_n w_n)$$

$$(v_1,\ldots,v_n)\otimes(w_1,\ldots,w_n) = (v_1 \otimes_1 w_1,\ldots,v_n \otimes_n w_n)$$

Show that $\left(\prod_{i=1}^{n} V_i, \oplus, \otimes, \mathbf{0}, \mathbf{1}\right)$ with $\oplus$ and $\otimes$ defined as above and $\mathbf{0} = (0_1,\ldots,0_n)$, $\mathbf{1} = (1_1,\ldots,1_n)$ is a semiring.

References

[1] A. A. Fraenkel, Y. Bar-Hillel, and A. Levy, *Foundations of Set Theory*, vol. 67. Amsterdam: Elsevier, 1973.

[2] T. Jech, *Set Theory*. Springer Science & Business Media, 2013.

[3] K. Kunen, *Set Theory: An Introduction to Independence Proofs*, vol. 102. Amsterdam: Elsevier, 2014.

[4] J. Rotman, *An Introduction to the Theory of Groups*, vol. 148. New York: Springer Science & Business Media, 2012.

[5] J. S. Golan, *Semirings and Their Applications*. New York: Springer Science & Business Media, 2013.

[6] B. S. Schröder, *Ordered Sets*. Berlin, Heidelberg: Springer, 2003.

[7] B. Dushnik and E. W. Miller, "Partially ordered sets," *American Journal of Mathematics*, vol. 63, no. 3, pp. 600–610, 1941.

[8] G. Grätzer, *General Lattice Theory*. New York: Springer Science & Business Media, 2002.

[9] B. A. Davey and H. A. Priestley, *Introduction to Lattices and Order*. Cambridge University Press, 2002.

[10] S. Givant and P. Halmos, *Introduction to Boolean Algebras*. New York: Springer Science & Business Media, 2008.

[11] P. Johnstone, "Conditions related to De Morgan's law," *Applications of Sheaves*, pp. 479–491, 1979.

[12] J. Kepner, W. Arcand, W. Bergeron, N. Bliss, R. Bond, C. Byun, G. Condon, K. Gregson, M. Hubbell, J. Kurz, A. McCabe, P. Michaleas, A. Prout, A. Reuther, A. Rosa, and C. Yee, "Dynamic Distributed Dimensional Data Model (D4M) database and computation system," in *2012 IEEE International Conference on Acoustics, Speech and Signal Processing (ICASSP)*, pp. 5349–5352, IEEE, 2012.

[13] J. Kepner, "D4M: Dynamic Distributed Dimensional Data Model." http://d4m.mit.edu.

10 Structural Properties of Associative Arrays

Summary

Associative arrays have a number of important properties that can be used to describe their structure. These properties include dimension, total number of elements, number of nonzero elements, density, sparsity, size, image, and rank. Of particular interest is how these properties behave under element-wise addition, element-wise multiplication, and array multiplication. The properties that are the most predictable are typically the properties that are most convenient to work with when using associative arrays in real applications. This chapter uses the formal definition of associative arrays to explore the properties of associative arrays with respect to these operations.

10.1 Estimating Structure

A key aspect of modern data processing systems is the required data storage for the inputs, intermediate results, and the outputs. Ideally, the data storage decreases after each step in a data processing system. Data processing that inadvertently causes the data storage requirements to increase significantly is a common source of error in real-world systems. For many data processing systems, understanding the necessary storage has become as importanct as numerical accuracy. Many data processing systems analyze *sparse* data that are dominated by unstored zero values. The storage of zero values can be eliminated because element-wise addition and element-wise multiplication with zero always produce the same results without requiring computation

$$a \oplus 0 = a \qquad a \otimes 0 = 0$$

Furthermore, the above calculations can be performed without loss of accuracy. Thus, a data processing system with mostly zero values might perform relatively few calculations that impact numerical accuracy. In contrast, the data storage requirements of sparse data can be dramatically impacted with relatively incidental calculations. For example, if $\oplus$ is equivalent to logical &, then for two sparse associative arrays $\mathbf{A}$ and $\mathbf{B}$, the associative array $\mathbf{C}$ resulting from their element-wise addition is dense because all common zero values will become one

$$\mathbf{C} = \mathbf{A} \oplus \mathbf{B}$$

Avoiding such operations that may dramatically increase the required data storage is critical to developing effective data processing systems.

Estimating the data storage required by an associative array can often be achieved by predicting the pattern of nonzero entries in the associative array or the number of edges in the corresponding graph. This pattern is often referred to as the structure of the associative array, matrix, or graph. The storage required is usually highly correlated with this structure. How the storage required depends upon the parameters of the system is often called the *memory complexity* or *storage complexity* [1, 2]. Historically, sparse matrix algorithms have often begun with a step that estimates the structure of the output from the structure of the inputs [3–6]. This structural step is then followed by a step that does the numerical computation with a static data structure. For a survey, see [7] and references therein. The structural step can also be used to optimize the numerical step on a parallel computer [8]. Structural prediction can also be significantly aided by graph theory [9–11].

This chapter begins by reiterating the mathematical definition of an associative array and provides additional definitions for structural terms. Wherever possible, associative array structural terms are defined in a manner that is consistent with their matrix equivalents. The rest of the chapter provides a variety of observations on the structural properties of associative arrays with respect to the three core operations: element-wise addition, element-wise multiplication, and array multiplication.

10.2 Associative Array Formal Definition

The previous chapters have motivated the use of semirings in the context of associative arrays. Semirings lead naturally to the following definition of an associative array as a mathematical object. Associative arrays can have d-dimensions. Without loss of generality, the $d = 2$ case is most common and will often be referred to as an associative array when there is no ambiguity. Two-dimensional associative arrays are used to describe most of the properties of associative arrays, but it can be assumed that all of these properties also hold for d-dimensional associative arrays unless otherwise noted.

Definition 10.1

Associative Array

An *associative array* is a map from a pair of strict totally ordered key sets K_1 and K_2 to a value set V with a semiring structure

$$\mathbf{A} : K_1 \times K_2 \to V$$

that has a finite number of nonzero elements and where $(V, \oplus, \otimes, 0, 1)$ is a semiring.

From here onward, whenever "array" is used, "associative array" is implicitly meant. Given a set of keys

$$K_1 \times K_2$$

and a commutative semiring

$$(V, \oplus, \otimes, 0, 1)$$

the set of all two-dimensional associative arrays over these keys and values is denoted

$$\mathbb{A}(K_1, K_2; V)$$

or, when understood by the context, simply $\mathbb{A}$.

Example 10.1

Familiar examples are the associative arrays encountered in Figure 4.1

$$\mathbf{A}, \mathbf{A}_1, \mathbf{A}_2 : K \times K \to V$$

where K is the set of all alphanumeric strings. V is the set of all alphanumeric strings representing the music data entries in the table $\mathbf{A}$ ordered lexicographically, so V has the max-min algebra. The nonzero values of $\mathbf{A}_1$ and $\mathbf{A}_2$ are what are presented in the figure. All other values are zero and are not displayed.

Example 10.2

$m \times n$ matrices over a field or more generally a ring $(V, \oplus, \otimes, 0, 1)$ are maps

$$\mathbf{A} : \{1, \ldots, m\} \times \{1, \ldots, n\} \to V$$

Letting $K_1 = \{1, \ldots, m\}$ and $K_2 = \{1, \ldots, n\}$ and recognizing that all rings are semirings, it follows that matrices over a ring are examples of two-dimensional associative arrays.

Example 10.3

The associative *zero array*

$$\mathbb{0} \in \mathbb{A}$$

is defined by

$$\mathbb{0}(k) = 0$$

for every key tuple $k \in K_1 \times K_2$.

Example 10.4

For $k' \in K_1 \times K_2$, the associative *unit arrays*

$$\mathbf{e}_{k'} \in \mathbb{A}$$

are defined by

$$\mathbf{e}_{k'}(k) = \begin{cases} 1 & \text{if } k = k' \\ 0 & \text{if } k \neq k' \end{cases}$$

Example 10.5

The *empty array* $\emptyset : \emptyset \to V$ is simply the empty set $\emptyset$ since there are no elements in $\emptyset \times V = \emptyset$. An array that is not equal to the empty array is said to be a *non-empty array*.

A common notation to define two-dimensional associative arrays $\mathbf{A} : K_1 \times K_2 \to V$ is to make use of matrix notation when the sizes of K_1 and K_2 are small finite numbers. Here, $\mathbf{A}$ is represented via an array with the rows labeled by the non-empty row elements of K_1, the columns labeled by the non-empty column elements of K_2, and the entry in row k_1 and column k_2 being the value of $\mathbf{A}(k_1, k_2)$.

Example 10.6

Let

$$K_1 = \{\text{fries}, \text{pizza}, \text{soda}\} \quad \text{and} \quad K_2 = \{\text{cost}, \text{quantity}\}$$

and consider the associative array

$$\mathbf{A} : K_1 \times K_2 \to \mathbb{R}_{\geq 0}$$

defined by

$$\mathbf{A} = \begin{array}{c} \\ \text{fries} \\ \text{pizza} \\ \text{soda} \end{array} \begin{array}{c} \begin{array}{cc} \text{cost} & \text{quantity} \end{array} \\ \left[\begin{array}{cc} 1.00 & 200 \\ 5.00 & 12 \\ 1.50 & 3 \end{array} \right] \end{array}$$

Specific entries in this above associative array can be written

$$\mathbf{A}(\text{pizza}, \text{quantity}) = 12 \quad \text{and} \quad \mathbf{A}(\text{soda}, \text{cost}) = 1.50$$

10.3 Padding Associative Arrays with Zeros

An associative array, when the domain is fixed, is determined by its nonzero values. For this reason, it would be expected that extending the domain and defining the extension to be zero at these added points would not affect the information that the original array conveys. Thus, *padding* an array with zeros should not present problems. Given two associative arrays

$$\mathbf{A} : K_1 \times K_2 \to V$$
$$\mathbf{B} : K_1' \times K_2' \to V$$

then $\mathbf{B}$ is said to be a *zero padding* of $\mathbf{A}$ if the key space of $\mathbf{A}$ is a subset of the key space of $\mathbf{B}$

$$K_1 \times K_2 \subset K_1' \times K_2'$$

and all the nonzero entries in $\mathbf{A}$ are also in $\mathbf{B}$. In other words

$$\mathbf{A}(k_1, k_2) = \mathbf{B}(k_1, k_2)$$

for all $k_1 \in K_1$ and $k_2 \in K_2$. Finally, all other entries in $\mathbf{B}$ are zero. That is

$$\mathbf{B}(k_1', k_2') = 0$$

for $k_1' \notin K_1$ and $k_2' \notin K_2$.

A zero padding of an associative array is an extension of the array that is zero for all the new key tuples. Because the corresponding values of the new key tuples are automatically decided, a zero padding is uniquely identified by its domain. Given

$$\mathbf{A} : K_1 \times K_2 \to V$$

and

$$K_1 \times K_2 \subset K_1' \times K_2'$$

the unique zero padding of $\mathbf{A}$ to the new key tuples $K_1' \times K_2$ is denoted

$$\mathbf{A}' = \text{pad}_{K_1' \times K_2'} \mathbf{A}$$

Example 10.7

Every associative array $\mathbf{A} : K_1 \times K_2 \to V$ is a zero padding of itself

$$\mathbf{A} = \text{pad}_{K_1 \times K_2} \mathbf{A}$$

Notationally, it is often convenient to simplify the key tuples of an associative array from $K_1 \times K_2$ to $K \times K$ to K^2. If K_1 and K_2 share a total ordering $\leq$, then this simplification of notation can be easily done with zero padding. Given

$$\mathbf{A} : K_1 \times K_2 \to V$$

where K_1 and K_2 have a compatible total order $\leq$ so that the total orders on K_1 and K_2 give the same ordering of elements in $K_1 \cap K_2$, then the key tuples of (A) can be made *square* by letting

$$K = K_1 \cup K_2$$

and zero padding $\mathbf{A}$

$$\mathbf{A} = \text{pad}_{K \times K} \mathbf{A}$$

The resulting associative array can now be written as

$$\mathbf{A} : K^2 \to V$$

By zero padding as necessary, the array operations $\oplus, \otimes, \otimes.\oplus$ become defined between any two arrays.

10.4 Zero, Null, Zero-Sum-Free

As a data structure that returns a value if given some number of keys, the associative array clearly constitutes a function between a set of key tuples and a value space. Furthermore, in practice every associative array is constructed from an empty associative array by adding and deleting nonzero values. Thus, it can be assumed that any associative array will only have values assigned to a finite number of key tuples and that the remaining keys will have some default *null* value 0. This assumption motivates including in the associative array definition a condition requiring that there be a finite number of nonzero entries.

Furthermore, as explained above, these maps should support operations that resemble those on ordinary vectors and matrices, such as matrix multiplication, matrix addition, and element-wise multiplication. Thus, the value space is assumed to have a commutative semiring structure. The default *null* value of the array is defined as algebraically equivalent to the 0 value of the semiring

Using 0 as the default for all unspecified values raises the issue of how to distinguish between a value that is explicitly 0 and a value that is unspecified. Some approaches to fixing this issue include adding a pseudo-identity value *null* to the semiring V with the property that for every $v \in V$

$$v \oplus \text{null} = \text{null} \oplus v = v$$

and

$$\text{null} \otimes \text{null} = \text{null} \otimes v = v \otimes \text{null} = \text{null} \oplus \text{null} = \text{null}$$

It can be checked that $V \cup \{\text{null}\}$ is a commutative semiring with additive identity null. Associative arrays do not make a distinction between null and 0, but acknowledge the possibility to correct the issue with this approach. The term null is standard in the relational database community where it is distinct from 0. In practice, in any specific application context, the distinction between null and 0 can almost always be addressed via practical workarounds.

The definition of a semiring does not require the existence of additive inverses

$$v \oplus \text{-}v = 0$$

However, semirings do not exclude the existence of additive inverses. There are contexts in which excluding additive inverses can be useful for analyzing associative array mathematics and optimizing their implementations. In a zero-sum-free semiring

$$u \oplus v = 0$$

implies $u = v = 0$ for all elements $u, v \in V$. Correspondingly, zero-sum-free semirings have the property that for $u, v \neq 0$

$$u \oplus v \neq 0$$

The above property is useful for predicting the number of nonzero entries in an associative array and can be helpful in implementing associative array algorithms.

10.5 Properties of Matrices and Associative Arrays

Matrices and two-dimensional associative arrays have many similarities and some subtle differences. It is informative to explore these properties. For the purpose of this comparison, let the matrix $\mathbf{A}$ be the mapping

$$\mathbf{A} : I \times J \to V$$

from the index sets

$$I = \{1, \ldots, m\}$$

and

$$J = \{1, \ldots, n\}$$

to the semiring V. The number of elements in I and J are denoted $m = |I|$ and $n = |J|$. In associative arrays, because of the large key space, it is often desirable to focus on rows and columns that have nonzero entries and ignore the rows and columns that are entirely zero. Associative array properties that deal with only the nonzero rows and columns are underlined. For example, let $\underline{I}$ be the indices of the non-empty rows and $\underline{J}$ be the indices of the non-empty columns. The number of elements in $\underline{I}$ and $\underline{J}$ are denoted $\underline{m} = |\underline{I}|$ and $\underline{n} = |\underline{J}|$.

Likewise, let the associative array $\mathbf{A}$ be the mapping

$$\mathbf{A} : K_1 \times K_2 \to V$$

from the strict totally ordered key sets K_1 and K_2 to the semiring V. The number of elements in the key sets K_1 and K_2 is denoted $m = |K_1|$ and $n = |K_2|$. Furthermore, let $\underline{K}_1$ be the keys of the non-empty rows and $\underline{K}_2$ be the keys of the non-empty columns. The number of elements in $\underline{K}_1$ and $\underline{K}_2$ are denoted $\underline{m} = |\underline{K}_1|$ and $\underline{n} = |\underline{K}_2|$.

The comparable properties of matrices and associative arrays that are of interest to explore include

dim — dimensions of the rows and columns

$$\text{dim}(\mathbf{A}) = (m,n)$$

total — number of values

$$\text{total}(\mathbf{A}) = mn$$

support — pairs of indices or keys corresponding to nonzero values

$$\text{support}(\mathbf{A})$$

nnz — number of nonzero values

$$\text{nnz}(\mathbf{A}) = |\text{support}(\mathbf{A})|$$

density — fraction of values that are nonzero

$$\text{density}(\mathbf{A}) = \frac{\text{nnz}(\mathbf{A})}{\text{total}(\mathbf{A})}$$

sparsity — fraction of values that are zero

$$\text{sparsity}(\mathbf{A}) = 1 - \text{density}(\mathbf{A})$$

size — number of non-empty rows and columns

$$\text{size}(\mathbf{A}) = \underline{m} \times \underline{n}$$

<u>total</u> — number values in the non-empty rows and columns

$$\underline{\text{total}}(\mathbf{A}) = \underline{m}\ \underline{n}$$

<u>density</u> — fraction of values in non-empty rows and columns that are nonzero

$$\underline{\text{density}}(\mathbf{A}) = \frac{\text{nnz}(\mathbf{A})}{\underline{\text{total}}(\mathbf{A})}$$

<u>sparsity</u> — fraction of values in non-empty rows and columns that are zero

$$\underline{\text{sparsity}}(\mathbf{A}) = 1 - \underline{\text{density}}(\mathbf{A})$$

Table 10.1
The impact of zero padding on the properties of matrices and associative arrays.

property	formula	zero padding
dim	$m \times n$	increases
total	mn	increases
support	support($\mathbf{A}$)	same
nnz	nnz($\mathbf{A}$)	same
density	nnz($\mathbf{A}$)/total($\mathbf{A}$)	decreases
sparsity	$1 -$ density($\mathbf{A}$)	increases
size	$\underline{m} \times \underline{n}$	same
$\underline{\text{total}}$	$\underline{m}\,\underline{n}$	same
$\underline{\text{density}}$	nnz($\mathbf{A}$)/$\underline{\text{total}}$($\mathbf{A}$)	same
$\underline{\text{sparsity}}$	$1 -$ $\underline{\text{density}}$($\mathbf{A}$)	same
image	$\mathbf{Av}$ for all $\mathbf{v}$	same
rank	rank($\mathbf{A}$)	same

image — all possible points that can be generated by multiplying $\mathbf{A}$ by any vector $\mathbf{v}$

$$\text{image}(\mathbf{A}) = \{\mathbf{w} \mid \text{there exists } \mathbf{v} \text{ with } \mathbf{w} = \mathbf{Av}\}$$

rank — the minimum number of linearly independent vectors needed to create image($\mathbf{A}$)

$$\text{rank}(\mathbf{A})$$

Usage of the term rank shall implicitly assume that it exists and that a linearly independent generating set of vectors also exists.

10.6 Properties of Zero Padding

As defined above, the properties of matrices and associative arrays are equivalent. One interesting aspect of these properties is how they behave under zero padding (see Table 10.1). The properties that remain under zero padding are often the properties that are most useful for associative arrays. The proofs of the zero padding properties summarized in Table 10.1 are provided in this section.

For each of the following proofs of the properties in Table 10.1, let

$$\mathbf{A} : K_1 \times K_2 \to V$$

be an array and

$$\mathbf{B} : K_1' \times K_2' \to V$$

be a zero padding of $\mathbf{A}$. Furthermore, let the dimensions of $\mathbf{A}$ and $\mathbf{B}$ be

$$m = |K_1| \quad n = |K_2| \quad m' = |K_1'| \quad n' = |K_2'|$$

The main fact used in the proofs below of the properties in Table 10.1 is that

$$\mathbf{B}(k_1,k_2) \neq 0$$

if and only if

$$(k_1,k_2) \in K_1 \times K_2$$

and

$$\mathbf{A}(k_1,k_2) \neq 0$$

Thus it is required that

$$\mathbf{B}(k_1,k_2) = 0$$

for all

$$(k_1,k_2) \notin K_1 \times K_2$$

for **B** to be a zero padding of **A**, and so

$$(k_1,k_2) \in K_1 \times K_2$$

if

$$\mathbf{B}(k_1,k_2) \neq 0$$

Moreover, **B** agrees with **A** on $K_1 \times K_2$, so

$$\mathbf{A}(k_1,k_2) = \mathbf{B}(k_1,k_2) \neq 0$$

if $(k_1,k_2) \in K_1 \times K_2$.

dim

By definition of a zero padding

$$K_1 \subset K_1' \quad \text{and} \quad K_2 \subset K_2'$$

Thus

$$m = |K_1| \leq |K_1'| = m'$$

and

$$n = |K_2| \leq |K_2'| = n'$$

showing that

$$(m,n) \leq (m',n')$$

total

From the argument for dim above, it is known that

$$m \leq m' \quad \text{and} \quad n \leq n'$$

Thus

$$mn \leq m'n'$$

support

As above

$$\mathbf{B}(k_1, k_2) \neq 0$$

if and only if

$$(k_1, k_2) \in K_1 \times K_2$$

and

$$\mathbf{A}(k_1, k_2) \neq 0$$

Thus, the support of **B** is exactly that of **A**.

nnz

In the zero padding **B**, every nonzero entry corresponds uniquely to a nonzero entry in **A**. Thus

$$\text{nnz}(\mathbf{A}) = \text{nnz}(\mathbf{B})$$

density

Because

$$\text{total}(\mathbf{A}) \leq \text{total}(\mathbf{B})$$

and

$$\text{nnz}(\mathbf{A}) = \text{nnz}(\mathbf{B})$$

then

$$\frac{\text{nnz}(\mathbf{A})}{\text{total}(\mathbf{A})} \geq \frac{\text{nnz}(\mathbf{B})}{\text{total}(\mathbf{B})}$$

sparsity

Because

$$\text{density}(\mathbf{A}) \geq \text{density}(\mathbf{B})$$

this implies that

$$\text{-density}(\mathbf{A}) \leq \text{-density}(\mathbf{B})$$

and thus that

$$1 - \text{density}(\mathbf{A}) \leq 1 - \text{density}(\mathbf{B})$$

size

Suppose that the row in $\mathbf{B}$ corresponding to the index $k_1 \in K_1'$ has a nonzero value in it, then there is a $k_2 \in K_2'$ such that

$$\mathbf{B}(k_1, k_2) \neq 0$$

but then

$$(k_1, k_2) \in K_1 \times K_2$$

and

$$\mathbf{A}(k_1, k_2) \neq 0$$

and so row k_1 in $\mathbf{A}$ also has a nonzero value in it. The above argument shows that

$$\underline{m}' \leq \underline{m}$$

The same argument can be repeated for the columns to show that

$$\underline{n}' \leq \underline{n}$$

Likewise, the reverse inequalities are also true

$$\underline{m} \leq \underline{m}'$$

and

$$\underline{n} \leq \underline{n}'$$

which follows from the fact that $\mathbf{B}$ extends $\mathbf{A}$, so if

$$\mathbf{A}(k_1, k_2) \neq 0$$

then

$$\mathbf{B}(k_1, k_2) \neq 0$$

Combining the inequalities results in

$$\underline{m} = \underline{m}'$$

and

$$\underline{n} = \underline{n}'$$

total

Since $\underline{m} = \underline{m}'$ and $\underline{n} = \underline{n}'$ then

$$\underline{mn} = \underline{m}'\underline{n}'$$

density

Because

$$\text{nnz}(\mathbf{A}) = \text{nnz}(\mathbf{B})$$

and

$$\underline{\text{total}}(\mathbf{A}) = \underline{\text{total}}(\mathbf{B})$$

it follows that their quotients are also equal, thus

$$\underline{\text{density}}(\mathbf{A}) = \underline{\text{density}}(\mathbf{B})$$

sparsity

Since $\underline{\text{density}}(\mathbf{A})$ and $\underline{\text{density}}(\mathbf{B})$ are equal

$$1 - \underline{\text{density}}(\mathbf{A}) = 1 - \underline{\text{density}}(\mathbf{B})$$

so that

$$\underline{\text{sparsity}}(\mathbf{A}) = \underline{\text{sparsity}}(\mathbf{B})$$

image

Here, zero padding does not literally keep the images the same, but they are closely related in that the image of **B** is determined by the image of **A** by simply padding the vectors with zeros. Recall that if

$$\mathbf{A}\mathbf{v} = \mathbf{u}$$

then $\mathbf{v}$ is an n-tuple (indexed by K_2) and $\mathbf{u}$ is an m-tuple indexed by K_1. If $\mathbf{v}$ is an n'-tuple (indexed by K_2'), then in the product $\mathbf{B}\mathbf{v}$

$$\mathbf{B}(k_1,k_2) \otimes \mathbf{v}(k_2) = 0$$

regardless of the value of $\mathbf{v}(k_2)$ as long as

$$(k_1,k_2) \notin K_1 \times K_2$$

Thus $\mathbf{w}(k_1) = 0$ whenever $k_1 \notin K_1$. When $k_1 \in K_1$

$$\begin{aligned}
\mathbf{v}(k_1) &= \bigoplus_{k_2 \in K_2'} \mathbf{B}(k_1,k_2) \otimes \mathbf{v}(k_2) \\
&= \bigoplus_{k_2 \in K_2} \mathbf{B}(k_1,k_2) \otimes \mathbf{v}(k_2) \\
&= \bigoplus_{k_2 \in K_2} \mathbf{A}(k_1,k_2) \otimes \mathbf{v}(k_2)
\end{aligned}$$

In other words, if $\mathbf{v}'$ is the result of removing those entries of $\mathbf{v}$ corresponding to indexes not in K_2, then $\mathbf{u}$ is $\mathbf{A}\mathbf{v}'$ padded with zeroes for the entries of K_1' not in K_1.

rank

The assignment process described above for turning the image of **A** into the image of **B** is actually a linear isomorphism of image(**A**) onto image(**B**). This isomorphism retains the dimensions of **A**. Thus, rank(**A**) = rank(**B**).

From the above descriptions and proofs, it is clear that dimension and size are closely related properties. Size is always less than or equal to dimension, so that

$$\underline{m} \leq m$$

and

$$\underline{n} \leq n$$

In practical settings, it is common for a matrix to have few empty rows or empty columns. The dimensions of a matrix are often similar to its size suggesting

$$\underline{m} \lesssim m$$

and

$$\underline{n} \lesssim n$$

For associative arrays, the number of row and column keys is usually far larger than the row or column size so that

$$\underline{m} \ll m$$

and

$$\underline{n} \ll n$$

For example, if the keys were "twenty-character strings," the key space might have

$$256^{20} = 2^{160} = 1461501637330902918203684832716283019655932542976$$

elements.

Total and $\underline{\text{total}}$ are also closely related and have similar relationships. It is always the case that for both matrices and associative arrays

$$\underline{m}\,\underline{n} \leq mn$$

It is common practice for matrices to be constructed so that

$$\underline{m}\,\underline{n} \lesssim mn$$

Likewise, for associative arrays

$$\underline{m}\,\underline{n} \ll mn$$

The number of nonzeros and the support are closely related since the number of nonzeros is equal to the number of entries in the support

$$\text{nnz}(\mathbf{A}) = |\text{support}(\mathbf{A})|$$

Both nnz and support are independent of zero padding, and the above relationship is always true for both matrices and associative arrays.

Density and sparsity depend on the total number of entries. An associative array may often have a very large total number of entries, in which case the density will be very low and the sparsity will be very high.

Density and sparsity depend on the total number of entries in the non-empty rows and columns, and so are similar for both matrices and associative arrays.

Both image and rank are unaffected by zero padding and are equivalent for matrices and associative arrays.

10.7 Support and Size

One of the defining properties of an associative array is that it has a finite number of nonzero entries. The set of key tuples in an associative array with nonzero values is referred to as the *support* of the associative array. Given an associative array

$$\mathbf{A} : K_1 \times K_2 \to V$$

the support($\mathbf{A}$) is the set of key tuples

$$\text{support}(\mathbf{A}) = \{(k_1, k_2) \in K_1 \times K_2 \mid \mathbf{A}(k_1, k_2) \neq 0\}$$

More generally, the support() function maps associative arrays to a set of key tuples

$$\text{support} : \mathbb{A} \to \mathcal{P}(K_1 \times K_2)$$

Projecting support($\mathbf{A}$) onto each of its dimensions produces the following notion of the *size* of an associative array.

Definition 10.2

Size of Array

Given a non-empty two-dimensional array $\mathbf{A} \in \mathbb{A}$, the size of the first dimension is the number of row keys corresponding to nonzero entries, and the size of the second dimension is the number of column keys corresponding to nonzero entries.

$$\text{size}(\mathbf{A}) = (\text{size}(\mathbf{A},1), \text{size}(\mathbf{A},2)) = \begin{pmatrix} \text{number of } k_1 & \text{number of } k_2 \\ \text{such that there is } k_2 \,, & \text{such that there is } k_1 \\ \text{with } \mathbf{A}(k_1,k_2) \neq 0 & \text{with } \mathbf{A}(k_1,k_2) \neq 0 \end{pmatrix}$$

Given a two-dimensional associative array $\mathbf{A} : K^2 \to V$, the first coordinate of size$(\mathbf{A}, 1)$ is called the *number of rows* while the second coordinate size$(\mathbf{A}, 2)$ is called the *number of columns*.

Example 10.8

Recall the arrays $\mathbf{A}, \mathbf{A}_1, \mathbf{A}_2 : K^2 \to V$ encountered in Figure 4.1

$$\begin{aligned}\text{size}(\mathbf{A}) &= (24, 9)\\ \text{size}(\mathbf{A}_1) &= (3, 2)\\ \text{size}(\mathbf{A}_2) &= (3, 3)\end{aligned}$$

Example 10.9

If $\mathbf{A}$ is an $m \times n$ matrix over a ring or an $m \times n$ matrix over a field, then size$(\mathbf{A}) = (m, n)$ if every row and column of $\mathbf{A}$ has at least one nonzero element.

Example 10.10

The *zero array* $\mathbb{0} \in \mathbb{A}$ has size$(\mathbb{0}) = (0, 0)$.

Example 10.11

The *unit arrays* $\mathbf{e}_k \in \mathbb{A}$, where $k \in K_1 \times K_2$, have size$(\mathbf{e}_k) = (1, 1)$.

10.8 Image and Rank

The rank of a matrix is the dimension of the column and row space of a matrix, or, equivalently, the dimension of the image. That is, rank$(\mathbf{A})$ is the number of unique vectors required to generate the row space, where the row space is all vectors produced by linear combinations of

$$\{\mathbf{A}(1, :), \mathbf{A}(2, :), \ldots, \mathbf{A}(m, :)\}$$

Likewise, the rank is also the number of unique vectors required to generate the column space. The column space is all vectors produced by linear combinations of

$$\{\mathbf{A}(:, 1), \mathbf{A}(:, 2), \ldots, \mathbf{A}(:, n)\}$$

For matrices, the rank has a number of interesting properties. For example, because the rank is the same for both the row space and the column space of a matrix, then for any matrix $\mathbf{A}$

$$\text{rank}(\mathbf{A}^{\mathsf{T}}) = \text{rank}(\mathbf{A})$$

A more important result is the rank-nullity theorem

Theorem 10.1

Rank-Nullity

[12–15] For any $m \times n$ matrix $\mathbf{A}$ with entries in a field, if X is the set of vectors $\mathbf{x}$ for which

$$\mathbf{A}\mathbf{x} = \mathbb{0}$$

then

$$\text{rank}(\mathbf{A}) + \dim(X) = n$$

The set X in Theorem 10.1 is called the *null space* of $\mathbf{A}$ and $\dim X$ is called the *nullity* of $\mathbf{A}$. If an $m \times n$ matrix $\mathbf{A}$ has m_0 empty rows and n_0 empty columns, then $\mathbf{A}$ has a row space of dimension at most

$$\underline{m} = m - m_0$$

and a column space of dimension at most

$$\underline{n} = n - n_0$$

Furthermore, since

$$\underline{m} = \text{size}(\mathbf{A}, 1)$$

and

$$\underline{n} = \text{size}(\mathbf{A}, 2)$$

then

$$\text{rank}(\mathbf{A}) \leq \min(\underline{m}, \underline{n})$$

For associative arrays, rank can be defined analogously. The rank of an associative array is the dimension of the column and row space of an associative array.

10.9 Example: Music

Figure 10.1 from Chapter 4 shows a sparse associative array $\mathbf{E}$ of music tracks and various features of each track. Each track is a row key and each feature is a column key.

The dimensions of this associative array are very large if the array includes all strings under a specific length. If the row keys are limited to 10-character printable ASCII strings and the column keys are limited to 28-character printable ASCII strings, then dimensions

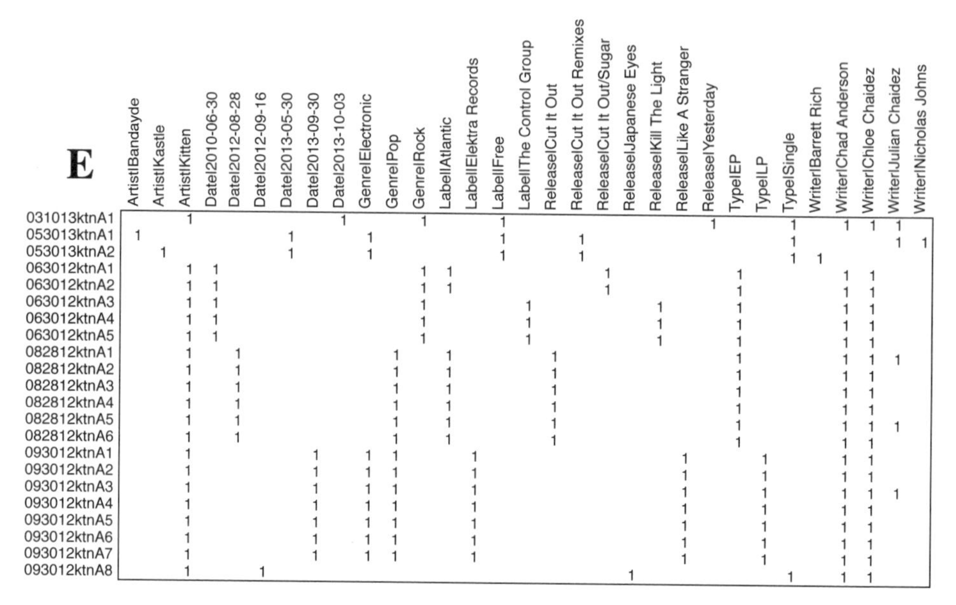

Figure 10.1
D4M sparse associative array **E** representation of a table of data from a music database. The column key and the value are concatenated with a separator symbol (in this case |) so that every unique pair of column and value has its own column in the sparse view. The new value is usually 1 to denote the existence of an entry. Column keys are an ordered set of database fields.

of **E** are

$$m = 95^{10} \approx 10^{19}$$
$$n = 95^{28} \approx 10^{55}$$

Likewise, the total number of values is also very large

$$mn = 95^{10} \times 95^{28} = 95^{38} \approx 10^{75}$$

The number of nonzero values in **E** is a much smaller number

$$\text{nnz}(\mathbf{E}) = 182$$

Correspondingly, the density of **E** is very small

$$\text{density}(\mathbf{E}) = 182/95^{38} \approx 10^{-73}$$

and the sparsity is effectively 1

$$\text{sparsity}(\mathbf{E}) = 1 - \text{density}(\mathbf{E}) \approx 1 - 10^{-73} \approx 1$$

In general, if the number of nonzero values is bounded, then the density will approach 0 and the sparsity will approach 1 as the square of the dimensions increases. The support of **E** is the set of 182 row and column key tuples corresponding to the nonzero values of **E**

$$\begin{aligned}\text{support}(\mathbf{E}) = \{&(\mathsf{053013ktnA1}, \mathsf{Artist|Bandayde}),\\ &(\mathsf{053013ktnA2}, \mathsf{Artist|Kastle}),\\ &(\mathsf{031013ktnA1}, \mathsf{Artist|Kitten}),\\ &\ldots\}\end{aligned}$$

Figure 10.1 only displays the non-empty rows and non-empty columns so the size of **E** is the number of rows and columns seen in the figure

$$\underline{m} = 22$$
$$\underline{n} = 31$$

The total number of values in the non-empty rows and columns is

$$\underline{\text{total}}(\mathbf{E}) = \underline{m}\,\underline{n} = 22 \times 31 = 682$$

The <u>density</u> of **E** is

$$\underline{\text{density}}(\mathbf{E}) = 182/682 \approx 0.27$$

which is far more informative than the density and indicates that approximately 27% of the values shown in Figure 10.1 are nonzero. Likewise, the <u>sparsity</u> of **E** is

$$\underline{\text{sparsity}}(\mathbf{E}) = 1 - 182/682 \approx 0.73$$

and shows that approximately 73% of the values displayed in Figure 10.1 are nonzero.

10.10 Example: Art

Figure 10.2 taken from Chapter 5 depicts a sparse associative array **E** of edges from a line art painting along with various features of each edge. Each edge is a row key and each edge feature is a column key.

The associative array dimensions are very large if the array has all strings under a specific length. If the row keys are limited to 2-character printable ASCII strings and the column keys are limited to 7-character printable ASCII strings, then dimensions of **E** are

$$m = 95^2 = 9025$$
$$n = 95^7 \approx 10^{14}$$

In addition, the total number of values is also very large

$$mn = 95^2 \times 95^7 = 95^9 \approx 10^{18}$$

	Color	Order	V01	V02	V03	V04	V05	V06	V07	V08	V09	V10	V11	V12	V13	V14	V15	V16	V17	V18	V19	V20
B1	blue	2	1	1	1																	
B2	blue	2				1	1	1	1	1												
G1	green	2	1	1	1																	
G2	green	2				1	1	1	1	1												
O1	orange	1	1	1	1																	
O2	orange	1	1	1	1																	
O3	orange	1				1	1	1	1	1												
O4	orange	1				1	1	1	1	1												
O5	orange	1		1				1			1		1					1				1
P1	pink	3	1	1	1																	
P2	pink	3				1	1	1	1	1												
P3	pink	3										1	1		1	1						
P4	pink	3		1								1										
P5	pink	3		1										1	1							
P6	pink	3															1	1	1			
P7	pink	3					1											1		1		
P8	pink	3							1									1			1	
S1	silver	2	1	1	1																	
S2	silver	2				1	1	1	1	1												

$$\mathbf{E}$$

Figure 10.2
Associative array **E** representation of the incidence matrix of the hyper-edges depicted in Figure 5.9.

The corresponding number of nonzero values in **E** is significantly less

$$\text{nnz}(\mathbf{E}) = 109$$

Likewise, the density of **E** is extremely small

$$\text{density}(\mathbf{E}) = 109/95^7 \approx 10^{-16}$$

and the sparsity is almost exactly 1

$$\text{sparsity}(\mathbf{E}) = 1 - \text{density}(\mathbf{E}) \approx 1 - 10^{-16} \approx 1$$

The support of **E** is the set of 109 row and column key tuples specifying the nonzero values of **E**

$$\begin{aligned} \text{support}(\mathbf{E}) = \{&(\mathsf{B1}, \mathsf{Color}), \\ &(\mathsf{B2}, \mathsf{Color}), \\ &(\mathsf{G1}, \mathsf{Green}), \\ &\ldots\} \end{aligned}$$

Figure 10.2 only shows non-empty rows and non-empty columns so the size of **E** is the number of rows and columns shown in the figure

$$\underline{m} = 19, \quad \underline{n} = 21$$

The total number of values in the non-empty rows and columns is

$$\underline{\text{total}}(\mathbf{E}) = \underline{m}\ \underline{n} = 19 \times 21 = 399$$

The density of $\mathbf{E}$ is

$$\underline{\text{density}}(\mathbf{E}) = 109/399 \approx 0.27$$

which provides more information than the density and shows that approximately 27% of the values in Figure 10.2 are nonzero. Likewise, the sparsity of $\mathbf{E}$ is

$$\underline{\text{sparsity}}(\mathbf{E}) = 1 - 109/399 \approx 0.73$$

and indicates that approximately 73% of the values displayed in Figure 10.2 are nonzero.

10.11 Properties of Element-Wise Addition

How the properties behave under element-wise addition is shown in Table 10.2. The properties that are the best understood after element-wise addition are naturally the easiest to analyze and use with associative arrays. The proofs of the element-wise addition properties given in Table 10.2 are as follows.

Definition 10.3

Element-Wise Addition

Given associative arrays

$$\mathbf{A}, \mathbf{B} : K_1 \times K_2 \to V$$

define

$$\mathbf{C} = \mathbf{A} \oplus \mathbf{B} : K_1 \times K_2 \to V \quad \text{by} \quad \mathbf{C}(k_1, k_2) = \mathbf{A}(k_1, k_2) \oplus \mathbf{B}(k_1, k_2)$$

dim

Element-wise addition does not change dimension $(|K_1|, |K_2|)$ since the arrays $\mathbf{A}, \mathbf{B}$ are assumed to have equal row and column key sets K_1, K_2, and the result $\mathbf{A} \oplus \mathbf{B}$ has the same key sets.

total

Element-wise addition does not change dimension $(|K_1|, |K_2|)$; therefore, total, which is the product of the dimensions $|K_1| \cdot |K_2|$, does not change.

Table 10.2
The implications of element-wise addition on the properties of matrices and associative arrays for values that are a semiring. It is assumed that the semiring is zero-sum-free.

property	formula	$\mathbf{C}$	$= \mathbf{A} \oplus \mathbf{B}$
dim	$m \times n$	$\mathrm{dim}(\mathbf{C})$	$= \mathrm{dim}(\mathbf{A}) = \mathrm{dim}(\mathbf{B})$
total	mn	$\mathrm{total}(\mathbf{C})$	$= \mathrm{total}(\mathbf{A}) = \mathrm{total}(\mathbf{B})$
support	$\mathrm{support}(\mathbf{C})$	$\mathrm{support}(\mathbf{C})$	$= \mathrm{support}(\mathbf{A}) \cup \mathrm{support}(\mathbf{B})$
nnz	$\mathrm{nnz}(\mathbf{C})$	$\mathrm{nnz}(\mathbf{C})$	$\geq \max(\mathrm{nnz}(\mathbf{A}), \mathrm{nnz}(\mathbf{B}))$
			$\leq \mathrm{nnz}(\mathbf{A}) + \mathrm{nnz}(\mathbf{B})$
density	$\mathrm{nnz}(\mathbf{C})/\mathrm{total}(\mathbf{C})$	$\mathrm{density}(\mathbf{C})$	$\geq \max(\mathrm{density}(\mathbf{A}), \mathrm{density}(\mathbf{B}))$
			$\leq \mathrm{density}(\mathbf{A}) + \mathrm{density}(\mathbf{A})$
sparsity	$1 - \mathrm{density}(\mathbf{C})$	$\mathrm{sparsity}(\mathbf{C})$	$\leq \min(\mathrm{sparsity}(\mathbf{A}), \mathrm{sparsity}(\mathbf{B}))$
			$\geq \mathrm{sparsity}(\mathbf{A}) + \mathrm{sparsity}(\mathbf{B})$
size	$\underline{m} \times \underline{n}$	$\mathrm{size}(\mathbf{C})$	$\geq \max(\mathrm{size}(\mathbf{A}), \mathrm{size}(\mathbf{B}))$
			$\leq \mathrm{size}(\mathbf{A}) + \mathrm{size}(\mathbf{B})$
$\underline{\mathrm{total}}$	$\underline{m}\,\underline{n}$	$\underline{\mathrm{total}}(\mathbf{C})$	$\geq \max(\underline{\mathrm{total}}(\mathbf{A}), \underline{\mathrm{total}}(\mathbf{B}))$
image	$\mathbf{Cv}$ for all $\mathbf{v}$	$\mathrm{image}(\mathbf{C})$	$\subset \mathrm{image}(\mathbf{A}) \oplus \mathrm{image}(\mathbf{B})$
rank	$\mathrm{rank}(\mathbf{C})$	$\mathrm{rank}(\mathbf{C})$	$\leq \mathrm{rank}(\mathbf{A}) + \mathrm{rank}(\mathbf{B})$

support

For any v, $v \oplus 0 = v$. However, in some semirings, there are nonzero u and v with $u \oplus v = 0$. Semirings in which this cannot occur are zero-sum-free.

Theorem 10.2

Support of Element-Wise Addition

For any matrices or associative arrays $\mathbf{A}$ and $\mathbf{B}$

$$\mathrm{support}(\mathbf{A} \oplus \mathbf{B}) \subset \mathrm{support}(\mathbf{A}) \cup \mathrm{support}(\mathbf{B})$$

If V is zero-sum-free, then

$$\mathrm{support}(\mathbf{A} \oplus \mathbf{B}) = \mathrm{support}(\mathbf{A}) \cup \mathrm{support}(\mathbf{B})$$

Proof. See Exercise 10.10. □

It is important to note that there are associative arrays $\mathbf{A}$ and $\mathbf{B}$ where

$$\mathrm{support}(\mathbf{A}) \cup \mathrm{support}(\mathbf{B}) \not\subset \mathrm{support}(\mathbf{A} \oplus \mathbf{B})$$

For example, if the entries in the associative array are characters, then it might make sense to consider them as living in the finite ring $\mathbb{Z}_{256}$, which under addition is the cyclic group

$\mathbb{Z}/256\mathbb{Z}$ [15]. In this setting, $100 + 156 = 0$, for example, so if **A** has all entries 100 and **B** has all entries 156, then $\mathbf{A} \oplus \mathbf{B} = \mathbb{0}$, and $\text{support}(\mathbf{A}) \cup \text{support}(\mathbf{B}) \not\subset \text{support}(\mathbf{A} \oplus \mathbf{B})$.

nnz

Since nnz is the cardinality of the support, the following theorem can be proven.

Theorem 10.3

NNZ of Element-Wise Addition

For any matrices or associative arrays **A** and **B**

$$\text{nnz}(\mathbf{A} \oplus \mathbf{B}) \leq \text{nnz}(\mathbf{A}) + \text{nnz}(\mathbf{B})$$

If V is zero-sum-free then, then

$$\max(\text{nnz}(\mathbf{A}), \text{nnz}(\mathbf{B})) \leq \text{nnz}(\mathbf{A} \oplus \mathbf{B})$$

Proof. See Exercise 10.11. □

density

Density is a straightforward transformation of nnz and total.

sparsity

Sparsity is a straightforward transformation of nnz and total.

size

Consider size in a manner similar to nnz.

Theorem 10.4

Size of Element-Wise Addition

If **A** and **B** are matrices or associative arrays, then

$$\text{size}(\mathbf{A} \oplus \mathbf{B}) \leq \text{size}(\mathbf{A}) + \text{size}(\mathbf{B})$$

Note that $+, \leq, \max$ are evaluated element-wise. If V is zero-sum-free, then

$$\max(\text{size}(\mathbf{A}), \text{size}(\mathbf{B})) \leq \text{size}(\mathbf{A} \oplus \mathbf{B})$$

Proof. See Exercise 10.12. □

total

Using size allows finding total by using its definition.

Theorem 10.5

Total of Element-Wise Addition

If $\mathbf{A}$ and $\mathbf{B}$ are matrices or associative arrays and V is zero-sum-free, then

$$\max(\underline{\text{total}}(\mathbf{A}), \underline{\text{total}}(\mathbf{B})) \leq \underline{\text{total}}(\mathbf{A} \oplus \mathbf{B})$$

Proof. See Exercise 10.13. □

image

Given two sets X and Y of vectors, write $X \oplus Y$ for the set $\{x \oplus y \mid x \in X, y \in Y\}$.

Theorem 10.6

Image of Element-Wise Addition

For any matrices or associative arrays $\mathbf{A}$ and $\mathbf{B}$

$$\text{image}(\mathbf{A} \oplus \mathbf{B}) \subset \text{image}(\mathbf{A}) \oplus \text{image}(\mathbf{B})$$

Proof. See Exercise 10.14. □

rank

The previous theorem can also be used to show that rank is subadditive.

Theorem 10.7

Rank of Element-Wise Addition

For any matrices or associative arrays $\mathbf{A}$ and $\mathbf{B}$

$$\text{rank}(\mathbf{A} \oplus \mathbf{B}) \leq \text{rank}(\mathbf{A}) + \text{rank}(\mathbf{B})$$

Proof. See Exercise 10.15. □

Table 10.3

The result of element-wise multiplication on the properties of matrices and associative arrays for values that are a semiring. Making the assumption that the semiring is zero-sum-free does not affect the below properties in this case. It will, however, be assumed that the semiring has no zero divisors so that if $a \otimes b = 0$, then $a = 0$ or $b = 0$.

property	formula	$\mathbf{C}$	$= \mathbf{A} \otimes \mathbf{B}$
dim	$m \times n$	$\text{dim}(\mathbf{C})$	$= \text{dim}(\mathbf{A}) = \text{dim}(\mathbf{B})$
total	mn	$\text{total}(\mathbf{C})$	$= \text{total}(\mathbf{A}) = \text{total}(\mathbf{B})$
support	$\text{support}(\mathbf{C})$	$\text{support}(\mathbf{C})$	$= \text{support}(\mathbf{A}) \cap \text{support}(\mathbf{B})$
nnz	$\text{nnz}(\mathbf{C})$	$\text{nnz}(\mathbf{C})$	$\leq \min(\text{nnz}(\mathbf{A}), \text{nnz}(\mathbf{B}))$
density	$\text{nnz}(\mathbf{C})/\text{total}(\mathbf{C})$	$\text{density}(\mathbf{C})$	$\leq \min(\text{density}(\mathbf{A}), \text{density}(\mathbf{B}))$
sparsity	$1 - \text{density}(\mathbf{C})$	$\text{sparsity}(\mathbf{C})$	$\geq \max(\text{sparsity}(\mathbf{A}), \text{sparsity}(\mathbf{B}))$
size	$\underline{m} \times \underline{n}$	$\text{size}(\mathbf{C})$	$\leq \min(\text{size}(\mathbf{A}), \text{size}(\mathbf{B}))$
$\underline{\text{total}}$	$\underline{m}\ \underline{n}$	$\underline{\text{total}}(\mathbf{C})$	$\leq \min(\underline{\text{total}}(\mathbf{A}), \underline{\text{total}}(\mathbf{B}))$
rank	$\text{rank}(\mathbf{C})$	$\text{rank}(\mathbf{C})$	$\leq \text{rank}(\mathbf{A})\text{rank}(\mathbf{B})$

10.12 Properties of Element-Wise Multiplication

How the properties are affected by element-wise multiplication is shown in Table 10.3. The properties that are the most well-defined after element-wise multiplication are often the easiest to use with associative arrays. The proofs of the element-wise multiplication properties listed in Table 10.3 are described in the rest of this section.

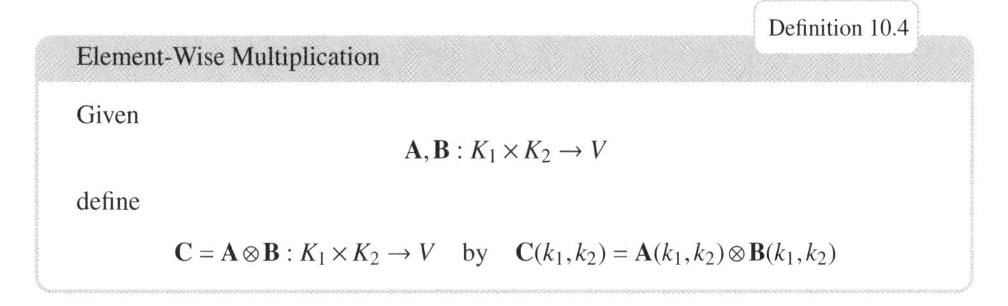

Definition 10.4

Element-Wise Multiplication

Given

$$\mathbf{A}, \mathbf{B} : K_1 \times K_2 \to V$$

define

$$\mathbf{C} = \mathbf{A} \otimes \mathbf{B} : K_1 \times K_2 \to V \quad \text{by} \quad \mathbf{C}(k_1, k_2) = \mathbf{A}(k_1, k_2) \otimes \mathbf{B}(k_1, k_2)$$

dim

Element-wise multiplication does not change dimension.

total

Element-wise multiplication does not change dimension; therefore, total, which is the product of the dimensions, does not change.

support

For any v, $v \otimes 0 = 0$. However, in some semirings, there are nonzero u and v with $u \otimes v = 0$, called zero divisors. This is not true for fields and gives the following result.

Theorem 10.8

Support of Element-Wise Multiplication

For any matrices or associative arrays **A** and **B**,

$$\text{support}(\mathbf{A} \otimes \mathbf{B}) \subset \text{support}(\mathbf{A}) \cap \text{support}(\mathbf{B})$$

If V has no zero divisors, then

$$\text{support}(\mathbf{A} \otimes \mathbf{B}) = \text{support}(\mathbf{A}) \cap \text{support}(\mathbf{B})$$

Proof. See Exercise 10.16. □

It is worth noting that there are associative arrays **A** and **B** that have

$$\text{support}(\mathbf{A} \otimes \mathbf{B}) \subsetneq \text{support}(\mathbf{A}) \cap \text{support}(\mathbf{B})$$

For example, if the entries in the associative array are characters, then it might make sense to consider them as living in the finite ring $\mathbb{Z}_{256}$. In this case, since $16^2 = 0$, an associative array with all nonzero entries 16 could have its element-wise multiplication square be all 0.

nnz

Since nnz is the cardinality of the support, it is possible to prove the following theorem about nnz.

Theorem 10.9

NNZ of Element-Wise Multiplication

For any matrices or associative arrays **A** and **B**

$$\text{nnz}(\mathbf{A} \otimes \mathbf{B}) \leq \min(\text{nnz}(\mathbf{A}), \text{nnz}(\mathbf{B}))$$

Proof. See Exercise 10.17. □

density

Density is a straightforward transformation of nnz and total.

sparsity

Sparsity is a straightforward transformation of nnz and total.

size

Size can be considered in a manner similar to nnz by using the set of nonzero rows and columns.

Theorem 10.10

Size of Element-Wise Multiplication

If **A** and **B** are matrices or associative arrays, then

$$\text{size}(\mathbf{A} \otimes \mathbf{B}) \leq \min(\text{size}(\mathbf{A}), \text{size}(\mathbf{B}))$$

Proof. See Exercise 10.18. □

total

Using size allows finding $\underline{\text{total}}$ by using its definition.

Theorem 10.11

$\underline{\text{Total}}$ of Element-Wise Multiplication

If **A** and **B** are matrices or associative arrays, then

$$\underline{\text{total}}(\mathbf{A} \otimes \mathbf{B}) \leq \min(\underline{\text{total}}(\mathbf{A}), \underline{\text{total}}(\mathbf{B}))$$

Proof. See Exercise 10.19. □

rank

The image of $\mathbf{A} \otimes \mathbf{B}$ is not as revealing, but its rank has the following property.

Theorem 10.12

Rank of Element-Wise Multiplication

For any matrices or associative arrays **A** and **B**

$$\text{rank}(\mathbf{A} \otimes \mathbf{B}) \leq \text{rank}(\mathbf{A})\, \text{rank}(\mathbf{B})$$

Proof. The proof will proceed in two steps. The first will be to show that the arrays **A** and **B** may be written as sums of rank 1 arrays, where the number of such rank 1 arrays is at most the rank of **A** and **B**, respectively. Let

$$\tilde{\mathbf{A}}(:,1),\ldots,\tilde{\mathbf{A}}(:,r)$$

be a basis for image(**A**), where $r = \text{rank}(\mathbf{A})$. Let $\mathbf{A}(:,j)$ be the j-th column of **A**. By the definition of a basis, each column $\mathbf{A}(:,j)$ is a linear combination of the basis vectors with coefficients $\hat{\mathbf{A}}(j,k) \in V$ such that

$$\mathbf{A}(:,j) = \bigoplus_{k=1}^{r} \hat{\mathbf{A}}(j,k) \otimes \tilde{\mathbf{A}}(:,k)$$

More specifically

$$\mathbf{A}(i,j) = \bigoplus_{k=1}^{r} \hat{\mathbf{A}}(j,k) \otimes \tilde{\mathbf{A}}(i,k)$$

or, because $\otimes$ is commutative

$$\mathbf{A}(i,j) = \bigoplus_{k=1}^{r} \tilde{\mathbf{A}}(i,k) \otimes \hat{\mathbf{A}}(j,k)$$

Thus, all of **A** can be constructed via the array multiplication

$$\mathbf{A} = \tilde{\mathbf{A}}\hat{\mathbf{A}}^{\mathsf{T}} = \bigoplus_{k=1}^{r} \tilde{\mathbf{A}}(:,k)\hat{\mathbf{A}}(:,k)^{\mathsf{T}}$$

Now

$$\tilde{\mathbf{A}}(:,k)\hat{\mathbf{A}}(:,k)^{\mathsf{T}}$$

is the product of a column vector and a row vector. If it can be shown that this product has rank 1, then the proof is complete. But this rank is immediate from the fact that $\tilde{\mathbf{A}}(:,k)$ is a basis for the column space of

$$\tilde{\mathbf{A}}(:,k)\hat{\mathbf{A}}(:,k)^{\mathsf{T}}$$

since the columns of this array are scalar multiples coming from $\hat{\mathbf{A}}(:,k)^{\mathsf{T}}$ of $\tilde{\mathbf{A}}(:,k)$. In general, this argument shows that if **v** and **w** are $n \times 1$ column vectors then $\mathbf{v}\mathbf{w}^{\mathsf{T}}$ is a rank 1 array.

Now let

$$\mathbf{A} = \bigoplus_{k=1}^{\text{rank}(\mathbf{A})} \tilde{\mathbf{A}}(:,k)\hat{\mathbf{A}}(:,k)^{\mathsf{T}}$$

$$\mathbf{B} = \bigoplus_{k'=1}^{\text{rank}(\mathbf{B})} \tilde{\mathbf{B}}(:,k')\hat{\mathbf{B}}(:,k')^{\mathsf{T}}$$

where the $\tilde{\mathbf{A}}(:,k)$, $\tilde{\mathbf{B}}(:,k')$ are column vectors and $\hat{\mathbf{A}}(:,k)^\mathsf{T}$, $\hat{\mathbf{B}}(:,k')^\mathsf{T}$ are row vectors. The element-wise product is linear, following from distributivity of $\otimes$ over $\oplus$ in V, and so

$$\begin{aligned}
\mathbf{A} \otimes \mathbf{B} &= \left(\bigoplus_{k=1}^{\mathrm{rank}(\mathbf{A})} \tilde{\mathbf{A}}(:,k)\hat{\mathbf{A}}(:,k)^\mathsf{T} \right) \otimes \left(\bigoplus_{k'=1}^{\mathrm{rank}(\mathbf{B})} \tilde{\mathbf{B}}(:,k')\hat{\mathbf{B}}(:,k')^\mathsf{T} \right) \\
&= \bigoplus_{k=1}^{\mathrm{rank}(\mathbf{A})} \bigoplus_{k'=1}^{\mathrm{rank}(\mathbf{B})} \left(\tilde{\mathbf{A}}(:,k)\hat{\mathbf{A}}(:,k)^\mathsf{T}\right) \otimes \left(\tilde{\mathbf{B}}(:,k')\hat{\mathbf{B}}(:,k')^\mathsf{T}\right)
\end{aligned}$$

Note that by direct computation

$$\left(\left(\tilde{\mathbf{A}}(:,k)\hat{\mathbf{A}}(:,k)^\mathsf{T}\right) \otimes \left(\tilde{\mathbf{B}}(:,k')\hat{\mathbf{B}}(:,k')^\mathsf{T}\right)\right)(i,j) = \tilde{\mathbf{A}}(i,k) \otimes \hat{\mathbf{A}}(k,j) \otimes \tilde{\mathbf{B}}(i,k') \otimes \hat{\mathbf{B}}(k',j)$$

The commutativity of $\otimes$ allows the terms in the above expression to be reordered and regrouped as

$$\left(\tilde{\mathbf{A}}(i,k) \otimes \tilde{\mathbf{B}}(i,k')\right) \otimes \left(\hat{\mathbf{A}}(j,k)^\mathsf{T} \otimes \hat{\mathbf{B}}(j,k')^\mathsf{T}\right)$$

Thus

$$\left(\tilde{\mathbf{A}}(:,k)\hat{\mathbf{A}}(:,k)^\mathsf{T}\right) \otimes \left(\tilde{\mathbf{B}}(:,k')\hat{\mathbf{B}}(:,k')^\mathsf{T}\right) = \left(\tilde{\mathbf{A}}(:,k) \otimes \tilde{\mathbf{B}}(:,k')\right)\left(\hat{\mathbf{A}}(:,k)^\mathsf{T} \otimes \hat{\mathbf{B}}(:,k')^\mathsf{T}\right)$$

and so

$$\begin{aligned}
\mathbf{A} \otimes \mathbf{B} &= \bigoplus_{k=1}^{\mathrm{rank}(\mathbf{A})} \bigoplus_{k'=1}^{\mathrm{rank}(\mathbf{B})} \left(\tilde{\mathbf{A}}(:,k) \otimes \tilde{\mathbf{B}}(:,k')\right)\left(\hat{\mathbf{A}}(:,k)^\mathsf{T} \otimes \hat{\mathbf{B}}(:,k')^\mathsf{T}\right) \\
&= \bigoplus_{k=1}^{\mathrm{rank}(\mathbf{A})} \bigoplus_{k'=1}^{\mathrm{rank}(\mathbf{B})} \left(\tilde{\mathbf{A}}(:,k) \otimes \tilde{\mathbf{B}}(:,k')\right)\left(\hat{\mathbf{B}}(:,k') \otimes \hat{\mathbf{A}}(:,k)\right)^\mathsf{T}
\end{aligned}$$

The above equation shows that $\mathbf{A} \otimes \mathbf{B}$ is a sum of rank($\mathbf{A}$)rank($\mathbf{B}$) rank 1 arrays, since both $\tilde{\mathbf{A}}(:,k) \otimes \tilde{\mathbf{B}}(:,k')$ and $\hat{\mathbf{B}}(:,k') \otimes \hat{\mathbf{A}}(:,k)$ are column vectors.

By the sub-additivity property of rank, it is confirmed that

$$\mathrm{rank}(\mathbf{A} \otimes \mathbf{B}) \leq \mathrm{rank}(\mathbf{A})\mathrm{rank}(\mathbf{B})$$

□

10.13 Array Multiplication

How the properties change under array multiplication is shown in Table 10.4. The properties that are the most consistent after array multiplication are usually the simplest to apply

Table 10.4

The impact of matrix-multiplication on the properties of matrices and associative arrays with semiring values. Assuming that the semiring is zero-sum-free or has no zero divisors has minimal effect on the properties below.

property	formula	$\mathbf{C}$	$= \mathbf{A}\,\mathbf{B} = \mathbf{A} \oplus.\otimes \mathbf{B}$
dim	$m \times n$	$\dim(\mathbf{C})$	$= (\dim(\mathbf{A},1), \dim(\mathbf{B},2))$
total	mn	$\text{total}(\mathbf{C})$	$= \dim(\mathbf{A},1)\,\dim(\mathbf{B},2)$
nnz	$\text{nnz}(\mathbf{C})$	$\text{nnz}(\mathbf{C})$	$\leq \text{total}(\mathbf{C})$
density	$\text{nnz}(\mathbf{C})/\text{total}(\mathbf{C})$	$\text{density}(\mathbf{C})$	≤ 1
sparsity	$1 - \text{density}(\mathbf{C})$	$\text{sparsity}(\mathbf{C})$	≤ 1
size	$\underline{m} \times \underline{n}$	$\text{size}(\mathbf{C})$	$\leq (\text{size}(\mathbf{A},1), \text{size}(\mathbf{B},2))$
rank	$\text{rank}(\mathbf{C})$	$\text{rank}(\mathbf{C})$	$\leq \min(\text{rank}(\mathbf{A}), \text{rank}(\mathbf{B}))$
			$\geq \text{rank}(\mathbf{A}) + \text{rank}(\mathbf{B}) - \dim(\mathbf{A},2)$

to associative arrays. The proofs of the array multiplication properties shown in Table 10.4 are described as follows.

Definition 10.5

Array Multiplication

Given

$$\mathbf{A} : K_1 \times K_3 \to V \qquad \mathbf{B} : K_3 \times K_2 \to V$$

define

$$\mathbf{C} = \mathbf{AB} = \mathbf{A} \oplus.\otimes \mathbf{B} : K_1 \times K_2 \to V \quad \text{by} \quad \mathbf{C}(k_1,k_2) = \bigoplus_{k_3 \in K_3} \mathbf{A}(k_1,k_3) \otimes \mathbf{B}(k_3,k_2)$$

dim

An $m \times n$ matrix or associative array times an $n \times m$ matrix or associative array produces an $n \times n$ matrix or associative array.

total

This is the same as dimension for matrices and follows naturally from size for associative arrays.

support

There is little to be observed about the support of $\mathbf{AB}$. It is possible to have

$$\text{support}(\mathbf{AB}) \subsetneq \text{support}(\mathbf{A}) \cap \text{support}(\mathbf{B})$$

or

$$\text{support}(\mathbf{AB}) \supsetneq \text{support}(\mathbf{A}) \cup \text{support}(\mathbf{B})$$

or anywhere in between. An example of the first case is

$$\mathbf{A} = \mathbf{B} = \begin{bmatrix} 0 & 1 \\ 0 & 0 \end{bmatrix}$$

with

$$\mathbf{AB} = \begin{bmatrix} 0 & 1 \\ 0 & 0 \end{bmatrix}^2 = \begin{bmatrix} 0 & 0 \\ 0 & 0 \end{bmatrix}$$

in which

$$\text{support}(\mathbf{A}) \cap \text{support}(\mathbf{B}) = \{(1,2)\}$$

and

$$\text{support}(\mathbf{AB}) = \emptyset$$

An example of the second case is

$$\mathbf{A} = \mathbf{B} = \begin{bmatrix} 0 & 1 \\ 1 & 1 \end{bmatrix}$$

with

$$\mathbf{AB} = \begin{bmatrix} 0 & 1 \\ 1 & 1 \end{bmatrix}^2 = \begin{bmatrix} 1 & 1 \\ 1 & 2 \end{bmatrix}$$

so that

$$\text{support}(\mathbf{A}) \cup \text{support}(\mathbf{B}) = \{(1,2),(2,1),(2,2)\}$$

and

$$\text{support}(\mathbf{AB}) = \{(1,1),(1,2),(2,1),(2,2)\}$$

Two examples of the third case include

$$\mathbf{A} = \mathbf{B} = \mathbf{AB} = \mathbb{I}$$

and

$$\mathbf{A} = \begin{bmatrix} 1 & 0 \\ 0 & 0 \end{bmatrix}, \mathbf{B} = \begin{bmatrix} 1 & 1 \\ 0 & 1 \end{bmatrix}$$

with

$$\mathbf{AB} = \begin{bmatrix} 1 & 1 \\ 0 & 0 \end{bmatrix}$$

in which

$$\text{support}(\mathbf{AB}) = \{(1,1),(1,2)\}$$

which is between

$$\text{support}(\mathbf{A}) \cap \text{support}(\mathbf{B}) = \{(1,1)\}$$

and

$$\text{support}(\mathbf{A}) \cup \text{support}(\mathbf{B}) = \{(1,1),(1,2),(2,1)\}$$

nnz

In the same way that there is not much to say about support, there also is not much to say about nnz. Indeed, the examples given to show that the support of $\mathbf{AB}$ can vary wildly also show that it is possible for $\text{nnz}(\mathbf{AB})$ to be larger than $\max(\text{nnz}(\mathbf{A}), \text{nnz}(\mathbf{B}))$, or to be smaller than $\min(\text{nnz}(\mathbf{A}), \text{nnz}(\mathbf{B}))$, or in between them.

density

Currently, little has been shown about nnz which limits what is known about density.

sparsity

Currently, little has been shown about nnz which limits what is known about sparsity.

size

The same example used to show that it is possible to have

$$\text{nnz}(\mathbf{AB}) < \min(\text{nnz}(\mathbf{A}), \text{nnz}(\mathbf{B}))$$

works for size as well. Let

$$\mathbf{A} = \mathbf{B} = \begin{bmatrix} 1 & 1 \\ -1 & -1 \end{bmatrix}$$

then

$$\begin{aligned}
\text{size}(\mathbf{AB}, 1) &= \text{size}\left(\begin{bmatrix} 1 & 1 \\ -1 & -1 \end{bmatrix}^2, 1\right) \\
&= \text{size}\left(\begin{bmatrix} 0 & 0 \\ 0 & 0 \end{bmatrix}, 1\right) \\
&= 0
\end{aligned}$$

but

$$\begin{aligned}
\min(\text{size}(\mathbf{A}, 1), \text{size}(\mathbf{B}, 1)) &= \min\left(\text{size}\left(\begin{bmatrix} 1 & 1 \\ -1 & -1 \end{bmatrix}, 1\right), \text{size}\left(\begin{bmatrix} 1 & 1 \\ -1 & -1 \end{bmatrix}, 1\right)\right) \\
&= 2
\end{aligned}$$

The same example also shows that it is possible to have

$$\text{size}(\mathbf{AB},2) < \min(\text{size}(\mathbf{A},2),\text{size}(\mathbf{B},2))$$

Nevertheless, there are some relations between size(**AB**) and size(**A**) and size(**B**).

Theorem 10.13

Size of Array Multiplication

If **A** and **B** are matrices or associative arrays, then

$$\text{size}(\mathbf{AB},1) \le \text{size}(\mathbf{A},1) \quad \text{and} \quad \text{size}(\mathbf{AB},2) \le \text{size}(\mathbf{B},2)$$

Consequently

$$\text{size}(\mathbf{AB}) \le \max(\text{size}(\mathbf{A}),\text{size}(\mathbf{B}))$$

Proof. See Exercise 10.20. □

Hopeful extensions to this would be

$$\text{size}(\mathbf{AB},2) \le \text{size}(\mathbf{A},2) \quad \text{and} \quad \text{size}(\mathbf{AB},1) \le \text{size}(\mathbf{B},1)$$

and hence the stronger upper bound

$$\text{size}(\mathbf{AB}) \le \min(\text{size}(\mathbf{A}),\text{size}(\mathbf{B}))$$

The above extensions, however, do not hold if

$$\mathbf{A} = \begin{bmatrix} 1 & 0 \\ 1 & 0 \end{bmatrix}, \quad \mathbf{B} = \begin{bmatrix} 1 & 1 \\ 1 & 1 \end{bmatrix}, \quad \mathbf{C} = \begin{bmatrix} 1 & 1 \\ 0 & 0 \end{bmatrix}$$

then

$$\mathbf{AB} = \mathbf{BC} = \begin{bmatrix} 1 & 1 \\ 1 & 1 \end{bmatrix}$$

but

$$\text{size}(\mathbf{AB},2) > \text{size}(\mathbf{A},2) \quad \text{and} \quad \text{size}(\mathbf{BC},1) > \text{size}(\mathbf{C},1)$$

rank

Rank obeys

$$\text{rank}(\mathbf{A}) + \text{rank}(\mathbf{B}) - n \le \text{rank}(\mathbf{AB}) \le \min(\text{rank}(\mathbf{A}),\text{rank}(\mathbf{B}))$$

The left inequality, known as Sylvester's inequality, is a special case of Frobenius's inequality.

Lemma 10.14

Frobenius's Inequality

[13] For any matrices $\mathbf{A}, \mathbf{B}$, and $\mathbf{C}$ with dimensions such that $\mathbf{ABC}$ is defined, then

$$\text{rank}(\mathbf{AB}) + \text{rank}(\mathbf{BC}) \leq \text{rank}(\mathbf{ABC}) + \text{rank}(\mathbf{B})$$

This above lemma leads to the following theorem.

Theorem 10.15

Sylvester's Inequality

For any two $n \times n$ matrices $\mathbf{A}$ and $\mathbf{B}$

$$\text{rank}(\mathbf{A}) + \text{rank}(\mathbf{B}) - n \leq \text{rank}(\mathbf{AB})$$

Proof. Applying Frobenius's inequality to $\mathbf{A}$, $\mathbb{I}$, and $\mathbf{B}$ gives

$$\begin{aligned}
\text{rank}(\mathbf{A}) + \text{rank}(\mathbf{B}) &= \text{rank}(\mathbf{A}\mathbb{I}) + \text{rank}(\mathbb{I}\mathbf{B}) \\
&\leq \text{rank}(\mathbf{A}\mathbb{I}\mathbf{B}) + \text{rank}(\mathbb{I}) \\
&= \text{rank}(\mathbf{AB}) + n
\end{aligned}$$

Rearranging the above expressions gives

$$\text{rank}(\mathbf{A}) + \text{rank}(\mathbf{B}) - n \leq \text{rank}(\mathbf{AB})$$

as desired. □

The above proof gives, as a special case, the following corollary.

Corollary 10.16

Rank n Matrices Closed under Array Multiplication

The set of $n \times n$ rank n matrices is closed under multiplication.

Proof. Suppose $\mathbf{A}$ and $\mathbf{B}$ are rank n, $n \times n$ matrices. Then $\mathbf{AB}$ is $n \times n$, so the rows of $\mathbf{AB}$ are

$$\{\mathbf{A}(1,:), \mathbf{A}(2,:), \ldots, \mathbf{A}(n,:)\}$$

The above are n row vectors and can be generated by n vectors through linear combination, so

$$\text{rank}(\mathbf{AB}) \leq n$$

Furthermore

$$\text{rank}(\mathbf{AB}) \geq \text{rank}(\mathbf{A}) + \text{rank}(\mathbf{B}) - n = n + n - n = n$$

so

$$\text{rank}(\mathbf{AB}) = n$$

□

The right equality follows straightforwardly from the rank-nullity theorem, and from the fact that rank does not change under transposition.

Theorem 10.17

Rank of Array Multiplication

For any $n \times n$ matrices $\mathbf{A}$ and $\mathbf{B}$

$$\text{rank}(\mathbf{A}) + \text{rank}(\mathbf{B}) - n \leq \text{rank}(\mathbf{AB}) \leq \min(\text{rank}(\mathbf{A}), \text{rank}(\mathbf{B}))$$

Proof. By Theorem 10.15, the left-hand side is

$$\text{rank}(\mathbf{A}) + \text{rank}(\mathbf{B}) - n \leq \text{rank}(\mathbf{AB})$$

Now, if

$$\mathbf{Bv} = \mathbb{0}$$

then

$$\mathbf{ABv} = \mathbf{A}\mathbb{0} = \mathbb{0}$$

so

$$\ker(\mathbf{B}) \subset \ker(\mathbf{AB})$$

where $\ker(\mathbf{B})$ is the set of all vectors $\mathbf{v}$ that satisfy

$$\mathbf{Bv} = \mathbb{0}$$

Likewise, where $\ker(\mathbf{AB})$ is the set of all vectors $\mathbf{v}$ that satisfy

$$\mathbf{ABv} = \mathbb{0}$$

Thus, if X is a set of vectors that generates via their linear combination $\ker(\mathbf{AB})$, X will also necessarily generate all the vectors in $\ker(\mathbf{B})$. Hence

$$\dim(\ker(\mathbf{B})) \leq \dim(\ker(\mathbf{AB}))$$

and so

$$\text{rank}(\mathbf{AB}) = n - \dim(\ker(\mathbf{AB})) \leq n - \dim(\ker(\mathbf{B})) = \text{rank}(\mathbf{B})$$

Since the row rank and column rank are the same, and transposition swaps rows and columns, for any matrix $\mathbf{M}$ (including $\mathbf{A}$ and $\mathbf{AB}$) it is known that

$$\text{rank}\left(\mathbf{M}^\mathsf{T}\right) = \text{rank}(\mathbf{M})$$

and since

$$(\mathbf{AB})^\mathsf{T} = \mathbf{B}^\mathsf{T}\mathbf{A}^\mathsf{T}$$

then

$$\text{rank}(\mathbf{AB}) = \text{rank}\left((\mathbf{AB})^\mathsf{T}\right) = \text{rank}\left(\mathbf{B}^\mathsf{T}\mathbf{A}^\mathsf{T}\right) \le \text{rank}\left(\mathbf{A}^\mathsf{T}\right) = \text{rank}(\mathbf{A})$$

This means that

$$\text{rank}(\mathbf{AB}) \le \text{rank}(\mathbf{B})$$

and

$$\text{rank}(\mathbf{AB} \le \text{rank}(\mathbf{A})$$

and so the right-hand side is

$$\text{rank}(\mathbf{AB}) \le \min(\mathbf{A}, \mathbf{B})$$

Thus

$$\text{rank}(\mathbf{A}) + \text{rank}(\mathbf{B}) - n \le \text{rank}(\mathbf{AB}) \le \min(\text{rank}(\mathbf{A}), \text{rank}(\mathbf{B}))$$

as desired. □

10.14 Closure of Operations between Arrays

One of the major motivations for the precise definition of an associative array given in Definition 10.1 was the collection of the operations of scalar multiplication, element-wise multiplication, element-wise addition, and array multiplication. However, it must still be shown that the definitions used to define these operations return associative arrays, namely that they fulfill the finite support condition.

Let

$$\mathbf{A}, \mathbf{B} : K_1 \times K_2 \to V$$

and

$$\mathbf{C} : K_2 \times K_3 \to V$$

be arrays. The element 0 in the semiring $(V, \oplus, \otimes, 0, 1)$ necessarily fulfills the identity

$$0 \otimes v = 0 = v \otimes 0$$

for every value $v \in V$ by the definition of a semiring. Thus

$$\text{size}(v\mathbf{A}, 1) \le \text{size}(\mathbf{A}, 1)$$

$$\text{size}(v\mathbf{A}, 2) \le \text{size}(\mathbf{A}, 2)$$

Additionally, for element-wise multiplication, there are the inequalities

$$\text{size}(\mathbf{A} \otimes \mathbf{B}, 1) \leq \min(\text{size}(\mathbf{A}, 1), \text{size}(\mathbf{B}, 1))$$

$$\text{size}(\mathbf{A} \otimes \mathbf{B}, 2) \leq \min(\text{size}(\mathbf{A}, 2), \text{size}(\mathbf{B}, 2))$$

For element-wise addition, there are the inequalities

$$\text{size}(\mathbf{A} \oplus \mathbf{B}, 1) \leq \text{size}(\mathbf{A}, 1) + \text{size}(\mathbf{B}, 1)$$

$$\text{size}(\mathbf{A} \oplus \mathbf{B}, 2) \leq \text{size}(\mathbf{A}, 2) + \text{size}(\mathbf{B}, 2)$$

Finally, in the case of array multiplication

$$\text{size}(\mathbf{AC}, 1) \leq \text{size}(\mathbf{A}, 1)$$

$$\text{size}(\mathbf{AC}, 2) \leq \text{size}(\mathbf{C}, 2)$$

The above observations show that the finite support condition holds for each of the resulting associative arrays.

10.15 Conclusions, Exercises, and References

There are a number of important properties that describe associative arrays. These properties include dimension, total number of elements, number of nonzero elements, density, sparsity, size, image, and rank. Of particular interest is how these properties behave under element-wise addition, element-wise multiplication, and array multiplication. The properties that are the most consistent are often the properties that are most useful to work with for associative arrays.

Exercises

Exercise 10.1 — Is the following map an associative array? Why or why not? If the map is an associative array, give its size.

$$\mathbf{A} : \{0, 1\} \times \{0, 1, 2\} \to \mathbb{R} \cup \{\text{-}\infty, \infty\}$$

given by

$$\mathbf{A} = \begin{array}{c} \\ 0 \\ 1 \end{array} \begin{array}{c} \begin{array}{ccc} 0 & 1 & 2 \end{array} \\ \left[\begin{array}{ccc} 0 & \infty & 1 \\ \text{-}\infty & 8 & \pi \end{array} \right] \end{array}$$

where $\mathbb{R} \cup \{\text{-}\infty, \infty\}$ is equipped with

(a) the min-max tropical algebra

(b) the max-plus algebra

Exercise 10.2 — Is the following map an associative array? Why or why not? If the map is an associative array, give its size.

$$\mathbf{B} : K_1 \times K_2 \to \mathbb{R}_+$$

with $\mathbf{B}(k_1, k_2) = 1$ for all $k_1 \in K_1$ and $k_2 \in K_2$.

Exercise 10.3 — Is the following map an associative array? Why or why not? If the map is an associative array, give its size.

(a) $$\mathbf{C} = \begin{array}{c} \\ 1 \\ 2 \end{array} \begin{array}{c} \begin{array}{ccc} 1 & 2 & 3 \end{array} \\ \begin{bmatrix} 7 & 2 & 1 \\ 0 & 3 & 3 \end{bmatrix} \end{array}$$

(b) $$\mathbf{D} = \begin{array}{c} \\ 1 \\ 2 \end{array} \begin{array}{c} \begin{array}{ccc} 1 & 2 & 3 \end{array} \\ \begin{bmatrix} 4 & 2 & 5 \\ 1 & 0 & 1 \end{bmatrix} \end{array}$$

(c) $$\mathbf{E} = \begin{array}{c} \\ 1 \\ 2 \\ 3 \end{array} \begin{array}{c} \begin{array}{cccc} 1 & 2 & 3 & 4 \end{array} \\ \begin{bmatrix} 3 & 2 & 1 & 0 \\ 0 & 1 & 2 & 3 \\ 0 & 1 & 0 & 1 \end{bmatrix} \end{array}$$

Exercise 10.4 — Is the following map an associative array? Why or why not? If the map is an associative array, give its size.

$$\mathbf{F} : \{1,2\} \times \{1,2\} \to \{\text{blue}, \text{red}\}$$

given by

$$\mathbf{F} = \begin{array}{c} \\ 1 \\ 2 \end{array} \begin{array}{c} \begin{array}{cc} 1 & 2 \end{array} \\ \begin{bmatrix} \text{blue} & \text{red} \\ \text{red} & \text{blue} \end{bmatrix} \end{array}$$

Exercise 10.5 — Is the following map an associative array? Why or why not? If the map is an associative array, give its size.

(a) $$\mathbf{G} = \begin{array}{c} \\ 1 \\ 2 \end{array} \begin{array}{c} \begin{array}{cc} 1 & 2 \end{array} \\ \begin{bmatrix} \emptyset & \{0\} \\ \{1\} & \emptyset \end{bmatrix} \end{array}$$

(b) $$\mathbf{H} = \begin{array}{cc} & \begin{array}{cc} 1 & 2 \end{array} \\ \begin{array}{c} 1 \\ 2 \end{array} & \begin{bmatrix} \{0,1\} & \{2\} \\ \emptyset & \{0\} \end{bmatrix} \end{array}$$

(c) $$\mathbf{I} = \begin{array}{cc} & \begin{array}{c} 1 \end{array} \\ \begin{array}{c} 1 \\ 2 \end{array} & \begin{bmatrix} \{0\} \\ \{0,2\} \end{bmatrix} \end{array}$$

Exercise 10.6 — Show that the following maps fulfill the inequalities in Section 10.14, where $\mathbf{A}, \mathbf{B}, \mathbf{C}$, and v are given by

$$\mathbf{A} = \begin{array}{cc} & \begin{array}{ccc} 1 & 2 & 3 \end{array} \\ \begin{array}{c} 1 \\ 2 \end{array} & \begin{bmatrix} 7 & 2 & 1 \\ 0 & 3 & 3 \end{bmatrix} \end{array}, \quad \mathbf{B} = \begin{array}{cc} & \begin{array}{ccc} 1 & 2 & 3 \end{array} \\ \begin{array}{c} 1 \\ 2 \end{array} & \begin{bmatrix} 4 & 2 & 5 \\ 1 & 0 & 1 \end{bmatrix} \end{array}, \quad \mathbf{C} = \begin{array}{cc} & \begin{array}{cccc} 1 & 2 & 3 & 4 \end{array} \\ \begin{array}{c} 1 \\ 2 \\ 3 \end{array} & \begin{bmatrix} 3 & 2 & 1 & 0 \\ 0 & 1 & 2 & 3 \\ 0 & 1 & 0 & 1 \end{bmatrix} \end{array}$$

and $v = 3$, where $V = \mathbb{R} \cup \{-\infty, \infty\}$ is equipped with the max-plus algebra.

Exercise 10.7 — Show that the following maps fulfill the inequalities in Section 10.14, where $\mathbf{A}, \mathbf{B}, \mathbf{C}$, and v are given by

$$\mathbf{A} = \begin{array}{cc} & \begin{array}{ccc} 1 & 2 & 3 \end{array} \\ \begin{array}{c} 1 \\ 2 \end{array} & \begin{bmatrix} 7 & 2 & 1 \\ 0 & 3 & 3 \end{bmatrix} \end{array}, \quad \mathbf{B} = \begin{array}{cc} & \begin{array}{ccc} 1 & 2 & 3 \end{array} \\ \begin{array}{c} 1 \\ 2 \end{array} & \begin{bmatrix} 4 & 2 & 5 \\ 1 & 0 & 1 \end{bmatrix} \end{array}, \quad \mathbf{C} = \begin{array}{cc} & \begin{array}{cccc} 1 & 2 & 3 & 4 \end{array} \\ \begin{array}{c} 1 \\ 2 \\ 3 \end{array} & \begin{bmatrix} 3 & 2 & 1 & 0 \\ 0 & 1 & 2 & 3 \\ 0 & 1 & 0 & 1 \end{bmatrix} \end{array}$$

and $v = 3$, where $V = \mathbb{R} \cup \{-\infty, \infty\}$ is equipped with the tropical max-min algebra.

Exercise 10.8 — Show that the following maps fulfill the inequalities in Section 10.14, where $\mathbf{A}, \mathbf{B}, \mathbf{C}$, and v are given by

$$\mathbf{A} = \begin{array}{cc} & \begin{array}{cc} 1 & 2 \end{array} \\ \begin{array}{c} 1 \\ 2 \end{array} & \begin{bmatrix} \emptyset & \{0\} \\ \{1\} & \emptyset \end{bmatrix} \end{array}, \quad \mathbf{B} = \begin{array}{cc} & \begin{array}{cc} 1 & 2 \end{array} \\ \begin{array}{c} 1 \\ 2 \end{array} & \begin{bmatrix} \{0,1\} & \{2\} \\ \emptyset & \{0\} \end{bmatrix} \end{array}, \quad \mathbf{C} = \begin{array}{cc} & \begin{array}{c} 1 \end{array} \\ \begin{array}{c} 1 \\ 2 \end{array} & \begin{bmatrix} \{0\} \\ \{0,2\} \end{bmatrix} \end{array}$$

and $v = \{0\}$, where $V = \mathcal{P}(\{0,1,2\})$ is equipped with the standard semiring structure.

Exercise 10.9 — Prove the inequalities in Section 10.14.

Exercise 10.10 — Prove Theorem 10.2.

Exercise 10.11 — Prove Theorem 10.3.

Exercise 10.12 — Prove Theorem 10.4.

Exercise 10.13 — Prove Theorem 10.5.

Exercise 10.14 — Prove Theorem 10.6.

Exercise 10.15 — Prove Theorem 10.7.

Exercise 10.16 — Prove Theorem 10.8.

Exercise 10.17 — Prove Theorem 10.9.

Exercise 10.18 — Prove Theorem 10.10.

Exercise 10.19 — Prove Theorem 10.11.

Exercise 10.20 — Prove Theorem 10.13.

Exercise 10.21 — Prove that if $K_1 \times K_2$ is infinite, then there does not exist an identity for element-wise multiplication.

Exercise 10.22 — Prove that if $K \times K$ is infinite, then there does not exist an identity for array multiplication.

Exercise 10.23 — Element-wise multiplication, element-wise addition, and array multiplication operations have been defined between two arrays

$$\mathbf{A} : K_1 \times K_2 \to V$$

and

$$\mathbf{B} : K_1' \times K_2' \to V$$

when there are additional conditions imposed on their domains. For element-wise multiplication and element-wise addition, it is necessary that

$$K_1 \times K_2 = K_1' \times K_2'$$

in order for

$$\mathbf{A} \otimes \mathbf{B}$$

and

$$\mathbf{A} \oplus \mathbf{B}$$

to be defined. For array multiplication, it is necessary that $K_2 = K_1'$ for $\mathbf{AB}$ to be defined.

(a) Expand the definitions of element-wise multiplication and element-wise addition to include the more general case where $K_1 \times K_2$ and $K_1' \times K_2'$ are possibly distinct, producing an associative array with domain

$$(K_1 \cup K_1') \times (K_2 \cup K_2')$$

(b) Do the same with array multiplication.

(c) Show that given any zero array

$$\mathbb{0} : K_1' \times K_2' \to V$$

the generalized array addition $\mathbf{A} \oplus \mathbb{0}$ is a zero padding of $\mathbf{A}$.

References

[1] A. Buluc and J. R. Gilbert, "On the representation and multiplication of hypersparse matrices," in *International Parallel and Distributed Processing Symposium Workshops(IPDPS)*, pp. 1–11, IEEE, 2008.

[2] A. Buluç and J. R. Gilbert, "Parallel sparse matrix-matrix multiplication and indexing: Implementation and experiments," *SIAM Journal on Scientific Computing*, vol. 34, no. 4, pp. C170–C191, 2012.

[3] A. George and E. Ng, "An implementation of Gaussian elimination with partial pivoting for sparse systems," *SIAM Journal on Scientific and Statistical Computing*, vol. 6, no. 2, pp. 390–409, 1985.

[4] T. F. Coleman and A. Pothen, "The null space problem II. algorithms," *SIAM Journal on Algebraic Discrete Methods*, vol. 8, no. 4, pp. 544–563, 1987.

[5] A. George and E. Ng, "On the complexity of sparse QR and LU factorization of finite-element matrices," *SIAM Journal on Scientific and Statistical Computing*, vol. 9, no. 5, pp. 849–861, 1988.

[6] J. R. Gilbert and M. T. Heath, "Computing a sparse basis for the null space," *SIAM Journal on Algebraic Discrete Methods*, vol. 8, no. 3, pp. 446–459, 1987.

[7] M. Benzi, "Preconditioning techniques for large linear systems: a survey," *Journal of Computational Physics*, vol. 182, no. 2, pp. 418–477, 2002.

[8] J. W. Liu, "Computational models and task scheduling for parallel sparse Cholesky factorization," *Parallel Computing*, vol. 3, no. 4, pp. 327–342, 1986.

[9] J. R. Gilbert, "Predicting structure in sparse matrix computations," *SIAM Journal on Matrix Analysis and Applications*, vol. 15, no. 1, pp. 62–79, 1994.

[10] E. Cohen, "Structure prediction and computation of sparse matrix products," *Journal of Combinatorial Optimization*, vol. 2, no. 4, pp. 307–332, 1998.

[11] L. Grady and E. L. Schwartz, "Faster graph-theoretic image processing via small-world and quadtree topologies," in *Conference on Computer Vision and Pattern Recognition*, vol. 2, pp. II–360, IEEE Computer Society, 2004.

[12] G. Strang, "The fundamental theorem of linear algebra," *The American Mathematical Monthly*, vol. 100, no. 9, pp. 848–855, 1993.

[13] C. D. Meyer, *Matrix Analysis and Applied Linear Algebra*, vol. 2. SIAM, 2000.

[14] G. Strang, *Introduction to Linear Algebra*, vol. 5. Wellesley, MA: Wellesley-Cambridge Press, 2009.

[15] M. Artin, *Algebra*. New York: Pearson, 2010.

11 Graph Construction and Graphical Patterns

Summary

Graph construction, a fundamental operation in a data processing system, is typically done by multiplying the incidence array representations of a graph, $\mathbf{E}_{\text{in}}$ and $\mathbf{E}_{\text{out}}$, to produce an adjacency array of the graph, $\mathbf{A}$, that can be processed with a variety of algorithms. Various mathematical criteria can be used to determine if the product $\mathbf{A} = \mathbf{E}_{\text{out}}^{\mathsf{T}}\mathbf{E}_{\text{in}}$ will be the adjacency array of the graph. An adjacency array of a graph can have certain patterns that are termed *special matrices*. Examples of useful patterns include the concepts of diagonal, off-diagonal, symmetric, skew-symmetric, upper and lower triangular, block, and block diagonal. Many of these patterns are also relevant when matrices are generalized to associative arrays. This chapter formally defines the relationships among adjacency arrays, incidence arrays, special matrices, and their corresponding graphs, all of which have properties that can be used to eliminate steps in a data processing pipeline.

11.1 Introduction

Adjacency arrays, typically denoted $\mathbf{A}$, have much in common with adjacency matrices. Likewise, incidence arrays or edge arrays, typically denoted $\mathbf{E}$, have much in common with incidence matrices [2–5], edge matrices [6], adjacency lists [7], and adjacency structures [8]. The powerful link between adjacency arrays and incidence arrays via array multiplication is the focus of the first part of this chapter.

The discussion of adjacency arrays and incidence arrays is readily illustrated with a variety of special matrices. The field of special matrices analyzes matrices with particular properties. These matrices have been the object of study since the first definition of matrices (see [9] and references therein). Among these properties are the structure of zero and nonzero elements in a matrix. These matrix structures come in many varieties, such as

Bidiagonal — nonzero along the main and an adjacent diagonal, and 0 elsewhere [10]
Block Diagonal — nonzero in blocks along the diagonal and 0 elsewhere [11, 12]
Convergent — becomes 0 when raised to higher powers [13]
Diagonal — 0 outside main diagonal [14]

Diagonalizable — transformable to a diagonal matrix [15, 16]
Exchange — 1 along main anti-diagonal and 0 elsewhere [17]
Hessenberg — 0 below (above) the lower (upper) main adjacent diagonal [18, 19]
Identity — 1 along main diagonal and 0 elsewhere [20]
Jacobi — nonzero along main diagonal and both adjacent diagonals, and 0 elsewhere [21]
Permutation — each row and each column contains a single 1 and is 0 elsewhere [22]
Shift — 1 on an adjacent main diagonal and 0 elsewhere [23]
Signature — ± 1 along the main diagonal and 0 elsewhere [24]
Triangular — 0 either above or below the main diagonal [25]
Zero — All values are 0 [20]

In each of the above special matrices, the pattern of zeroes and nonzero elements implies a different graph. However, the graph properties are not independent of the ordering of the rows and columns in the matrix. In the case of associative arrays, the ordering of the rows and columns is determined by the order function of their respective row and column key sets. Thus, special matrices, graphs, and order functions on row and column key sets are all closely related and are discussed in detail in the second half of the chapter.

11.2 Adjacency and Incidence Array Definitions

Associative arrays derive much of their power from their ability to represent data intuitively in easily understandable tables. Two properties of associative arrays in particular are different from other two-dimensional arrangements of data. First, each row and column key in an array is unique and sortable, allowing rows and columns to be queried efficiently. Secondly, because associative array implementations store no rows or columns that are entirely empty, insertion, selection, and deletion of data can be performed by element-wise addition, element-wise multiplication, and array multiplication.

Associative arrays can represent graphs through both incidence and adjacency arrays.

Definition 11.1

Adjacency Array

Let G be a (directed, weighted) graph G with vertex set $K_{\text{out}} \cup K_{\text{in}}$, where K_{out} and K_{in} are the sets of vertices with outgoing and incoming edges, respectively.
The array

$$\mathbf{A} : K_{\text{out}} \times K_{\text{in}} \to V$$

is an *adjacency array* of G if there is an edge from k_{out} to k_{in} if and only if

$$\mathbf{A}(k_{\text{out}}, k_{\text{in}}) \neq 0$$

Definition 11.2

Standard Adjacency Array

If G is a weighted (directed) graph then

$$\mathbf{A} : K_{\text{out}} \times K_{\text{in}} \to V$$

is the *standard adjacency array* if $\mathbf{A}(k_{\text{out}}, k_{\text{in}})$ is equal to the weight of the edge from k_{out} to k_{in}, if there is such an edge, and 0 otherwise.
If G is not weighted, then the standard adjacency array is defined in the same way, with weights assumed to be 1.

There can be multiple adjacency arrays for a given graph. Whenever *the* adjacency array is spoken of, it is assumed to mean the *standard* adjacency array.

Definition 11.3

Incidence Array

Let G be a (directed, weighted, multi, hyper) graph with edge set K and vertex set $K_{\text{out}} \cup K_{\text{in}}$, where K_{out} and K_{in} are the sets of vertices with outgoing and incoming edges, respectively.
Arrays

$$\mathbf{E}_{\text{out}} : K \times K_{\text{out}} \to V \quad \text{and} \quad \mathbf{E}_{\text{in}} : K \times K_{\text{in}} \to V$$

are *out-vertex* and *in-vertex incidence arrays* of G, respectively, if

$$\mathbf{E}_{\text{out}}(k, k_{\text{out}}) \neq 0 \quad \text{and} \quad \mathbf{E}_{\text{in}}(k, k_{\text{in}})$$

if and only if edge k is directed outward from vertex k_{out} and inward to vertex k_{in}.

There can be multiple distinct out-vertex (in-vertex) incidence arrays for a given graph. Graphs have been discussed in the preceding text, but to supplement the formal definition of an associative array, it is useful to formally define what it means to be a graph.

Definition 11.4

Undirected Graph

An *undirected graph*

$$G = (K, E)$$

consists of a set K whose elements are called *vertices* and a set E whose elements are unordered pairs of vertices $\{k, k'\}$ called *edges*.

An undirected graph can be thought of geometrically as a set of vertices in K, where each edge

$$\{k, k'\} \in E$$

represents an edge between vertex k and k'. In addition, the standard adjacency array $\mathbf{A}$ of an undirected graph obeys the transpose identity

$$\mathbf{A} = \mathbf{A}^{\mathsf{T}}$$

Example 11.1

An undirected line graph $G = (K, E)$ can be constructed with vertice and eges

$$K = \{1, \ldots, n\} \quad \text{and} \quad E = \{\{k, k+1\} \mid 1 \leq k < n\}$$

This graph forms a sequential, undirected line through all of the vertices. A graph drawing of this line graph for $n = 4$ and its standard adjacency array are

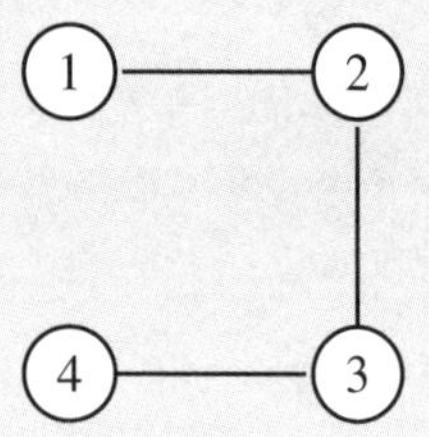

$$\mathbf{A} = \begin{array}{c} \\ 1 \\ 2 \\ 3 \\ 4 \end{array} \begin{array}{c} \begin{array}{cccc} 1 & 2 & 3 & 4 \end{array} \\ \left[\begin{array}{cccc} & 1 & & \\ 1 & & 1 & \\ & 1 & & 1 \\ & & 1 & \end{array} \right] \end{array}$$

Example 11.2

An undirected loop graph $G = (K, E)$ can be constructed with vertices and edges

$$K = \{1, \ldots, n\} \quad \text{and} \quad E = \{\{k, k+1\} \mid 1 \leq k < n\} \cup \{(n, 1)\}$$

This graph forms a sequential, undirected loop through all of the vertices. A graph drawing of this graph for $n = 4$ and its standard adjacency array are

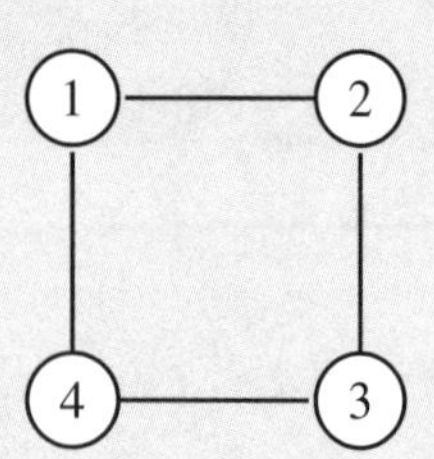

$$\mathbf{A} = \begin{array}{c} \\ 1 \\ 2 \\ 3 \\ 4 \end{array} \begin{array}{c} \begin{array}{cccc} 1 & 2 & 3 & 4 \end{array} \\ \left[\begin{array}{cccc} & 1 & & 1 \\ 1 & & 1 & \\ & 1 & & 1 \\ 1 & & 1 & \end{array} \right] \end{array}$$

Definition 11.5

Directed Graph

A *directed graph*

$$G = (K, E)$$

consists of a set K whose elements are called *vertices* and a subset E of $K \times K$ whose elements $(k_{\text{out}}, k_{\text{in}})$ are called *edges*.

A directed graph is a set of vertices where an edge

$$(k_{\text{out}}, k_{\text{in}}) \in E$$

is an arrow directed from vertex k_{out} to vertex k_{in}.

By convention, every vertex will either be an *out-vertex* with an edge directed outward from that vertex, or an *in-vertex* with an edge directed inward to that vertex. Hence, it is assumed that $K = K_{\text{out}} \cup K_{\text{in}}$ where K_{out} and K_{in} are the sets of out-vertices and in-vertices. In terms of the adjacency arrays, this convention is merely a restatement of the fact that zero rows and columns can be assumed to have been removed.

Example 11.3

A directed line graph $G = (K_{\text{out}} \cup K_{\text{in}}, E)$ can be constructed with vertices

$$K_{\text{out}} = \{1, \ldots, n-1\} \quad \text{and} \quad K_{\text{in}} = \{2, \ldots, n\}$$

and edges

$$E = \{(k, k+1) \mid 1 \leq k < n\}$$

This (directed) graph forms a sequential, directed line through all of the vertices. A graph drawing of this line graph for $n = 4$ and its standard adjacency array are

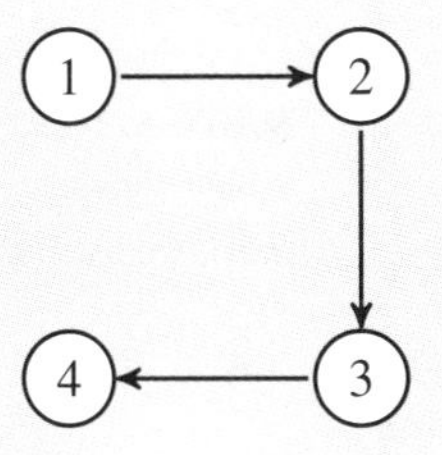

$$\mathbf{A} = \begin{array}{c} \\ 1 \\ 2 \\ 3 \end{array} \begin{array}{c} \begin{array}{ccc} 2 & 3 & 4 \end{array} \\ \left[\begin{array}{ccc} 1 & & \\ & 1 & \\ & & 1 \end{array} \right] \end{array}$$

Example 11.4

A directed loop graph $G = (K_{\text{out}} \cup K_{\text{in}}, E)$ can be constructed with vertices

$$K_{\text{out}} = \{1,\ldots,n\} \quad \text{and} \quad K_{\text{in}} = \{1,\ldots,n\}$$

and edges

$$E = \{(k, k+1) \mid 1 \leq k < n\} \cup \{(n,1)\}$$

This graph produces a sequential, directed loop through all of the vertices. A graph drawing of this graph for $n = 4$ is

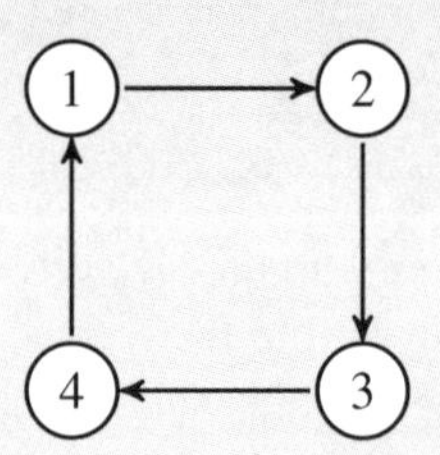

$$\mathbf{A} = \begin{array}{c} \\ 1 \\ 2 \\ 3 \\ 4 \end{array} \begin{array}{c} \begin{array}{cccc} 1 & 2 & 3 & 4 \end{array} \\ \left[\begin{array}{cccc} & 1 & & \\ & & 1 & \\ & & & 1 \\ 1 & & & \end{array}\right] \end{array}$$

Definition 11.6

Weighted Directed Graph

A *weighted directed graph*

$$G = (K_{\text{out}} \cup K_{\text{in}}, E, \mathrm{W} : E \to V)$$

is a directed graph $(K_{\text{out}} \cup K_{\text{in}}, E)$ along with a *weight function*

$$\mathrm{W} : E \to V$$

which assigns to each edge a nonzero value in the semiring $(V, \oplus, \otimes, 0, 1)$.

A weighted directed graph can be interpreted visually as a set of vertices where each edge

$$(k_{\text{out}}, k_{\text{in}}) \in K_{\text{out}} \times K_{\text{in}}$$

represents an arrow pointing from vertex k_{out} to vertex k_{in} that has a nonzero value $v \in V$ assigned to the edge.

The standard adjacency array of a weighted graph G is the extension of the weight function to all pairs in $K_{\text{out}} \times K_{\text{in}}$. In other words, the weight function is the restriction of the standard adjacency array to the set of edges.

In a graph drawing, the weight function is represented by labeling the arrow representing the edge with the weight of the edge.

Example 11.5

A weighted directed graph $G = (K_{\text{out}} \cup K_{\text{in}}, E, V)$ can be constructed with vertices

$$K_{\text{out}} = K_{\text{in}} = \{\text{Boston}, \text{Cambridge}, \text{Salem}\}$$

and edges

$$E = \{(\text{Boston}, \text{Cambridge}), (\text{Cambridge}, \text{Boston}),\\ (\text{Cambridge}, \text{Salem}), (\text{Salem}, \text{Cambridge})\}$$

where the weights are the shortest driving distances between these cities This graph has the graph drawing

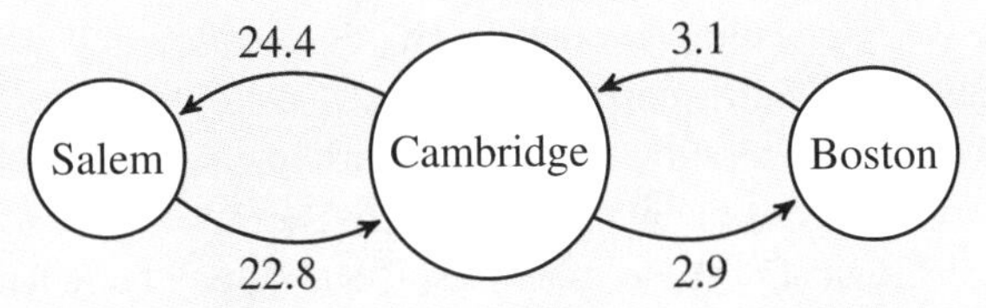

The adjacency array for this weighted directed graph is

$$\mathbf{A} = \begin{array}{c} \\ \text{Boston} \\ \text{Cambridge} \\ \text{Salem} \end{array} \begin{array}{c} \begin{array}{ccc} \text{Boston} & \text{Cambridge} & \text{Salem} \end{array} \\ \left[\begin{array}{ccc} & 3.1 & \\ 2.9 & & 24.4 \\ & 22.8 & \end{array}\right] \end{array}$$

Example 11.6

The graph of a 3×3 matrix over some polynomial ring

$$\mathbf{A} = \begin{array}{c} \\ 1 \\ 2 \\ 3 \end{array} \begin{array}{c} \begin{array}{ccc} 1 & 2 & 3 \end{array} \\ \left[\begin{array}{ccc} 1 & x & \\ & x & \\ x^2 & & \end{array}\right] \end{array}$$

has the vertices $G = K = \{1, 2, 3\}$ and edges $E = \{(1,1), (1,2), (2,2), (3,1)\}$ with weight function given by the restriction of $\mathbf{A}$ to E.

Example 11.7

The graph of the unit array $\mathbf{e}_{(k_1,k_2)} : K^2 \to V$ again has K as the set of vertices, but with a single edge connecting k_1 and k_2.

11.3 Adjacency Array Construction

Incidence arrays are often readily obtained from raw data. In many cases, an associative array representing a spreadsheet or database table is already in the form of an incidence array. However, to analyze a graph, it is often convenient to represent the graph as an adjacency array. Constructing an adjacency array from data stored in an incidence array via array multiplication is one of the most common and important steps in a data processing system.

Given a graph G with vertex set $K_{\text{out}} \cup K_{\text{in}}$ and edge set K, the construction of adjacency arrays for G relies on the assumption that $\mathbf{E}_{\text{out}}^{\mathsf{T}}\mathbf{E}_{\text{in}}$ is an adjacency array of G. This assumption is certainly true in the most common case, where the value set is composed of non-negative reals and the operations $\oplus$ and $\otimes$ are arithmetic plus $+$ and arithmetic times $\times$, respectively. However, one hallmark of associative arrays is their ability to contain as values nontraditional data. For these value sets, $\oplus$ and $\otimes$ may be redefined to operate on non-numerical values. For example, for the value of all alphanumeric strings, with

$$\oplus \equiv \max$$
$$\otimes \equiv \min$$

it is not immediately apparent in this case whether $\mathbf{E}_{\text{out}}^{\mathsf{T}}\mathbf{E}_{\text{in}}$ is an adjacency array of the graph whose set of vertices is $K_{\text{out}} \cup K_{\text{in}}$. In the subsequent sections, the criteria on the value set V and the operations $\oplus$ and $\otimes$ are presented so that

$$\mathbf{A} = \mathbf{E}_{\text{out}}^{\mathsf{T}}\mathbf{E}_{\text{in}}$$

always produces an adjacency array [1, 26].

For the purpose of establishing the minimum criteria for graph construction, this section will *not* assume a full commutative semiring structure for the value set. All that will be required is that the value set V be closed under two binary operations $\oplus$ and $\otimes$ and that V contains the identity elements of these operations, denoted 0 and 1, respectively. Associativity, commutativity, and distributivity are not required. However, in most practical applications these properties will hold.

Moreover, the graphs dealt with can be even more general by allowing for multiple directed edges between two vertices.

Definition 11.7

Directed Multi-Graph

A *directed multi-graph*

$$G = (K_{\text{out}} \cup K_{\text{in}}, K)$$

consists of a set of vertices $K_{\text{out}} \cup K_{\text{in}}$ and a set K whose elements are called *edges*, where each edge is assigned a *source* in K_{out} and a *target* in K_{in}.

Directed multi-graphs allow for multiple directed edges between two vertices. This has no effect on the definitions of incidence arrays or adjacency arrays.

Theorem 11.1

Constructing Adjacency Array from Incidence Arrays (Directed Multi-Graphs)

Let V be a set with closed binary operations $\oplus, \otimes$ with identities $0, 1 \in V$. Then the following are equivalent:

(i) $\oplus$ and $\otimes$ satisfy the properties

  (a) Zero-sum-free: $a \oplus b = 0$ if and only if $a = b = 0$,

  (b) Zero-divisor-free: $a \otimes b = 0$ if and only if $a = 0$ or $b = 0$, and

  (c) 0 is onnihilator for $\otimes$: $a \otimes 0 = 0 \otimes a = 0$.

(ii) If G is a directed multi-graph with out-vertex and in-vertex incidence arrays $\mathbf{E}_{\text{out}} : K \times K_{\text{out}} \to V$ and $\mathbf{E}_{\text{in}} : K \times K_{\text{out}} \to V$, then $\mathbf{E}_{\text{out}}^{\mathsf{T}} \mathbf{E}_{\text{in}}$ is an adjacency array for G.

Proof. For $\mathbf{E}_{\text{out}}^{\mathsf{T}} \mathbf{E}_{\text{in}}$ to be the adjacency array of G, the entry $\mathbf{A}(k_{\text{out}}, k_{\text{in}})$ must be nonzero if and only if there is an edge from k_{out} to k_{in}, which is equivalent to saying that the entry must be nonzero if and only if there is a $k \in K$ such that

$$\mathbf{E}_{\text{out}}^{\mathsf{T}}(k_{\text{out}}, k) \neq 0$$
$$\mathbf{E}_{\text{in}}(k, k_{\text{in}}) \neq 0$$

Taken altogether, the above pair of equations implies

$$\bigoplus_{k \in K} \mathbf{E}^{\mathsf{T}}_{\text{out}}(k_{\text{out}}, k) \otimes \mathbf{E}_{\text{in}}(k, k_{\text{in}}) \neq 0$$

$$\text{if and only if}$$

$$\mathbf{E}^{\mathsf{T}}_{\text{out}}(k_{\text{out}}, k) \neq 0 \quad \text{and} \quad \mathbf{E}_{\text{in}}(k, k_{\text{in}}) \neq 0 \tag{11.1}$$

This equivalence is shown to imply the desired algebraic properties of $\oplus, \otimes$. It is equivalent to the statement

$$\bigoplus_{k \in K} \mathbf{E}_{\text{out}}(k, x) \otimes \mathbf{E}_{\text{in}}(k, y) = 0$$

$$\text{if and only if}$$

$$\nexists k \in K \quad \text{such that} \quad \mathbf{E}_{\text{out}}(k, x) \neq 0 \quad \text{and} \quad \mathbf{E}_{\text{in}}(k, y) \neq 0$$

The above statement in turn is equivalent to

$$\bigoplus_{k \in K} \mathbf{E}_{\text{out}}(k, x) \otimes \mathbf{E}_{\text{in}}(k, y) = 0$$

$$\text{if and only if}$$

$$\mathbf{E}_{\text{out}}(k, x) = 0 \quad \text{or} \quad \mathbf{E}_{\text{in}}(k, y) = 0$$

This expression may be split up into two conditional statements. The first condition is that for all $k \in K$

$$\bigoplus_{k \in K} \mathbf{E}_{\text{out}}(k, x) \otimes \mathbf{E}_{\text{in}}(k, y) = 0$$

$$\text{implies}$$

$$\mathbf{E}_{\text{out}}(k, x) = 0 \quad \text{or} \quad \mathbf{E}_{\text{in}}(k, y) = 0 \tag{11.2}$$

The second condition is that for all $k \in K$

$$\mathbf{E}_{\text{out}}(k, x) = 0 \quad \text{or} \quad \mathbf{E}_{\text{in}}(k, y) = 0$$

$$\text{implies}$$

$$\bigoplus_{k \in K} \mathbf{E}_{\text{out}}(k, x) \otimes \mathbf{E}_{\text{in}}(k, y) = 0 \tag{11.3}$$

Lemma 11.2

Zero-Sum-Free is Necessary

Equation 11.2 implies that V is zero-sum-free.

Proof. Suppose there exist nonzero $v, w \in V$ such that $v \oplus w = 0$, or that nontrivial additive inverses exist. Then consider the graph

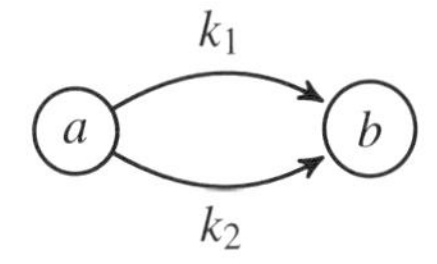

with incidence arrays

$$\mathbf{E}_{\text{out}} = \begin{array}{c} \\ k_1 \\ k_2 \end{array} \begin{array}{c} a \\ \left[\begin{array}{c} v \\ w \end{array}\right] \end{array} \quad \mathbf{E}_{\text{in}} = \begin{array}{c} \\ k_1 \\ k_2 \end{array} \begin{array}{c} b \\ \left[\begin{array}{c} 1 \\ 1 \end{array}\right] \end{array}$$

It is the case that

$$\begin{aligned} \mathbf{E}_{\text{out}}^{\mathsf{T}} \mathbf{E}_{\text{in}}(b,a) &= (v \otimes 1) \oplus (w \otimes 1) \\ &= v \oplus w \\ &= 0 \end{aligned}$$

which contradicts Equation 11.2. Therefore, no such nonzero v and w may be present in V, meaning it is necessary that V be zero-sum-free. □

Lemma 11.3

Zero-Divisor-Free is Necessary

Equation 11.2 implies that V is zero-divisor-free.

Proof. Suppose there exist nonzero $v, w \in V$ such that $v \otimes w = 0$. Then consider the graph

$$a \xrightarrow{\;k\;} b$$

with incidence arrays

$$\mathbf{E}_{\text{out}} = k \begin{array}{c} a \\ \left[\begin{array}{c} v \end{array}\right] \end{array} \quad \mathbf{E}_{\text{in}} = k \begin{array}{c} b \\ \left[\begin{array}{c} w \end{array}\right] \end{array}$$

It is the case that

$$\begin{aligned}\mathbf{E}_{\text{out}}^{\mathsf{T}}\mathbf{E}_{\text{in}}(a,b) &= \mathbf{E}_{\text{out}}(k,a)\otimes\mathbf{E}_{\text{in}}(k,b)\\ &= v\otimes w\\ &= 0\end{aligned}$$

which contradicts Equation 11.2. Therefore, no such v and w may be present in V, so V is zero-divisor-free. □

Lemma 11.4

0 Annihilates is Necessary

Equation 11.3 implies that 0 annihilates V under $\otimes$.

Proof. A first proof uses the result of Lemma 11.2. Consider the graph

$$a \xrightarrow{k_1} b \xrightarrow{k_2} c \xrightarrow{k_3} d$$

with incidence arrays

$$\mathbf{E}_{\text{out}} = \begin{matrix} & a & b & c \\ k_1 & v & 0 & 0 \\ k_2 & 0 & v & 0 \\ k_3 & 0 & 0 & v \end{matrix} \qquad \mathbf{E}_{\text{in}} = \begin{matrix} & b & c & d \\ k_1 & v & 0 & 0 \\ k_2 & 0 & v & 0 \\ k_3 & 0 & 0 & v \end{matrix}$$

It is the case that

$$\mathbf{E}_{\text{out}}^{\mathsf{T}}\mathbf{E}_{\text{in}}(a,c) = (v\otimes 0)\oplus(0\otimes v)\oplus(0\otimes 0) = 0 \tag{11.4}$$

Then, by Lemma 11.2, it follows that $v\otimes 0 = 0\times v = 0\otimes 0 = 0$.

A second proof avoids the use of Lemma 11.2. First consider the graph

$$a \xrightarrow{k_1} b \xrightarrow{k_2} c$$

with incidence arrays

$$\mathbf{E}_{\text{out}} = \begin{matrix} & a & b \\ k_1 & v & 0 \\ k_2 & 0 & v \end{matrix} \qquad \mathbf{E}_{\text{in}} = \begin{matrix} & b & c \\ k_1 & v & 0 \\ k_2 & 0 & v \end{matrix}$$

It is the case that

$$\mathbf{E}_{\text{out}}^{\mathsf{T}}\mathbf{E}_{\text{in}}(a,c) = (v\otimes 0)\oplus(0\otimes v) = 0 \tag{11.5}$$

Then using Equation 11.4 with Equation 11.5, it follows that $0 \otimes 0 = 0$.

Next consider the graph

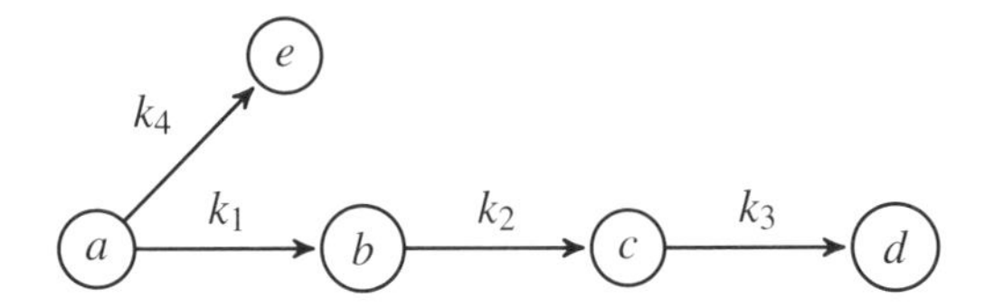

with incidence arrays

$$
\mathbf{E}_{\text{out}} = \begin{array}{c} \\ k_1 \\ k_2 \\ k_3 \\ k_4 \end{array} \begin{array}{c} \begin{array}{ccc} a & b & c \end{array} \\ \left[\begin{array}{ccc} v & 0 & 0 \\ 0 & v & 0 \\ 0 & 0 & v \\ v & 0 & 0 \end{array}\right] \end{array}
\qquad
\mathbf{E}_{\text{in}} = \begin{array}{c} \\ k_1 \\ k_2 \\ k_3 \\ k_4 \end{array} \begin{array}{c} \begin{array}{cccc} b & c & d & e \end{array} \\ \left[\begin{array}{cccc} v & 0 & 0 & 0 \\ 0 & v & 0 & 0 \\ 0 & 0 & v & 0 \\ 0 & 0 & 0 & v \end{array}\right] \end{array}
$$

It is the case that

$$\mathbf{E}_{\text{out}}^{\mathsf{T}} \mathbf{E}_{\text{in}}(a, c) = (v \otimes 0) \oplus (0 \otimes v) \oplus (0 \otimes 0) \oplus (v \otimes 0) = 0 \tag{11.6}$$

Then using Equation 11.4 with Equation 11.6, it follows that $v \otimes 0 = 0$. Using this in Equation 11.5 gives the final desired equality $0 \otimes v = 0$.

□

Conversely, Theorem 11.1(i) is sufficient for Theorem 11.1(ii) to hold. Assume that zero is an annihilator, V is zero-sum-free, and V is zero-divisors-free. Zero-sum-freeness and the nonexistence of zero divisors means that there exists $k \in K$ such that

$$\mathbf{E}_{\text{out}}(k, x) \neq 0 \quad \text{and} \quad \mathbf{E}_{\text{in}}(k, y) \neq 0$$

$$\text{implies}$$

$$\bigoplus_{k \in K} \mathbf{E}_{\text{out}}(k, x) \otimes \mathbf{E}_{\text{in}}(k, y) \neq 0$$

which is the contrapositive of Equation 11.2. In addition, since zero is an annihilator, it is the case that for all $k \in K$

$$\mathbf{E}_{\text{out}}(e, x) = 0 \quad \text{or} \quad \mathbf{E}_{\text{in}}(e, y) = 0$$

$$\text{implies}$$

$$\bigoplus_{k \in} \mathbf{E}_{\text{out}}(k, x) \otimes \mathbf{E}_{\text{in}}(k, y) = 0$$

which is Equation 11.3. As Equation 11.2 and Equation 11.3 combine to form Equation 11.1, it is established that the conditions are sufficient for Equation 11.1. □

The remaining product of the incidence arrays that is defined is $\mathbf{E}_{\text{in}}^{\mathsf{T}}\mathbf{E}_{\text{out}}$. The above requirements will now be shown to be necessary and sufficient for the remaining product to be the adjacency array of the reverse of the graph. Recall that the reverse of G is the graph $\bar{G}$ in which all the arrows in G have been reversed. Let G be a graph with incidence matrices $\mathbf{E}_{\text{out}}$ and $\mathbf{E}_{\text{in}}$.

Corollary 11.5

Constructing Adjacency Array of Reverse from Incidence Arrays

Condition (i) in Theorem 11.1 are necessary and sufficient so that $\mathbf{E}_{\text{in}}^{\mathsf{T}}\mathbf{E}_{\text{out}}$ is an adjacency matrix of the reverse of G.

Proof. See Exercise 11.4. □

Among those structures that satisfy the algebraic requirements of Theorem 11.1 are the semirings that are both zero-sum-free and have no zero divisors. For example, $\mathbb{N}$ and $\mathbb{R}_{\geq 0}$ with standard addition and multiplication or any bounded linearly ordered set with supremum and infimum.

There are also structures that are not semirings but nevertheless satisfy the requirements of Theorem 11.1, such as $\mathbb{N}$ with the operations

$$n \oplus m = \begin{cases} m & \text{if } n = 0 \text{ or } m = 0 \\ n+1 & \text{otherwise} \end{cases} \quad \text{and} \quad n \otimes m = \begin{cases} 0 & \text{if } n = 0 \\ n & \text{if } m = 1 \\ m & \text{if } n = 1 \\ m^2 & \text{otherwise} \end{cases}$$

Then $\oplus$ has identity 0, $n \oplus m = 0$ implies $n = m = 0$, $\otimes$ has identity 1, $n \otimes m = 0$ implies at least one of n, m is 0, and $0 \otimes n = n \otimes 0 = 0$. However, neither $\oplus$ nor $\otimes$ is associative or commutative, and $\otimes$ does not distribute over $\oplus$, as the following computations show.

$$\begin{aligned} (1 \oplus 1) \oplus 1 &= 3 & 1 \oplus (1 \oplus 1) &= 2 \\ 1 \oplus 2 &= 2 & 2 \oplus 1 &= 3 \\ (2 \otimes 3) \otimes 4 &= 4 & (2 \otimes 3) \otimes 4 &= 16 \\ 2 \otimes 3 &= 4 & 3 \otimes 2 &= 9 \\ 2 \otimes (1 \oplus 1) &= 4 & (2 \otimes 1) \oplus (2 \otimes 1) &= 3 \\ (1 \oplus 1) \otimes 2 &= 4 & (1 \otimes 2) \oplus (1 \otimes 2) &= 3 \end{aligned}$$

Non-examples, however, include the max-plus algebra and power set algebras (assuming more than one element).

Restricting the class of the graphs involved removes the necessity of being zero-sum-free.

Corollary 11.6

Constructing Adjacency Array from Incidence Arrays (Directed Graphs)

Let V be a set with closed binary operations $\oplus, \otimes$ with additive identity $0 \in V$. Then the following are equivalent:

(i) $\oplus$ and $\otimes$ satisfy the properties

(a) Zero-divisor-free: $a \otimes b = 0$ if and only if $a = 0$ or $b = 0$, and

(b) 0 is annihilator for $\otimes$: $a \otimes 0 = 0 \otimes a = 0$.

(ii) If G is a directed graph with out-vertex and in-vertex incidence arrays $\mathbf{E}_{\text{out}} : K \times K_{\text{out}} \to V$ and $\mathbf{E}_{\text{in}} : K \times K_{\text{out}} \to V$, then $\mathbf{E}_{\text{out}}^{\mathsf{T}} \mathbf{E}_{\text{in}}$ is an adjacency array for G.

Proof. The implication $(ii) \implies (i)$ follows from the proof of $(i) \implies (ii)$ given in Theorem 11.1 since the graphs constructed in the proofs of Lemma 11.3 and Lemma 11.4 are directed graphs.

As in the proof of Theorem 11.1, in proving $(i) \implies (ii)$ it is enough to show that Equation 11.2

$$\bigoplus_{k \in K} \mathbf{E}_{\text{out}}(k,x) \otimes \mathbf{E}_{\text{in}}(k,y) = 0$$

$$\text{implies}$$

$$\mathbf{E}_{\text{out}}(k,x) = 0 \quad \text{or} \quad \mathbf{E}_{\text{in}}(k,y) = 0$$

and Equation 11.3

$$\mathbf{E}_{\text{out}}(k,x) = 0 \quad \text{or} \quad \mathbf{E}_{\text{in}}(k,y) = 0$$

$$\text{implies}$$

$$\bigoplus_{k \in K} \mathbf{E}_{\text{out}}(k,x) \otimes \mathbf{E}_{\text{in}}(k,y) = 0$$

in a directed graph G.

Since there can be at most a single edge directed from x to y, there is at most a single edge k such that

$$\mathbf{E}_{\text{out}}(k,x) \neq 0 \quad \text{and} \quad \mathbf{E}_{\text{in}} \neq 0$$

with

$$\mathbf{E}_{\text{out}}(k',x) = 0 \quad \text{or} \quad \mathbf{E}_{\text{in}}(k',y) = 0$$

for all other edges k'. Since 0 is an annihilator, each of the terms

$$\mathbf{E}_{\text{out}}(k',x)\mathbf{E}_{\text{in}}(k',y)$$

is equal to 0, so

$$0 = \mathbf{E}_{\text{out}}(k,x)\mathbf{E}_{\text{in}}(k,y)$$

Since V is zero-divisor-free, it follows that either

$$\mathbf{E}_{\text{out}}(k,x) = 0 \quad \text{or} \quad \mathbf{E}_{\text{in}}(k,y) = 0$$

This proves that Equation 11.2 holds.

Conversely, the proof that Equation 11.3 holds follows exactly as in the proof of Theorem 11.1. □

The criteria guarantee an accurate adjacency array for any data set that satisfies them, regardless of value distribution in the incidence arrays. However, if the incidence arrays are known to possess a certain structure, it is possible to circumvent some of the conditions and still always produce adjacency arrays. For example, if each key set of an undirected incidence array $\mathbf{E}$ is a list of documents and the array entries are sets of words shared by documents, then it is necessary that a word in $\mathbf{E}(i,j)$ and $\mathbf{E}(m,n)$ has to be in $\mathbf{E}(i,n)$ and $\mathbf{E}(m,j)$. This structure means that when multiplying $\mathbf{E}^{\mathsf{T}}\mathbf{E}$ using $\oplus = \cup$ and $\otimes = \cap$, a non-empty set will never be "multiplied" by or intersected with a disjoint non-empty set. This condition eliminates the need for the zero-product property to be satisfied, as every multiplication of non-empty sets is already guaranteed to produce a non-empty set. The array produced will contain as entries a list of words shared by those two documents.

Though the criteria ensure that the product of incidence arrays will be an adjacency array, they do not ensure that certain matrix properties hold. For example, the property

$$(\mathbf{AB})^{\mathsf{T}} = \mathbf{B}^{\mathsf{T}}\mathbf{A}^{\mathsf{T}}$$

may be violated under these criteria, as $(\mathbf{E}_{\text{out}}^{\mathsf{T}}\mathbf{E}_{\text{in}})^{\mathsf{T}}$ is not necessarily equal to $\mathbf{E}_{\text{in}}^{\mathsf{T}}\mathbf{E}_{\text{out}}$. (For this matrix transpose property to always hold, the operation $\otimes$ would have to be commutative.)

11.4 Graph Construction with Different Semirings

The ability to change $\oplus$ and $\otimes$ operations allows different graph adjacency arrays to be constructed with the same element-wise addition, element-wise multiplication, and array multiplication syntax. Specific pairs of operations are best suited for constructing certain types of adjacency arrays. The pattern of edges resulting from array multiplication of

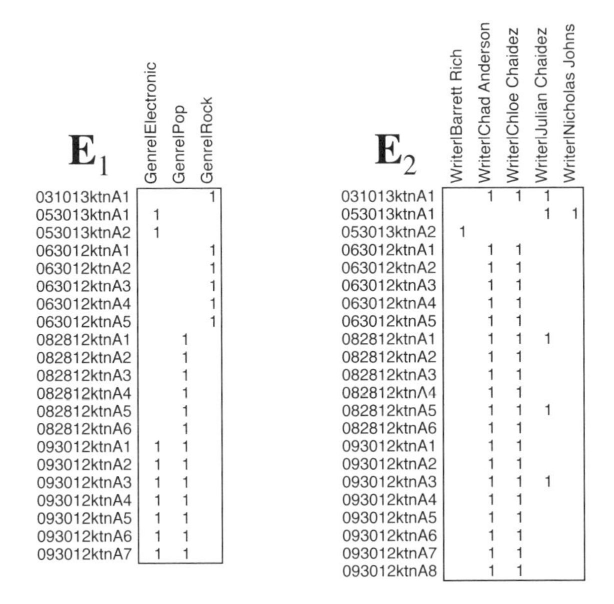

$\mathbf{E}_1$	Genre|Electronic	Genre|Pop	Genre|Rock
031013ktnA1			1
053013ktnA1	1		
053013ktnA2	1		
063012ktnA1			1
063012ktnA2			1
063012ktnA3			1
063012ktnA4			1
063012ktnA5			1
082812ktnA1		1	
082812ktnA2		1	
082812ktnA3		1	
082812ktnA4		1	
082812ktnA5		1	
082812ktnA6		1	
093012ktnA1	1	1	
093012ktnA2	1	1	
093012ktnA3	1	1	
093012ktnA4	1	1	
093012ktnA5	1	1	
093012ktnA6	1	1	
093012ktnA7	1	1	

$\mathbf{E}_2$	Writer|Barrett Rich	Writer|Chad Anderson	Writer|Chloe Chaidez	Writer|Julian Chaidez	Writer|Nicholas Johns
031013ktnA1		1	1	1	
053013ktnA1				1	1
053013ktnA2	1				
063012ktnA1		1	1		
063012ktnA2		1	1		
063012ktnA3		1	1		
063012ktnA4		1	1		
063012ktnA5		1	1		
082812ktnA1		1	1	1	
082812ktnA2		1	1		
082812ktnA3		1	1		
082812ktnA4		1	1		
082812ktnA5		1	1	1	
082812ktnA6		1	1		
093012ktnA1		1	1		
093012ktnA2		1	1		
093012ktnA3		1	1	1	
093012ktnA4		1	1		
093012ktnA5		1	1		
093012ktnA6		1	1		
093012ktnA7		1	1		
093012ktnA8		1	1		

Figure 11.1
Incidence arrays of music writers and music genres $\mathbf{E}_1$ and $\mathbf{E}_2$ as defined in Figure 4.4 for different tracks of music.

incidence arrays is generally preserved for various semirings. However, the nonzero values assigned to the edges can be very different and enable the construction different graphs.

For example, constructing an adjacency array of the graph of music writers connected to music genres from Figure 4.4 begins with selecting the incidence sub-arrays $\mathbf{E}_1$ and $\mathbf{E}_2$ as shown in Figure 11.1. Array multiplication of $\mathbf{E}_1^\mathsf{T}$ with $\mathbf{E}_2$ produces the desired adjacency array of the graph. Figure 11.2 illustrates this array multiplication for different operator pairs $\oplus$ and $\otimes$. The pattern of edges among vertices in the adjacency arrays shown in Figure 11.2 are the same for the different operator pairs, but the edge weights differ. All the nonzero values in $\mathbf{E}_1$ and $\mathbf{E}_2$ are 1. All the $\otimes$ operators in Figure 11.2 have the property

$$0 \otimes 1 = 1 \otimes 0 = 0$$

for their respective values of zero, be it 0, $-\infty$, or ∞. Likewise, all the $\otimes$ operators in Figure 11.2 also have the property

$$1 \otimes 1 = 1$$

except where $\otimes = +$, in which case

$$1 \otimes 1 = 2$$

The differences in the adjacency array weights are less pronounced than if the values of $\mathbf{E}_1$ and $\mathbf{E}_2$ were more diverse. The most apparent difference is between the $+.\times$ semiring

$\mathbf{E}_1^{\mathsf{T}} \; +.\times \; \mathbf{E}_2 =$

	Writer|Barrett Rich	Writer|Chad Anderson	Writer|Chloe Chaidez	Writer|Julian Chaidez	Writer|Nicholas Johns
Genre|Electronic	1	7	7	2	1
Genre|Pop		13	13	3	
Genre|Rock		6	6	1	

$\mathbf{E}_1^{\mathsf{T}} \; \substack{\max.+ \\ \min.+} \; \mathbf{E}_2 =$

	Writer|Barrett Rich	Writer|Chad Anderson	Writer|Chloe Chaidez	Writer|Julian Chaidez	Writer|Nicholas Johns
Genre|Electronic	2	2	2	2	2
Genre|Pop		2	2	2	
Genre|Rock		2	2	2	

$\mathbf{E}_1^{\mathsf{T}} \; \max.\min \; \mathbf{E}_2 =$

	Writer|Barrett Rich	Writer|Chad Anderson	Writer|Chloe Chaidez	Writer|Julian Chaidez	Writer|Nicholas Johns
Genre|Electronic	1	1	1	1	1
Genre|Pop		1	1	1	
Genre|Rock		1	1	1	

$\mathbf{E}_1^{\mathsf{T}} \; \min.\max \; \mathbf{E}_2 =$

	Writer|Barrett Rich	Writer|Chad Anderson	Writer|Chloe Chaidez	Writer|Julian Chaidez	Writer|Nicholas Johns
Genre|Electronic	1	1	1	1	1
Genre|Pop		1	1	1	
Genre|Rock		1	1	1	

$\mathbf{E}_1^{\mathsf{T}} \; \substack{\max.\times \\ \min.\times} \; \mathbf{E}_2 =$

	Writer|Barrett Rich	Writer|Chad Anderson	Writer|Chloe Chaidez	Writer|Julian Chaidez	Writer|Nicholas Johns
Genre|Electronic	1	1	1	1	1
Genre|Pop		1	1	1	
Genre|Rock		1	1	1	

Figure 11.2

Creating a graph of music writers related to music genres can be computed by multiplying $\mathbf{E}_1$ and $\mathbf{E}_2$ as defined in Figure 11.1. This correlation is performed using the transpose operation ^{T} and the array multiplication operation $\oplus.\otimes$. The resulting associative array has row keys taken from the column keys of $\mathbf{E}_1$ and column keys taken from the column keys of $\mathbf{E}_2$. The values represent the weights on the edges between the vertices of the graph. Different pairs of operations $\oplus$ and $\otimes$ produce different results. For display convenience, operator pairs that produce the same values in this specific example are stacked.

and the other semirings in Figure 11.2. In the case of $+.\times$ semiring, the $\oplus$ operation $+$ aggregates values from all the edges between two vertices. Additional positive edges will increase the overall weight in the adjacency array. In the other pairs of operations, the $\oplus$ operator is either max or min, which effectively selects only one edge weight to use for assigning the overall weight. Additional edges will only impact the edge weight in the adjacency array if the new edge is an appropriate maximum or minimum value. Thus, $+.\times$ constructs adjacency arrays that aggregate all the edges. The other semirings construct adjacency arrays that select extremal edges. Each can be useful for constructing graph adjacency arrays in an appropriate context.

$\mathbf{E}_1$

	Genre|Electronic	Genre|Pop	Genre|Rock
031013ktnA1			3
053013ktnA1	1		
053013ktnA2	1		
063012ktnA1			3
063012ktnA2			3
063012ktnA3			3
063012ktnA4			3
063012ktnA5			3
082812ktnA1		2	
082812ktnA2		2	
082812ktnA3		2	
082812ktnA4		2	
082812ktnA5		2	
082812ktnA6		2	
093012ktnA1	1	2	
093012ktnA2	1	2	
093012ktnA3	1	2	
093012ktnA4	1	2	
093012ktnA5	1	2	
093012ktnA6	1	2	
093012ktnA7	1	2	

$\mathbf{E}_2$

	Writer|Barrett Rich	Writer|Chad Anderson	Writer|Chloe Chaidez	Writer|Julian Chaidez	Writer|Nicholas Johns
031013ktnA1		1	1	1	
053013ktnA1				1	1
053013ktnA2	1				
063012ktnA1		1	1		
063012ktnA2		1	1		
063012ktnA3		1	1		
063012ktnA4		1	1		
063012ktnA5		1	1		
082812ktnA1		1	1	1	
082812ktnA2		1	1		
082812ktnA3		1	1		
082812ktnA4		1	1		
082812ktnA5		1	1	1	
082812ktnA6		1	1		
093012ktnA1		1	1		
093012ktnA2		1	1		
093012ktnA3		1	1	1	
093012ktnA4		1	1		
093012ktnA5		1	1		
093012ktnA6		1	1		
093012ktnA7		1	1		
093012ktnA8		1	1		

Figure 11.3
Incidence arrays from Figure 11.1 modified so that the nonzero values of $\mathbf{E}_1$ take on the values 1, 2, and 3.

The impact of different semirings on the graph adjacency array weights is more pronounced if the values of $\mathbf{E}_1$ and $\mathbf{E}_2$ are more diverse. Figure 11.3 modifies $\mathbf{E}_1$ so that a value of 2 is given to the nonzero values in the column Genre|Pop and a value of 3 is given to the nonzero values in the column Genre|Rock.

Figure 11.4 shows the results of constructing adjacency arrays with $\mathbf{E}_1$ and $\mathbf{E}_2$ using different semirings. The impact of changing the values in $\mathbf{E}_1$ can be seen by comparing Figure 11.2 with Figure 11.4. For the +.× semiring, the values in the adjacency array rows Genre|Pop and Genre|Rock are multiplied by 2 and 3. The increased adjacency array values for these rows are a result of the ⊗ operator being arithmetic multiplication × so that

$$2 \otimes 1 = 2 \times 1 = 2$$

$$3 \otimes 1 = 3 \times 1 = 3$$

For the max.+ and min.+ semirings, the values in the adjacency array rows Genre|Pop and Genre|Rock are larger by 1 and 2. The larger values in the adjacency array of these

$\mathbf{E}_1^{\mathsf{T}}$ +.× $\mathbf{E}_2$ =

	Writer|Barrett Rich	Writer|Chad Anderson	Writer|Chloe Chaidez	Writer|Julian Chaidez	Writer|Nicholas Johns
Genre|Electronic	1	7	7	2	1
Genre|Pop		26	26	6	
Genre|Rock		18	18	3	

$\mathbf{E}_1^{\mathsf{T}}$ max.+ / min.+ $\mathbf{E}_2$ =

	Writer|Barrett Rich	Writer|Chad Anderson	Writer|Chloe Chaidez	Writer|Julian Chaidez	Writer|Nicholas Johns
Genre|Electronic	2	2	2	2	2
Genre|Pop		3	3	3	
Genre|Rock		4	4	4	

$\mathbf{E}_1^{\mathsf{T}}$ max.min $\mathbf{E}_2$ =

	Writer|Barrett Rich	Writer|Chad Anderson	Writer|Chloe Chaidez	Writer|Julian Chaidez	Writer|Nicholas Johns
Genre|Electronic	1	1	1	1	1
Genre|Pop		1	1	1	
Genre|Rock		1	1	1	

$\mathbf{E}_1^{\mathsf{T}}$ min.max $\mathbf{E}_2$ =

	Writer|Barrett Rich	Writer|Chad Anderson	Writer|Chloe Chaidez	Writer|Julian Chaidez	Writer|Nicholas Johns
Genre|Electronic	1	1	1	1	1
Genre|Pop		2	2	2	
Genre|Rock		3	3	3	

$\mathbf{E}_1^{\mathsf{T}}$ max.× / min.× $\mathbf{E}_2$ =

	Writer|Barrett Rich	Writer|Chad Anderson	Writer|Chloe Chaidez	Writer|Julian Chaidez	Writer|Nicholas Johns
Genre|Electronic	1	1	1	1	1
Genre|Pop		2	2	2	
Genre|Rock		3	3	3	

Figure 11.4

Building a graph of music writers connected with the music genres can be accomplished by multiplying $\mathbf{E}_1$ and $\mathbf{E}_2$ as defined in Figure 11.3. The correlation is computed with the transpose operation ^{T} and array multiplication $\oplus.\otimes$. The resulting associative array has row keys taken from the column keys of $\mathbf{E}_1$ and column keys taken from the column keys of $\mathbf{E}_2$. The values represent the weights on the edges between the vertices of the graph. Different pairs of operations $\oplus$ and $\otimes$ produce different results. For display convenience, operator pairs that produce the same values in this specific example are stacked.

rows are due to the $\otimes$ operator being arithmetic addition + resulting in

$$2 \otimes 1 = 2 + 1 = 3$$
$$3 \otimes 1 = 3 + 1 = 4$$

For the max.min semiring, Figure 11.2 and Figure 11.4 have the same adjacency array because $\mathbf{E}_2$ is unchanged. The $\otimes$ operator corresponding to the minimum value function continues to select the smaller nonzero values from $\mathbf{E}_2$

$$2 \otimes 1 = \min(2, 1) = 1$$
$$3 \otimes 1 = \min(3, 1) = 1$$

In contrast, for the min.max semiring, the values in the adjacency array rows Genre|Pop and Genre|Rock are larger by 1 and 2. The increase in adjacency array values for these rows is a result of the ⊗ operator selecting the larger nonzero values from $\mathbf{E}_1$

$$2 \otimes 1 = \max(2,1) = 2$$
$$3 \otimes 1 = \max(3,1) = 3$$

Finally, for the max.× and min.× semirings, the values in the adjacency array rows Genre|Pop and Genre|Rock are increased by 1 and 2. Similar to the +.× semiring, the larger adjacency array values for these rows are a result of the ⊗ operator being arithmetic multiplication × resulting in

$$2 \otimes 1 = 2 \times 1 = 2$$
$$3 \otimes 1 = 3 \times 1 = 3$$

Figures 11.2 and 11.4 show that a wide range of graph adjacency arrays can be constructed via array multiplication of incidence arrays over different semirings. A synopsis of the graph constructions illustrated in Figures 11.2 and 11.4 is as follows

+.× — sum of products of edge weights connecting two vertices; computes the strength of all connections between two connected vertices

max.× — maximum of products of edge weights connecting two vertices; selects the edge with largest weighted product of all the edges connecting two vertices

min.× — minimum of products of edge weights connecting two vertices; selects the edge with smallest weighted product of all the edges connecting two vertices

max.+ — maximum of sum of edge weights connecting two vertices; selects the edge with largest weighted sum of all the edges connecting two vertices

min.+ — minimum of sum of edge weights connecting two vertices; selects the edge with smallest weighted sum of all the edges connecting two vertices

max.min — maximum of the minimum of weights connecting two vertices; selects the largest of all the shortest connections between two vertices

min.max — minimum of the maximum of weights connecting two vertices; selects the smallest of all the largest connections between two vertices

11.5 Special Arrays and Graphs

In the study and applications of matrices, certain structures that a given matrix exhibits are important; these matrices are typically termed special matrices. For example, the ideas corresponding to diagonal, off-diagonal, symmetric, skew-symmetric, upper and lower triangular, block, and block diagonal are all useful to consider over an appropriate matrix.

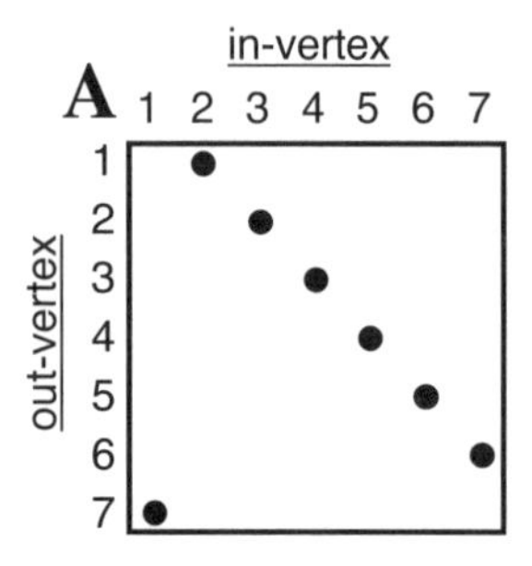

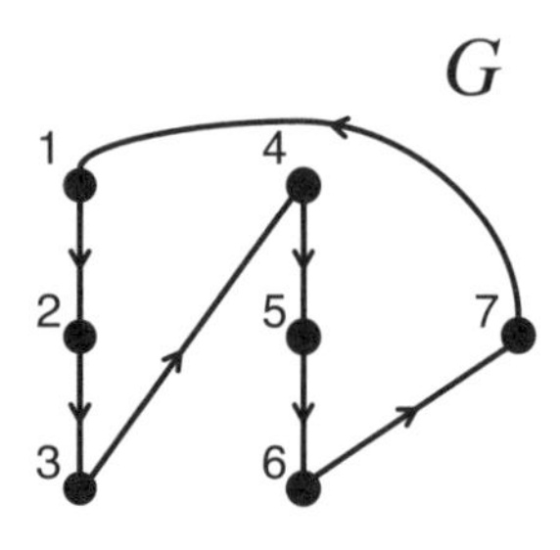

Figure 11.5
Permutation adjacency matrix **A** and corresponding graph G consisting of vertices with exactly one in-edge from another vertex and exactly one out-edge to another vertex.

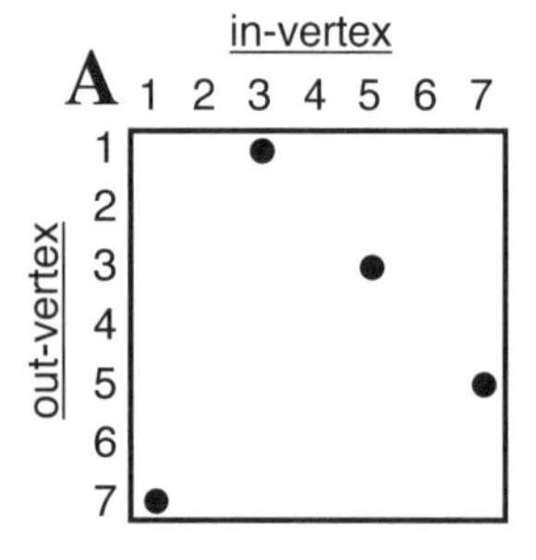

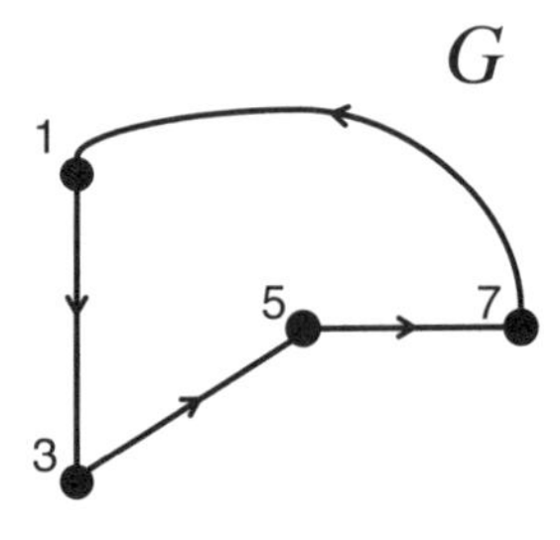

Figure 11.6
Partial permutation adjacency array **A** and corresponding graph G consisting of a subset of vertices with exactly one in-edge from another vertex and exactly one out-edge to another vertex.

With the generalization to associative arrays, it is worth examining which of these structures can be generalized to arrays.

The various structures on a matrix can be grouped into *order relations*, *algebraic relations*, and *subobject relations*; the first includes such notions as diagonal, off-diagonal, and upper and lower triangular; the second includes symmetric, skew-symmetric, and Hermitian; and the last includes block matrices and block diagonal.

An additional perspective on this concept is the adjacency matrix of a graph as defined and described in previous chapters, and the definition of a graph from a matrix. This correspondence can be extended to the more general scenario of a two-dimensional associative array. Looking at the graph of such an array provides visual intuition as to the meaning of these structures.

The structure of the nonzero entries in an associative array can be impacted by the ordering of the row keys and the column keys. A square array is *diagonal* (see Figure 11.7) if

$$\mathbf{A}(k_1, k_2) \neq 0 \quad \text{if and only if} \quad k_1 = k_2$$

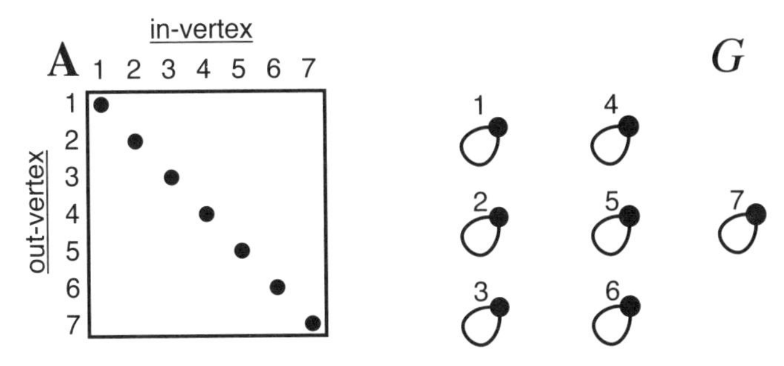

Figure 11.7
Diagonal adjacency array **A** and corresponding graph G consisting of vertices with self-edges.

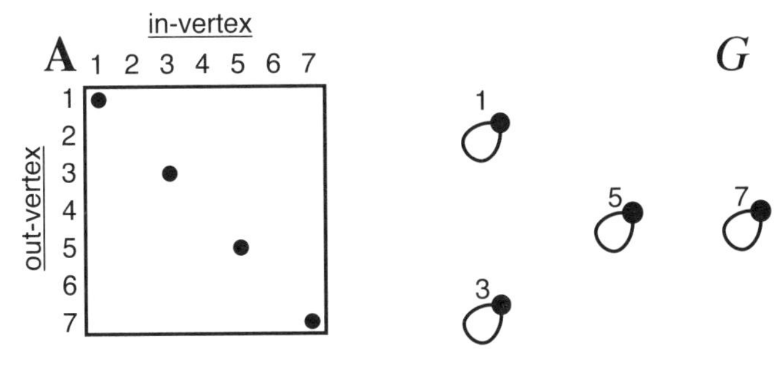

Figure 11.8
Partial diagonal adjacency array **A** and corresponding graph G consisting of a subset of vertices with self-edges.

and is a *partial diagonal* array (see Figure 11.8) if

$$\mathbf{A}(k_1, k_2) \neq 0 \quad \text{implies} \quad k_1 = k_2$$

The graph equivalent of diagonality is that a two-dimensional array

$$\mathbf{A} : K^2 \to V$$

is diagonal if and only if the graph of **A** consists of only self-loops, whereas off-diagonality means that the graph of **A** is a line graph.

Example 11.8

The array

$$\mathbf{E}(\text{031013ktnA1 : 053013ktnA2, ArtistlBandayde : ArtistlKitten})$$

in Figure 4.4 is partially diagonal.

A square array is a *permutation array* (see Figure 11.5) if each row and column has exactly one nonzero entry that has a value of 1

$$\mathbf{A}(k_1,k_2) = 1 \quad \text{if and only if} \quad \begin{array}{l} \mathbf{A}(k_1, K_2 \setminus \{k_2\}) = 0 \quad \text{and} \\ \mathbf{A}(K_1 \setminus \{k_1\}, k_2) = 0 \end{array}$$

Likewise, an array is a *partial permutation array* (see Figure 11.6) if each row and column has no more than one nonzero entry that has a value of 1

$$\mathbf{A}(k_1,k_2) = 1 \quad \text{implies} \quad \begin{array}{l} \mathbf{A}(k_1, K_2 \setminus \{k_2\}) = 0 \quad \text{and} \\ \mathbf{A}(K_1 \setminus \{k_1\}, k_2) = 0 \end{array}$$

11.6 Key Ordering

In the definition of an associative array $\mathbf{A}$, row keys and column keys are orderable, but the ordering function is left unspecified. To specify the order, let an associative array $\mathbf{A}$ have a pair of linear orders $<_1$ and $<_2$ on K_1 and K_2 such that for distinct

$$k_1, k_1' \in K_1 \quad \text{and} \quad k_2, k_2' \in K_2$$

then

$$k_1 <_1 k_1' \quad \text{or} \quad k_1' <_1 k_1$$

and

$$k_2 <_2 k_2' \quad \text{or} \quad k_2' <_2 k_2$$

Also, assume that K_1 and K_2 are finite.

For the same reason that it is possible to permute just the rows or just the columns of a matrix, there can be distinct orderings on the row space and the column space; in other words $<_1$ is not the same as $<_2$ in general even if $K_1 = K_2$.

There are order isomorphisms

$$f : K_1 \to \{1,\ldots,m\} \quad \text{and} \quad f' : K_2 \to \{1,\ldots,n\}$$

Lemma 11.7

Finite Linearly Ordered Sets are Order Isomorphic to Initial Segments of $\mathbb{N}$

Suppose K is a finite set with $<$ a strict total order on K. Then there exists an order isomorphism

$$f : \{1,\ldots,n\} \to K$$

Proof. See Exercise 11.10. □

With this order isomorphism in mind, each of K_1 and K_2 can be assumed to be of the form $\{1,\ldots,m\}$ and $\{1,\ldots,n\}$ with the standard orderings and an array

$$\mathbf{A} : K_1 \times K_2 \to V$$

can be replaced with an array

$$\{1,\ldots,m\} \times \{1,\ldots,n\} \to V$$

by composing $\mathbf{A}$ with an order isomorphism

$$\{1,\ldots,m\} \times \{1,\ldots,n\} \to K_1 \times K_2$$

The resulting array is called the *matrix associated with* $\mathbf{A}$ (with respect to $<_1$ and $<_2$).

Ordering the keys leads to a new notion of diagonality.

Definition 11.8

Ordered Diagonal

Suppose $\mathbf{A} : K_1 \times K_2 \to V$ is an array with K_1, K_2 finite and strict totally ordered by $<_1, <_2$, respectively. Let $\mathbf{B} : \{1,\ldots,m\} \times \{1,\ldots,n\} \to V$ be the matrix associated with $\mathbf{A}$ with respect to $<_1, <_2$.
Then $\mathbf{A}$ is *ordered (partially) diagonal* if and only if $\mathbf{B}$ is (partially) diagonal.

An array with specified orderings may be diagonal but not ordered diagonal, or ordered diagonal but not diagonal.

The notions of triangularity can also be defined by passing to the associated matrix. This necessitates a definition in the case of a matrix. $\mathbf{B} : \{1,\ldots,m\} \times \{1,\ldots,n\} \to V$ is *upper triangular* (see Figure 11.9 left) if

$$\mathbf{B}(k_1,k_2) \neq 0 \quad \text{if and only if} \quad k_1 \leq k_2$$

and is *partially upper triangular* (see Figure 11.9 right) if

$$\mathbf{B}(k_1,k_2) \neq 0 \quad \text{implies} \quad k_1 \leq k_2$$

Likewise, $\mathbf{B}$ is *lower triangular* (see Figure 11.10 left) if

$$\mathbf{B}(k_1,k_2) \neq 0 \quad \text{if and only if} \quad k_2 \leq k_1$$

and is *partially lower triangular* (see Figure 11.10 right) if

$$\mathbf{B}(k_1,k_2) \neq 0 \quad \text{implies} \quad k_2 \leq k_1$$

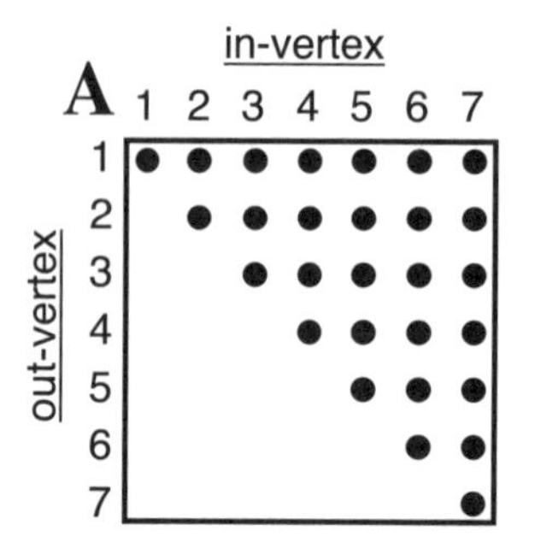

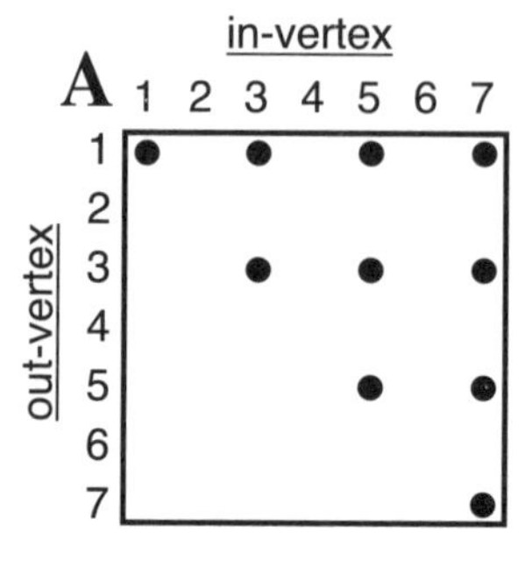

Figure 11.9
(Left) upper triangular adjacency array where every vertex has at least one edge with every other vertex. (Right) partially upper triangular adjacency array where a subset of vertices has at least one edge with every other vertex in the subset.

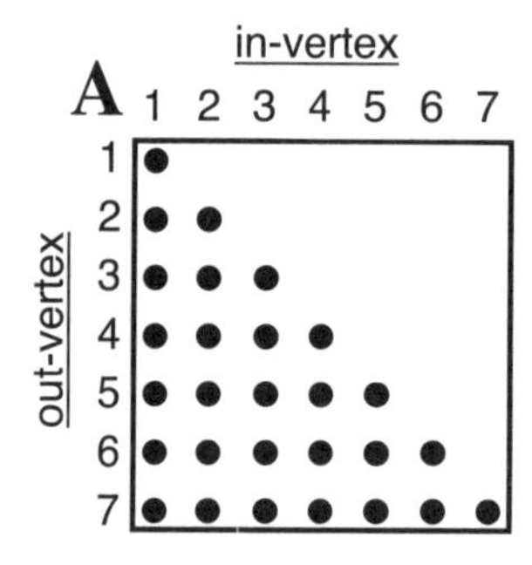

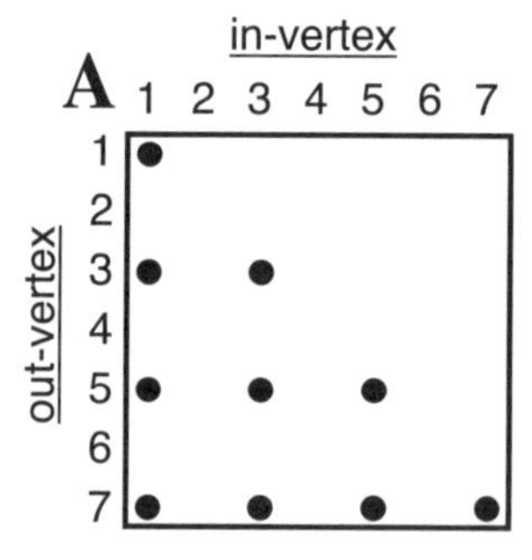

Figure 11.10
(Left) lower-triangular adjacency array where every vertex has at least one edge with every other vertex. (Right) partially lower triangular adjacency array where a subset of vertices has at least one edge with every other vertex in the subset.

Definition 11.9

Ordered Triangular

Suppose $\mathbf{A} : K_1 \times K_2 \to V$ is an array with K_1, K_2 finite and strict totally ordered by $<_1, <_2$, respectively. Let $\mathbf{B} : \{1,\ldots,m\} \times \{1,\ldots,n\} \to V$ be the matrix associated with $\mathbf{A}$ with respect to $<_1, <_2$.
Then $\mathbf{A}$ is *ordered (partially) upper/lower triangular* if $\mathbf{B}$ is (partially) upper/lower triangular.

The use of the prefix ordered is to ensure that the role of the orderings $<_1$ and $<_2$ is explicit and highlights the fact that K_1 and K_2 are being identified using those orderings. A diagonal array satisfies $\mathbf{A}(k_1, k_2) \neq 0$ if and only if $k_1 = k_2$, whereas being an ordered diagonal array would instead only say that k_1 and k_2 have the same placement in K_1 and K_2

relative to $<_1$ and $<_2$, respectively. When the role of the orderings is understood, "ordered" can be removed.

Example 11.9

If $\mathbf{A}$ is ordered diagonal under some orderings of K_1 and K_2, then it is also partially upper and lower triangular.

In practice, a more useful notion is that of a *weakly triangular* array. Informally, for arrays weak triangularity provides the ability to either push the elements in each column or row upward or to the right and permute the columns or rows to create an upper or lower triangular matrix.

The definition of weakly triangular makes use of two functions that indicate how many rows and columns have a size of at least a certain value. Specifically, the number of rows with more than $\underline{n}$ nonzero entries is given by

$$\text{row}_{\mathbf{A}}(\underline{n}) = |\{k_1 \mid \text{size}(\mathbf{A}(k_1,:),2) > \underline{n}\}$$

Likewise, the number of rows with more than $\underline{m}$ nonzero entries is given by

$$\text{col}_{\mathbf{A}}(\underline{m}) = |\{k_2 \mid \text{size}(\mathbf{A}(:,k_2),1) > \underline{m}\}$$

The array $\mathbf{A}$ with $\underline{m}$ non-empty rows and $\underline{n}$ non-empty columns is said to be *weakly upper triangular* if for

$$N = \max(\underline{m},\underline{n})$$

and

$$p \in \{0,\ldots,N\}$$

there are at most

$$p+1-\text{col}_{\mathbf{A}}(N-p+1)$$

elements of K_1 such that for each of these elements k the set

$$\text{supp}(\mathbf{A}) \cap (\{k\} \times K_2)$$

has size $N-p$. $\mathbf{A}$ is *weakly lower triangular* if for

$$p \in \{0,\ldots,N\}$$

there are at most

$$p+1-\text{col}_{\mathbf{A}}(N-p+1)$$

elements of K_2 such that for each of these elements k the set

$$\text{supp}(\mathbf{A}) \cap (K_1 \times \{k\})$$

has size $N-p$.

The graph-theoretic interpretation of weak triangularity for an associative array $\mathbf{A} : K^2 \to V$ is to say that if

$$N = \max\{\text{size}(\mathbf{A},1), \text{size}(\mathbf{A},2)\}$$

with row as above, then $\mathbf{A}$ is weakly upper triangular if for each p such that

$$0 \le p \le N$$

there are at most

$$1 + p - \text{col}_{\mathbf{A}}(N-p+1)$$

elements of K with exactly $N-p$ edges *out of* those elements, and weakly lower triangular if for each p such that

$$0 \le p \le N$$

there are at most

$$1 + p - \text{row}_{\mathbf{A}}(N-p+1)$$

elements of K with exactly $N-p$ edges *into* those elements.

Example 11.10

The array $\mathbf{A} : \{1,3,5,7\} \times \{2,4\} \to \mathbb{N}$ defined by

$$\mathbf{A} = \begin{array}{c} \\ 1 \\ 3 \\ 5 \\ 7 \end{array} \begin{array}{c} \begin{array}{cc} 2 & 4 \end{array} \\ \left[\begin{array}{cc} 1 & 2 \\ & 1 \\ & 1 \\ & 1 \end{array}\right] \end{array}$$

is weakly lower triangular because there is at most

$\max\{0, 0+1-\text{row}(4-0+1)\} = 1$ row of size $4-0=4$
$\max\{0, 1+1-\text{row}(4-1+1)\} = 1$ row of size $4-1=3$
$\max\{0, 2+1-\text{row}(4-2+1)\} = 2$ rows of size $4-2=2$
$\max\{0, 3+1-\text{row}(4-3+1)\} = 3$ rows of size $4-3=1$
$\max\{0, 4+1-\text{row}(4-4+1)\} = 1$ row of size 0

Example 11.11

The arrays $\mathbf{E}_1$ and $\mathbf{E}_2$ in Figure 4.4 are weakly lower triangular, but neither is weakly upper triangular.

Example 11.12

The matrices

$$\begin{array}{c} \\ 1 \\ 2 \\ 3 \end{array} \begin{array}{c} \begin{array}{cccc} 1 & 2 & 3 & 4 \end{array} \\ \begin{bmatrix} \{0\} & \emptyset & \{1,2\} & \{0,1\} \\ \emptyset & \{0,1\} & \{2\} & \{0,2\} \\ \{0,1,2\} & \emptyset & \emptyset & \{1\} \end{bmatrix} \end{array} \quad \text{and} \quad \begin{array}{c} \\ 1 \\ 2 \\ 3 \end{array} \begin{array}{c} \begin{array}{cc} 1 & 2 \end{array} \\ \begin{bmatrix} \pi & \infty \\ -\infty & -\infty \\ 0 & -\infty \end{bmatrix} \end{array}$$

are both weakly lower triangular, with the second also weakly upper triangular.

There is a close relationship between weak upper triangularity and weak lower triangularity.

Lemma 11.8

Transpose Flips Weakly Upper and Lower Triangularity

A matrix **A** is weakly upper triangular if and only if its transpose $\mathbf{A}^\mathsf{T}$ is weakly lower triangular.

Proof. See Exercise 11.11. □

11.7 Algebraic Properties

Algebraic relational structures refers to structures that are defined by the algebraic properties that a square two-dimensional array $\mathbf{A} : K^2 \to V$ has as a binary operation on K taking values in V. Commutativity is symmetry of the matrix and anti-commutativity (V is a ring) is that of skew-symmetry of the matrix. A two-dimensional square associative array **A** is symmetric if

$$\mathbf{A}(k_1, k_2) = \mathbf{A}(k_2, k_1)$$

Example 11.13

Every symmetric matrix is an example of a symmetric two-dimensional array, as is every diagonal array.

Skew-symmetry and Hermitian properties of a matrix in general cannot be extended to associative arrays since they depend upon the codomain V having additive inverses. The possible nonexistence of inverses is the distinguishing quality of semirings from rings. Similarly, issues exist with the notion of complex conjugation. For this reason, to expand these notions to associative arrays, additional assumptions would need to be made on V. For example, if V is a ring, then the associative array $\mathbf{A}$ is said to be *skew-symmetric* if

$$\mathbf{A}(k_1,k_2) = \text{-}\mathbf{A}(k_2,k_1)$$

Similarly, when V is a sub-semiring of $\mathbb{C}$ closed under complex conjugation, then the associative array $\mathbf{A}$ is said to be *Hermitian* if

$$\mathbf{A}(k_1,k_2) = \overline{\mathbf{A}(k_2,k_1)}$$

where $\overline{z} = x - y\sqrt{-1}$ is the complex conjugate of the complex number $z = x + y\sqrt{-1}$.

11.8 Subobject Properties

Subobject relational structures implies the ability to represent a matrix as an array of sub-matrices possibly satisfying one of the order-relational or algebraic-relational structures above. The first step is to define what the analog of a sub-matrix is in the context of associative arrays.

Definition 11.10

Sub-Array

Given an associative array

$$\mathbf{A} : K_1 \times K_2 \to V$$

the associative array

$$\mathbf{B} : K_1' \times K_2' \to V$$

is said to be a *sub-array* of $\mathbf{A}$ if

$$K_1' \times K_2' \subset K_1 \times K_2$$

and $\mathbf{B}$ agrees with $\mathbf{A}$ on $K_1' \times K_2'$.

Because the block structure of a matrix is typically used in reference to its structure, allowing arbitrary sub-arrays would prevent the blocks from being meaningfully ordered, even when K_1 and K_2 are explicitly ordered. For this reason, K_1 and K_2 are required to be partitioned into closed intervals $[k,k'] = \{k'' \mid k \leq k'' \leq k'\}$. Writing

$$K_1 = [k_1,k_2] \cup \cdots \cup [k_{2p-1},k_{2p}]$$
$$K_2 = [k_1',k_2'] \cup \cdots \cup [k_{2q-1}',k_{2q}']$$

where

$$k_1 \leq k_2 \leq_1 \cdots \leq_1 k_{2p-1} \leq k_{2p}$$
$$k'_1 \leq k'_2 \leq_2 \cdots \leq_2 k'_{2q-1} \leq k'_{2q}$$

then the partitions

$$\mathcal{K}_1 = \{[k_1, k_2], \ldots, [k_{2p-1}, k_{2p}]\}$$
$$\mathcal{K}_2 = \{[k'_1, k'_2], \ldots, [k'_{2q-1}, k'_{2q}]\}$$

can be ordered by

$$[k_1, k_2] \leq_1 \cdots \leq_1 [k_{2p-1}, k_{2p}]$$
$$[k'_1, k'_2] \leq_2 \cdots \leq_2 [k'_{2q-1}, k'_{2q}]$$

Definition 11.11

Block Associative Array and Associated Block Structure Map

Suppose K_1 and K_2 are finite totally ordered sets. A (two-dimensional) *block associative array* is an associative array

$$\mathbf{A} : K_1 \times K_2 \to V$$

coupled with a pair of partitions $\mathcal{K}_1, \mathcal{K}_2$ of K_1 and K_2 into closed intervals.
The *associated block structure map* is the associative array

$$\mathbf{A}' : \mathcal{K}_1 \times \mathcal{K}_2 \to \mathbb{A}$$

defined by

$$\mathbf{A}'([k_1, k'_1], [k_2, k'_2]) = \text{pad}_{K_1 \times K_2} \mathbf{A}|_{[k_1, k'_1] \times [k_2, k'_2]}$$

or that $([k_1, k'_1], [k_2, k'_2])$ is sent to the sub-array of $\mathbf{A}$ determined by $[k_1, k'_1] \times [k_2, k'_2]$ which has been zero padded.

The orderings on K_1 and K_2 give orderings on $\mathcal{K}_1$ and $\mathcal{K}_2$, respectively. As such, it makes sense to speak of the "n-th element" of $\mathcal{K}_i$.

Because K_1 and K_2 are finite, $\mathbb{A}(K_1, K_2; V)$ forms a semiring itself under element-wise (Hadamard) multiplication, and so the associated block structure map

$$\mathbf{A}' : \mathcal{K}_1 \times \mathcal{K}_2 \to \mathbb{A}$$

is a well-defined associative array. This is the reason the sub-arrays $\mathbf{A}|_{[k_1, k'_1] \times [k_2, k'_2]}$ are zero padded.

By combining the notions of the associated block structure map and the matrix associated with the key orderings, it is possible to define the notions of block diagonality and block triangularity.

Definition 11.12

Block Diagonal, Triangular

A block associative array

$$\mathbf{A} : K_1 \times K_2 \to V$$

with the partitions $(\mathcal{K}_1, \mathcal{K}_2)$ is *block (partially) diagonal* or *block (partially) upper/lower triangular* if the associated block structure map

$$\mathbf{A}' : \mathcal{K}_1 \times \mathcal{K}_2 \to \mathbb{A}$$

is ordered (partially) diagonal or ordered (partially) upper/lower triangular, respectively.

11.9 Conclusions, Exercises, and References

Graph construction is a key operation for data processing and is typically performed by multiplying the incidence array representations of a graph $\mathbf{E}_{\text{in}}$ and $\mathbf{E}_{\text{out}}$. The result is an adjacency matrix of the graph $\mathbf{A}$ that can be further processed with a variety of techniques. Various mathematical criteria ensure the product $\mathbf{A} = \mathbf{E}_{\text{out}}^{\mathsf{T}} \mathbf{E}_{\text{in}}$ is the adjacency array of the graph. An adjacency matrix of a graph can have certain patterns that are typically termed special matrices. The concepts of diagonal, off-diagonal, symmetric, skew-symmetric, upper and lower triangular, block, and block diagonal are all examples of useful patterns. These patterns are also relevant when matrices are generalized to associative arrays.

Exercises

Exercise 11.1 — Consider Example 1. Is $\mathbf{A}$ symmetric? If so, why?

Exercise 11.2 — Consider Example 3 and Example 4. Describe and explain the differences in their adjacency arrays.

Exercise 11.3 — Consider the conditions in (i)(a), (i)(b), and (i)(c) in Theorem 11.1. For each condition, explain why it is necessary for the construction of adjacency arrays from incidence arrays.

(a) Zero-sum-free

(b) No zero divisors

(c) 0 annihilator

Exercise 11.4 — Prove Corollary 11.5.

Exercise 11.5 — Compute the size, $\underline{m}$ and $\underline{n}$ and number of nonzeros (nnz) in

(a) Figure 11.1 $\mathbf{E}_1$

(b) Figure 11.1 $\mathbf{E}_2$

(c) Figure 11.2 $\mathbf{E}_1 \ {+}.{\times} \ \mathbf{E}_1^\mathsf{T}$

(d) Comment on any similarities among these quantities

Exercise 11.6 — For each of the array multiplications in Figure 11.2, state the value represented by the blank spaces in the result

(a) $\mathbf{E}_1 \ {+}.{\times} \ \mathbf{E}_1^\mathsf{T}$

(b) $\mathbf{E}_1 \ \max.{+} \ \mathbf{E}_1^\mathsf{T}$

(c) $\mathbf{E}_1 \ \min.{+} \ \mathbf{E}_1^\mathsf{T}$

(d) $\mathbf{E}_1 \ \max.\min \ \mathbf{E}_1^\mathsf{T}$

(e) $\mathbf{E}_1 \ \min.\max \ \mathbf{E}_1^\mathsf{T}$

(f) $\mathbf{E}_1 \ \max.{\times} \ \mathbf{E}_1^\mathsf{T}$

(g) $\mathbf{E}_1 \ \min.{\times} \ \mathbf{E}_1^\mathsf{T}$

Exercise 11.7 — Take a graph from your own experience and write down its adjacency array and its incidence arrays.

Exercise 11.8 — Using the adjacency array from the first exercise, compute the nearest neighbors of a vertex by using array multiplication. Select different functions for $\otimes$ and $\oplus$ and see how they affect the result.

Exercise 11.9 — Using the incidence arrays from the previous exercise, compute the adjacency array by using array multiplication. Select different functions for $\otimes$ and $\oplus$ and see how they affect the result.

Exercise 11.10 — Prove Lemma 11.7.

Exercise 11.11 — Prove Lemma 11.8.

References

[1] H. Jananthan, K. Dibert, and J. Kepner, "Constructing adjacency arrays from incidence arrays," in *International Parallel and Distributed Processing Symposium Workshops (IPDPSW)*, IEEE, 2017.

[2] R. H. Bruck and H. J. Ryser, "The nonexistence of certain finite projective planes," *Canadian Journal of Mathematics*, vol. 1, no. 191, p. 9, 1949.

[3] D. Fulkerson and O. Gross, "Incidence matrices and interval graphs," *Pacific Journal of Mathematics*, vol. 15, no. 3, pp. 835–855, 1965.

[4] D. Fulkerson and O. Gross, "Incidence matrices and interval graphs," *Pacific Journal of Mathematics*, vol. 15, no. 3, pp. 835–855, 1965.

[5] G. Fisher and O. Wing, "Computer recognition and extraction of planar graphs from the incidence matrix," *IEEE Transactions on Circuit Theory*, vol. 13, no. 2, pp. 154–163, 1966.

[6] F. Freudenstein and L. Dobjansky, "Some applications of graph theory to the structural analysis of mechanisms," *ASME Transaction Journal of Engineering for Industry*, vol. 89, no. 1, pp. 153–158, 1967.

[7] L. D. Bodin and S. J. Kursh, "A detailed description of a computer system for the routing and scheduling of street sweepers," *Computers & Operations Research*, vol. 6, no. 4, pp. 181–198, 1979.

[8] R. Tarjan, "Depth-first search and linear graph algorithms," *SIAM Journal on Computing*, vol. 1, no. 2, pp. 146–160, 1972.

[9] M. Fiedler, *Special Matrices and Their Applications in Numerical Mathematics*. Courier Corporation, 2008.

[10] M. Gu and S. C. Eisenstat, "A divide-and-conquer algorithm for the bidiagonal svd," *SIAM Journal on Matrix Analysis and Applications*, vol. 16, no. 1, pp. 79–92, 1995.

[11] H. D. Patterson and R. Thompson, "Recovery of inter-block information when block sizes are unequal," *Biometrika*, pp. 545–554, 1971.

[12] A. Y. Ng, M. I. Jordan, and Y. Weiss, "On spectral clustering: Analysis and an algorithm," in *NIPS*, vol. 14.2, pp. 849–856, 2001.

[13] C. D. Meyer, Jr and R. J. Plemmons, "Convergent powers of a matrix with applications to iterative methods for singular linear systems," *SIAM Journal on Numerical Analysis*, vol. 14, no. 4, pp. 699–705, 1977.

[14] R. A. Brualdi, S. V. Parter, and H. Schneider, "The diagonal equivalence of a nonnegative matrix to a stochastic matrix," *Journal of Mathematical Analysis and Applications*, vol. 16, no. 1, pp. 31–50, 1966.

[15] L. N. Trefethen, "Pseudospectra of matrices," *Numerical Analysis*, vol. 91, pp. 234–266, 1991.

[16] A. Ziehe, P. Laskov, G. Nolte, and K.-R. MÃžller, "A fast algorithm for joint diagonalization with non-orthogonal transformations and its application to blind source separation," *Journal of Machine Learning Research*, vol. 5, no. Jul, pp. 777–800, 2004.

[17] C. Ballantine, "Products of involutory matrices. I," *Linear and Multilinear Algebra*, vol. 5, no. 1, pp. 53–62, 1977.

[18] J. H. Wilkinson, *The Algebraic Eigenvalue Problem*, vol. 87. Oxford, U.K.: Clarendon Press, 1965.

[19] Z. Bai and J. Demmel, "On a block implementation of Hessenberg multishift QR iteration," *International Journal of High Speed Computing*, vol. 1, no. 01, pp. 97–112, 1989.

[20] A. Cayley, "A memoir on the theory of matrices," *Philosophical Transactions of the Royal Society of London*, vol. 148, pp. 17–37, 1858.

[21] C. De Boor and G. H. Golub, "The numerically stable reconstruction of a jacobi matrix from spectral data," *Linear Algebra and Its Applications*, vol. 21, no. 3, pp. 245–260, 1978.

[22] J. L. Stuart and J. R. Weaver, "Matrices that commute with a permutation matrix," *Linear Algebra and Its Applications*, vol. 150, pp. 255–265, 1991.

[23] H.-L. Gau, M.-C. Tsai, and H.-C. Wang, "Weighted shift matrices: Unitary equivalence, reducibility and numerical ranges," *Linear Algebra and its Applications*, vol. 438, no. 1, pp. 498–513, 2013.

[24] N. J. Higham, "J-orthogonal matrices: Properties and generation," *SIAM Review*, vol. 45, no. 3, pp. 504–519, 2003.

[25] W. Givens, "Computation of plain unitary rotations transforming a general matrix to triangular form," *Journal of the Society for Industrial and Applied Mathematics*, vol. 6, no. 1, pp. 26–50, 1958.

[26] K. Dibert, H. Jansen, and J. Kepner, "Algebraic conditions for generating accurate adjacency arrays," in *MIT Undergraduate Research Technology Conference (URTC)*, IEEE, 2015.

III LINEAR SYSTEMS

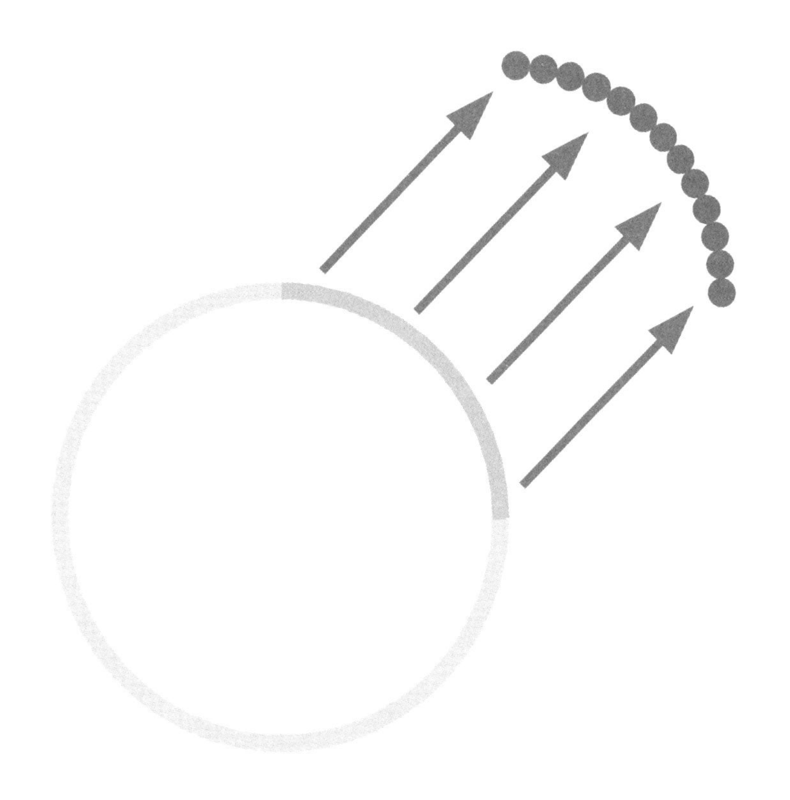

12 Survey of Common Transformations

Summary

A central feature of array multiplication is its ability to transform the values of an array. The ability to use a variety of $\oplus$ and $\otimes$ operations in different combinations significantly extends the flexibility of array transformations. Visualizing and deriving 2×2 transformations of the unit circle over various semirings provides valuable intuition on these transformations. Most general transformations can be viewed as a composition of identity, contraction, expansion, stretching, and rotation transformations. The diversity of shapes produced by these transformations reveals the many interesting patterns, which are even more impressive given that these transformations obey the properties of semirings, such as associativity, distributivity, and identities. The diversity of these behaviors under simple array transformation also serves to motivate the reexamination of other matrix concepts in subsequent chapters. This chapter visually depicts the enormous diversity of "linear" behavior that can be exhibited by diverse semirings and sets the stage for exploring these behaviors more rigorously in Part III of the book.

12.1 Array Transformations

The ability to change $\oplus$ and $\otimes$ operations allows different graph algorithms to be implemented using the same element-wise addition, element-wise multiplication, and array multiplication operations. Different semirings are best suited for certain classes of graph algorithms. The breadth-first-search behavior illustrated in Figure 6.7 is generally preserved for various semirings, but the values of the nonzero values assigned to the edges and vertices can be very different (see Figures 6.11 and 6.12). Likewise, the pattern of edges in a graph adjacency array produced by multiplying two incidence arrays is usually preserved for different semirings, but the exact values associated with those edges vary (see Figures 11.1, 11.2, 11.3, and 11.4).

The matrix multiplication properties of max.min, min.max, max.+, and min.+ semirings have received extensive attention (see [1–6] and references therein). Prior work on these semirings has focused on developing their mathematical properties. Depicting or visualizing the actual transformation performed by matrix multiplication in these semirings has not occurred. Visualizing matrix transformations provides important intuition for guiding their practical application [7–10]. This chapter depicts these transformations, for the first time, in order to provide this intuition.

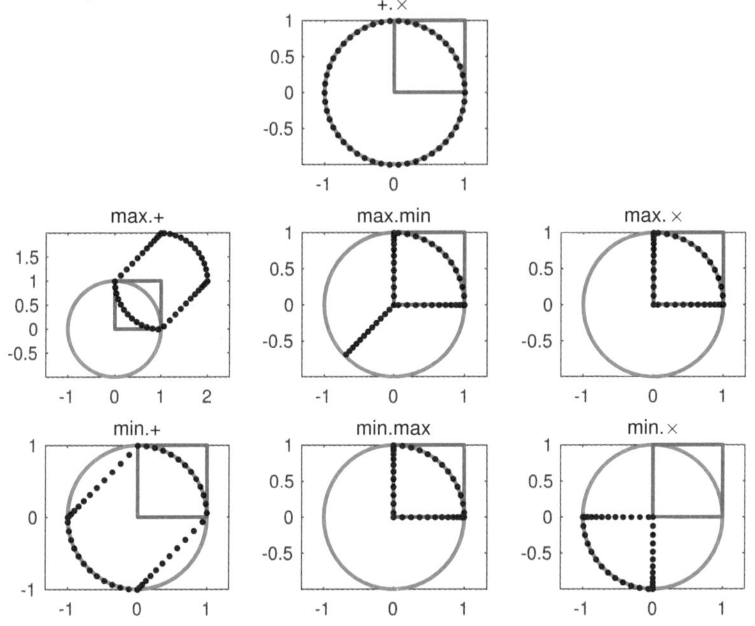

Figure 12.1
Transformation of points on the unit circle (green circle) via array multiplication by a 2×2 associative array (blue square), resulting in a new set of points (black dots). The title of each plot lists the $\oplus$ and $\otimes$ operators used in the array multiplication. The array values $\begin{bmatrix}1 & 0\\ 0 & 1\end{bmatrix}$ correspond to the identity transformation in the $+.\times$ semiring.

A useful way to see the impact of changing semirings is by plotting the transformation of the unit circle by various 2×2 arrays $\mathbf{A}$. Let the unit circle be represented by the $2\times n$ unit circle array $\mathbf{U}$

$$\mathbf{U} = \begin{array}{c} \\ x \\ y \end{array} \begin{array}{c} \begin{array}{ccccc} 1 & 2 & 3 & \cdots & n \end{array} \\ \begin{bmatrix} 1 & \cos(\phi) & \cos(2\phi) & \cdots & \cos(2\pi-\phi) \\ 0 & \sin(\phi) & \sin(2\phi) & \cdots & \sin(2\pi-\phi) \end{bmatrix} \end{array}$$

where $\phi = 2\pi/n$ is the angular offset between two neighboring points on the unit circle, $\mathbf{U}(x,:)$ is the row vector of the x-coordinates of the points, and $\mathbf{U}(y,:)$ is the row vector of all y-coordinates of the points.

Perhaps the simplest transformation is by the 2×2 identity array of the $+.\times$ semiring

$$\mathbf{A} = \mathbb{I}_{+.\times} = \begin{array}{c} \\ x \\ y \end{array} \begin{array}{c} \begin{array}{cc} x & y \end{array} \\ \left[\begin{array}{cc} 1 & 0 \\ 0 & 1 \end{array}\right] \end{array}$$

In this instance, array multiplication of $\mathbf{U}$ by $\mathbf{A}$ results in no change to $\mathbf{U}$

$$\begin{aligned} \mathbf{AU} &= \mathbf{A} \oplus.\otimes \mathbf{U} \\ &= \mathbb{I}_{+.\times} \; +.\times \; \mathbf{U} \\ &= \mathbf{U} \end{aligned}$$

The above transformation is depicted in Figure 12.1 (top, center). Array multiplication by $\mathbb{I}_{+.\times}$ results in no change to $\mathbf{U}$ when the $+.\times$ semiring is used. Figure 12.1 shows there is a significant change to $\mathbf{U}$ when $\mathbb{I}_{+.\times}$ is used with other semirings. These transformations are just a small sample of the enormous flexibility offered by combining different pairs of operations. What is most impressive is that these transformations all obey the properties of semirings, such as associativity, distributivity, and identities, which allow them to be combined and scaled to create a wide range of algorithms.

From the perspective of graphs, Figure 12.1 illustrates how the values of two edges leading into the same vertex are transformed by traversing these edges with a given semiring and with values specified by $\mathbb{I}_{+.\times}$. Using $\mathbb{I}_{+.\times}$ to traverse these edges in the $+.\times$ semiring leaves these edges untouched. Likewise, using $\mathbb{I}_{+.\times}$ to traverse these edges in the max.+ semiring adds a constant to the resulting edge values and then selects the maximum value. For a given application, it is necessary to understand the analytic techniques to select the semiring and the values so as to achieve a desired application goal.

Up to this point, the focus of the discussion of semirings has been on their properties as they pertain to associative arrays, with less attention given to the actual arithmetic of semiring calculations. It is worthwhile to review this arithmetic in the context of some simple examples to illustrate the detailed analytic calculations that can be utilized to determine the behavior of a given transformation. For a given semiring, the transformations of the unit circle are often consistent within quadrants of the plane. These quadrants can be defined as sets I, II, III, and IV of column keys of the unit circle array $\mathbf{U}$ as follows

Quadrant I — (positive, upper right, northeast quadrant)

$$\mathrm{I} = \{j \mid 0 \leq \mathbf{U}(x,j) \leq 1 \text{ and } 0 \leq \mathbf{U}(y,j) \leq 1\}$$

For the unit circle, these points correspond to angles in the range

$$0 \leq \phi \leq 0.5\pi$$

Quadrant II — (upper left, northwest quadrant)

$$\mathrm{II} = \{j \mid -1 \leq \mathbf{U}(x, j) \leq 0 \text{ and } 0 \leq \mathbf{U}(y, j) \leq 1\}$$

For the unit circle, these points correspond to angles in the range

$$0.5\pi \leq \phi \leq \pi$$

Quadrant III — (negative quadrant, lower left, southwest quadrant)

$$\mathrm{III} = \{j \mid -1 \leq \mathbf{U}(x, j) \leq 0 \text{ and } -1 \leq \mathbf{U}(y, j) \leq 0\}$$

For the unit circle, these points correspond to angles in the range

$$\pi \leq \phi \leq 1.5\pi$$

Quadrant IV — (lower right quadrant, southeast quadrant)

$$\mathrm{IV} = \{j \mid 0 \leq \mathbf{U}(x, j) \leq 1 \text{ and } -1 \leq \mathbf{U}(y, j) \leq 0\}$$

For the unit circle, these points correspond to angles in the range

$$1.5\pi \leq \phi \leq 2\pi$$

The subsequent discussion enumerates, derives, and visualizes these transformations of the unit circle for each semiring and for each quadrant. These analyses can be used to deduce the behavior of semiring transformations in many contexts.

12.2 Identity

As previously mentioned, array multiplication by $\mathbb{I}_{+.\times}$ results in no change to $\mathbf{U}$ when the $+.\times$ semiring is used, but this is not expected to be the case when $\mathbb{I}_{+.\times}$ is used over other semirings. To preserve the identity transformation

$$\mathbf{AU} = \mathbf{U}$$

would have required setting $\mathbf{A}$ to the identity of each semiring as follows

$$\mathbb{I}_{\max.+} = \begin{array}{c} \\ x \\ y \end{array} \begin{array}{c} \begin{array}{cc} x & y \end{array} \\ \begin{bmatrix} 0 & -\infty \\ -\infty & 0 \end{bmatrix} \end{array} \quad \mathbb{I}_{\max.\min} = \begin{array}{c} \\ x \\ y \end{array} \begin{array}{c} \begin{array}{cc} x & y \end{array} \\ \begin{bmatrix} \infty & -\infty \\ -\infty & \infty \end{bmatrix} \end{array} \quad \mathbb{I}_{\max.\times} = \begin{array}{c} \\ x \\ y \end{array} \begin{array}{c} \begin{array}{cc} x & y \end{array} \\ \begin{bmatrix} 1 & 0 \\ 0 & 1 \end{bmatrix} \end{array}$$

$$\mathbb{I}_{\min.+} = \begin{array}{c} \\ x \\ y \end{array} \begin{array}{c} \begin{array}{cc} x & y \end{array} \\ \begin{bmatrix} 0 & \infty \\ \infty & 0 \end{bmatrix} \end{array} \quad \mathbb{I}_{\min.\max} = \begin{array}{c} \\ x \\ y \end{array} \begin{array}{c} \begin{array}{cc} x & y \end{array} \\ \begin{bmatrix} -\infty & \infty \\ \infty & -\infty \end{bmatrix} \end{array} \quad \mathbb{I}_{\min.\times} = \begin{array}{c} \\ x \\ y \end{array} \begin{array}{c} \begin{array}{cc} x & y \end{array} \\ \begin{bmatrix} 1 & \infty \\ \infty & 1 \end{bmatrix} \end{array}$$

The above identity arrays are a direct result of the definitions of the 0 and 1 elements of their corresponding semirings. Analysis of these identity arrays provides some insight into the broader behavior of these semirings. The $\mathbb{I}_{+.\times}$ has the advantage that it is easily computed in a wide variety of semirings. Thus, additional insight can be obtained by analyzing the behavior of semirings when multiplying by the $\mathbb{I}_{+.\times}$ identity array. The subsequent sections carefully derive the transformation $\mathbb{I}_{+.\times}$ for each quadrant to explain the resulting visual depiction of the transformation. This consistent enumeration provides the reader with the necessary tools to analyze more complicated transformations. As a result of the analysis, many symmetries emerge whereby certain quadrants have similarities. These similarities are useful in applying these concepts to real data as they can allow for significant simplifications. For example, certain transformations in certain semirings will always project a quadrant to a well-defined point or set of points, and can be used in a real data processing system to simplify or eliminate steps entirely.

Multiplication Over max.+

Multiplying the unit circle **U** by $\mathbb{I}_{+.\times}$ in the max.+ semiring is shown in Figure 12.1 (middle, left) and performs the following transformation

$$\begin{array}{c} \\ x \\ y \end{array} \overset{\begin{array}{cc} x & y \end{array}}{\begin{bmatrix} 1 & 0 \\ 0 & 1 \end{bmatrix}} \text{ max.+ } \mathbf{U} = \begin{array}{c} \\ x \\ y \end{array} \overset{\begin{array}{ccc} 1 & \cdots & n \end{array}}{\begin{bmatrix} \max(\mathbf{U}(x,:)+1, \mathbf{U}(y,:)+0) \\ \max(\mathbf{U}(x,:)+0, \mathbf{U}(y,:)+1) \end{bmatrix}}$$

In the positive quadrant (I), the center of the unit circle is shifted to (1,1) because $(x+1, y+1)$ are always greater than (x, y), so that

$$\max(\mathbf{U}(x,\text{I})+1, \mathbf{U}(y,\text{I})+0) = \mathbf{U}(x,\text{I})+1$$

and

$$\max(\mathbf{U}(x,\text{I})+0, \mathbf{U}(y,\text{I})+1) = \mathbf{U}(y,\text{I})+1$$

resulting in

$$\mathbb{I}_{+.\times} \text{ max.+ } \mathbf{U}(:,\text{I}) = \begin{array}{c} \\ x \\ y \end{array} \overset{\text{I}}{\begin{bmatrix} \mathbf{U}(x,\text{I})+1 \\ \mathbf{U}(y,\text{I})+1 \end{bmatrix}}$$

The quadrant I max.+ transformation is visualized as a simple shift of the points up and to the right of their starting positions

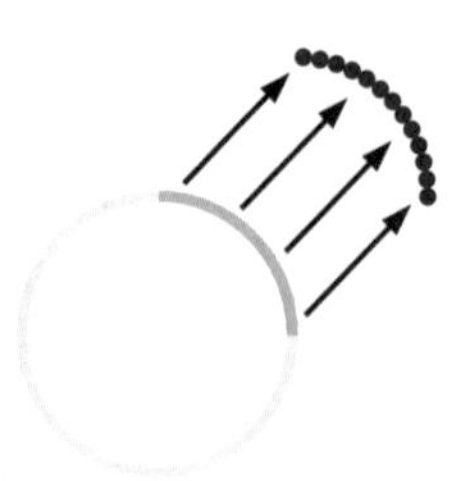

In the upper left quadrant (II), the unit circle projects itself onto its y values because the y values are all greater than their corresponding x values

$$\max(\mathbf{U}(x,\mathrm{II})+1, \mathbf{U}(y,\mathrm{II})+0) = \mathbf{U}(y,\mathrm{II})$$

since

$$\sin(\phi) \geq \cos(\phi)+1$$

for

$$0.5\pi \leq \phi \leq \pi$$

The y values are simply offset by 1 because

$$\max(\mathbf{U}(x,\mathrm{II})+0, \mathbf{U}(y,\mathrm{II})+1) = \mathbf{U}(y,\mathrm{II})+1$$

resulting in

$$\mathbb{I}_{+.\times} \ \mathrm{max.+} \ \mathbf{U}(:,\mathrm{II}) = \begin{matrix} \\ x \\ y \end{matrix} \begin{matrix} \mathrm{II} \\ \begin{bmatrix} \mathbf{U}(y,\mathrm{II}) \\ \mathbf{U}(y,\mathrm{II})+1 \end{bmatrix} \end{matrix}$$

The quadrant II max.+ transformation is thus an offset projection of the unit circle points onto the line $y = x+1$

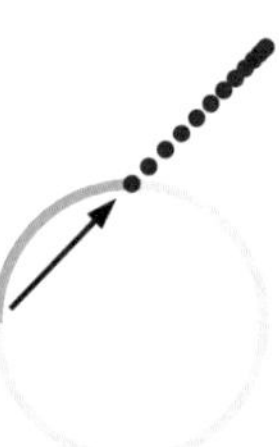

In the negative quadrant (III), the transformation is similar to the transformation in the positive quadrant (I) since the center point of the unit circle is moved to (1,1) because $(x+1, y+1)$ is always greater than (x,y), so that

$$\max(\mathbf{U}(x,\mathrm{III})+1, \mathbf{U}(y,\mathrm{III})+0) = \mathbf{U}(x,\mathrm{III})+1$$

and

$$\max(\mathbf{U}(x, \mathrm{III}) + 0, \mathbf{U}(y, \mathrm{III}) + 1) = \mathbf{U}(y, \mathrm{III}) + 1$$

resulting in

$$\mathbb{I}_{+.\times} \text{ max.+ } \mathbf{U}(:, \mathrm{III}) = \begin{matrix} \\ x \\ y \end{matrix} \begin{matrix} \mathrm{III} \\ \begin{bmatrix} \mathbf{U}(x, \mathrm{III}) + 1 \\ \mathbf{U}(y, \mathrm{III}) + 1 \end{bmatrix} \end{matrix}$$

Thus, the quadrant III max.+ transformation is similar to the quadrant I transformation and appears as linear shifting of the points to the upper right of their starting position by a fixed amount

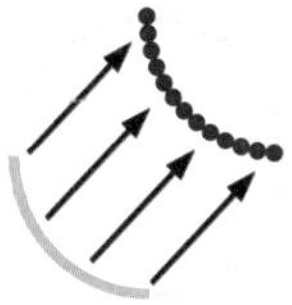

In the lower right quadrant (IV), the transformation is like the transformation in the upper left quadrant (II), with the difference that the unit circle projects itself onto its x values because the x values are offset by 1

$$\max(\mathbf{U}(x, \mathrm{IV}) + 1, \mathbf{U}(y, \mathrm{IV}) + 0) = \mathbf{U}(x, \mathrm{IV}) + 1$$

and the y values are all swapped with their corresponding x values

$$\max(\mathbf{U}(x, \mathrm{IV}) + 0, \mathbf{U}(y, \mathrm{IV}) + 1) = \mathbf{U}(x, \mathrm{IV})$$

since

$$\cos(\phi) \geq \sin(\phi) + 1$$

for

$$1.5\pi \leq \phi \leq 2\pi$$

resulting in

$$\mathbb{I}_{+.\times} \text{ max.min } \mathbf{U}(:, \mathrm{IV}) = \begin{matrix} \\ x \\ y \end{matrix} \begin{matrix} \mathrm{IV} \\ \begin{bmatrix} \mathbf{U}(x, \mathrm{IV}) + 1 \\ \mathbf{U}(x, \mathrm{IV}) \end{bmatrix} \end{matrix}$$

The quadrant IV max.+ transformation is depicted as the unit circle being translated and projected to lie on the line $y = x - 1$

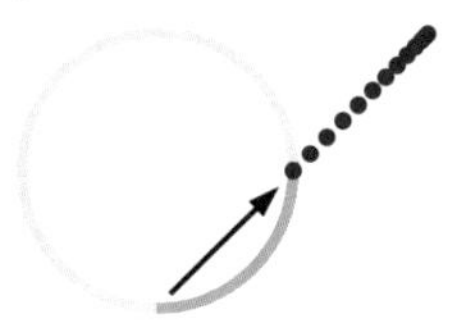

Multiplication Over min.+

Multiplying the unit circle **U** by $\mathbb{I}_{+.\times}$ in the min.+ semiring is shown in Figure 12.1 (bottom, left) and performs the following transformation

$$\begin{matrix} & x & y \\ x & 1 & 0 \\ y & 0 & 1 \end{matrix} \text{ min.+ } \mathbf{U} = \begin{matrix} & 1 \quad \cdots \quad n \\ x & \min(\mathbf{U}(x,:)+1, \mathbf{U}(y,:)+0) \\ y & \min(\mathbf{U}(x,:)+0, \mathbf{U}(y,:)+1) \end{matrix}$$

In the positive quadrant (I), the transformation replaces the x values with their corresponding y values

$$\min(\mathbf{U}(x,\text{I})+1, \mathbf{U}(y,\text{I})+0) = \mathbf{U}(y,\text{I})$$

Likewise, the y values are replaced with their corresponding x values

$$\max(\mathbf{U}(x,\text{I})+0, \mathbf{U}(y,\text{I})+1) = \mathbf{U}(x,\text{I})$$

resulting in

$$\mathbb{I}_{+.\times} \text{ max.+ } \mathbf{U}(:,\text{I}) = \begin{matrix} & \text{I} \\ x & \mathbf{U}(y,\text{I}) \\ y & \mathbf{U}(x,\text{I}) \end{matrix}$$

The quadrant I min.+ transformation appears as the unit circle unchanged but its points are reflected about the line $y = x$

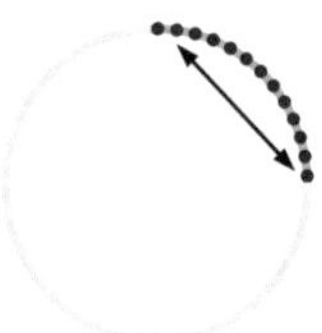

In the upper left quadrant (II), all the x values are projected onto their x values offset by 1 because

$$\min(\mathbf{U}(x,\text{II})+1, \mathbf{U}(y,\text{II})+0) = \mathbf{U}(x,\text{II})+1$$

since

$$\sin(\phi) \geq \cos(\phi) + 1$$

for

$$0.5\pi \leq \phi \leq \pi$$

Likewise, the y values are substituted with their corresponding x values

$$\min(\mathbf{U}(x,\mathrm{II}) + 0, \mathbf{U}(y,\mathrm{II}) + 1) = \mathbf{U}(x,\mathrm{II})$$

resulting in

$$\mathbb{I}_{+.\times} \text{ max.+ } \mathbf{U}(:,\mathrm{II}) = \begin{array}{c} \\ x \\ y \end{array} \begin{array}{c} \mathrm{II} \\ \left[\begin{array}{c} \mathbf{U}(x,\mathrm{II}) + 1 \\ \mathbf{U}(x,\mathrm{II}) \end{array}\right] \end{array}$$

The quadrant II min.+ transformation becomes the unit circle offset and moved to the line $y = x - 1$

In the negative quadrant (III), the transformation has an effect similar to the transformation in the positive quadrant (I) since the transformation swaps the x and y values of the unit circle because

$$\min(\mathbf{U}(x,\mathrm{III}) + 1, \mathbf{U}(y,\mathrm{III}) + 0) = \mathbf{U}(y,\mathrm{III})$$

and

$$\min(\mathbf{U}(x,\mathrm{III}) + 0, \mathbf{U}(y,\mathrm{III}) + 1) = \mathbf{U}(x,\mathrm{III})$$

resulting in

$$\mathbb{I}_{+.\times} \text{ max.+ } \mathbf{U}(:,\mathrm{III}) = \begin{array}{c} \\ x \\ y \end{array} \begin{array}{c} \mathrm{III} \\ \left[\begin{array}{c} \mathbf{U}(y,\mathrm{III}) \\ \mathbf{U}(x,\mathrm{III}) \end{array}\right] \end{array}$$

The quadrant III min.+ transformation can be visualized in the same manner as the quadrant I transformation. The unit circle appears unchanged but with its points being reflected about the line $y = x$

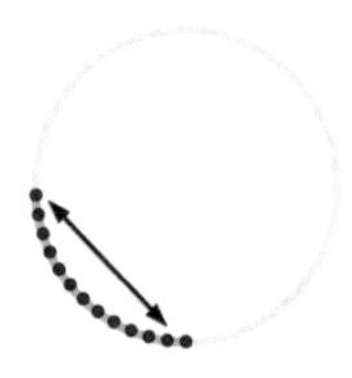

In the lower right quadrant (IV), the transformation has a similar impact as the transformation in the upper left quadrant (II), with the difference that the x values are set to their y values because

$$\min(\mathbf{U}(x,\mathrm{IV})+1, \mathbf{U}(y,\mathrm{IV})+0) = \mathbf{U}(y,\mathrm{IV})$$

Likewise, the y values are shifted by 1

$$\min(\mathbf{U}(x,\mathrm{IV})+0, \mathbf{U}(y,\mathrm{IV})+1) = \mathbf{U}(y,\mathrm{IV})+1$$

since

$$\cos(\phi) \geq \sin(\phi)+1$$

for

$$1.5\pi \leq \phi \leq 2\pi$$

resulting in

$$\mathbb{I}_{+.\times} \text{ max.min } \mathbf{U}(:,\mathrm{IV}) = \begin{matrix} \\ x \\ y \end{matrix} \begin{matrix} \mathrm{IV} \\ \begin{bmatrix} \mathbf{U}(y,\mathrm{IV}) \\ \mathbf{U}(y,\mathrm{IV})+1 \end{bmatrix} \end{matrix}$$

The quadrant IV min.+ transformation is thus similar to the quadrant II transformation and can be represented as the unit circle shifted and plotted onto the line given by the equation $y = x+1$

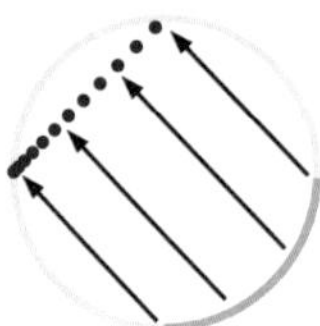

Multiplication Over max.min

Multiplying the unit circle $\mathbf{U}$ by $\mathbb{I}_{+.\times}$ in the max.min semiring is shown in Figure 12.1 (middle, center) and results in the following transformation

$$\begin{matrix} & x & y \\ x & 1 & 0 \\ y & 0 & 1 \end{matrix} \text{ max.min } \mathbf{U} = \begin{matrix} \\ x \\ y \end{matrix} \begin{bmatrix} \max(\min(\mathbf{U}(x,:),1), \min(\mathbf{U}(y,:),0)) \\ \max(\min(\mathbf{U}(x,:),0), \min(\mathbf{U}(y,:),1)) \end{bmatrix}$$

(columns labelled $1 \; \cdots \; n$)

In the positive quadrant (I), the x values are unchanged because

$$\max(\min(\mathbf{U}(x,\mathrm{I}),1),\min(\mathbf{U}(y,\mathrm{I}),0)) = \max(\mathbf{U}(x,\mathrm{I}),0) = \mathbf{U}(x,\mathrm{I})$$

Likewise, the y values are also unchanged

$$\max(\min(\mathbf{U}(x,\mathrm{I}),0),\min(\mathbf{U}(y,\mathrm{I}),1)) = \max(0,\mathbf{U}(y,\mathrm{I})) = \mathbf{U}(y,I)$$

resulting in

$$\mathbb{I}_{+.\times} \ \text{max.min} \ \mathbf{U}(:,\mathrm{I}) = \begin{array}{c} \\ x \\ y \end{array} \begin{array}{c} \mathrm{I} \\ \left[\begin{array}{c} \mathbf{U}(x,\mathrm{I}) \\ \mathbf{U}(y,\mathrm{I}) \end{array} \right] \end{array}$$

Thus, in quadrant I, the max.min transformation acts as the identity transformation with no shift in the unit circle points.

In the upper left quadrant (II), the x values all map onto 0 because

$$\max(\min(\mathbf{U}(x,\mathrm{II}),1),\min(\mathbf{U}(y,\mathrm{II}),0)) = \max(\mathbf{U}(x,\mathrm{II}),0) = 0$$

and the y values are unchanged because

$$\max(\min(\mathbf{U}(x,\mathrm{II}),0),\min(\mathbf{U}(y,\mathrm{II}),1)) = \max(0,\mathbf{U}(y,\mathrm{II})) = \mathbf{U}(y,\mathrm{II})$$

resulting in

$$\mathbb{I}_{+.\times} \ \text{max.min} \ \mathbf{U}(:,\mathrm{II}) = \begin{array}{c} \\ x \\ y \end{array} \begin{array}{c} \mathrm{II} \\ \left[\begin{array}{c} 0 \\ \mathbf{U}(y,\mathrm{II}) \end{array} \right] \end{array}$$

The quadrant II max.min transformation can be depicted as the unit circle directly projected onto the line $x = 0$

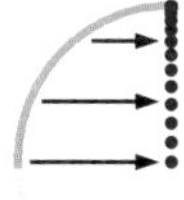

In the negative quadrant (III), the behavior is different from that of all the other quadrants.

Both the x and y values are transformed to the same value

$$\max(\min(\mathbf{U}(x,\mathrm{III}),1),\min(\mathbf{U}(y,\mathrm{III}),0)) = \max(\mathbf{U}(x,\mathrm{III}),\mathbf{U}(y,\mathrm{III}))$$

and

$$\max(\min(\mathbf{U}(x,\mathrm{III}),0),\min(\mathbf{U}(y,\mathrm{III}),1)) = \max(\mathbf{U}(x,\mathrm{III}),\mathbf{U}(y,\mathrm{III}))$$

resulting in

$$\mathbb{I}_{+.\times} \text{ max.min } \mathbf{U}(:,\mathrm{III}) = \begin{matrix} \\ x \\ y \end{matrix} \begin{matrix} \mathrm{III} \\ \begin{bmatrix} \max(\mathbf{U}(x,\mathrm{III}),\mathbf{U}(y,\mathrm{III})) \\ \max(\mathbf{U}(x,\mathrm{III}),\mathbf{U}(y,\mathrm{III})) \end{bmatrix} \end{matrix}$$

Thus, the quandrant III max.min transformation directly projects the unit circle onto the line $y = x$

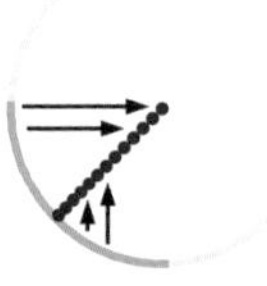

In the lower right quadrant (IV), the transformation appears similar to the quadrant II transformation. The x values remain unchanged because

$$\max(\min(\mathbf{U}(x,\mathrm{IV}),1),\min(\mathbf{U}(y,\mathrm{IV}),0)) = \max(\mathbf{U}(x,:),\mathbf{U}(y,\mathrm{IV})) = \mathbf{U}(x,\mathrm{IV})$$

The y values all become 0 because in this quadrant they are all less than 0

$$\max(\min(\mathbf{U}(x,\mathrm{IV}),0),\min(\mathbf{U}(y,\mathrm{IV}),1)) = \max(0,\mathbf{U}(y,\mathrm{IV})) = 0$$

resulting in

$$\mathbb{I}_{+.\times} \text{ max.min } \mathbf{U}(:,\mathrm{IV}) = \begin{matrix} \\ x \\ y \end{matrix} \begin{matrix} \mathrm{IV} \\ \begin{bmatrix} \mathbf{U}(x,\mathrm{IV}) \\ 0 \end{bmatrix} \end{matrix}$$

The quadrant IV max.min transformation can be viewed as a direct projection of the unit circle onto the line $y = 0$

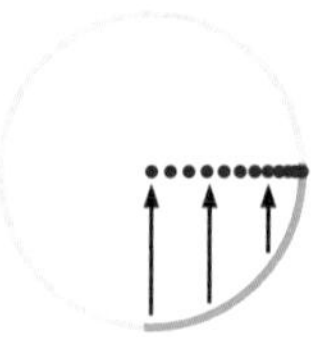

Multiplication Over min.max

Multiplying the unit circle $\mathbf{U}$ by $\mathbb{I}_{+.\times}$ in the min.max semiring is shown in Figure 12.1 (bottom, center) and performs the following transformation

$$\begin{array}{c} \\ x \\ y \end{array}\begin{array}{c} \begin{array}{cc} x & y \end{array} \\ \left[\begin{array}{cc} 1 & 0 \\ 0 & 1 \end{array}\right] \end{array} \text{min.max } \mathbf{U} = \begin{array}{c} \\ x \\ y \end{array}\begin{array}{c} 1 \quad \cdots \quad n \\ \left[\begin{array}{c} \min(\max(\mathbf{U}(x,:),1),\max(\mathbf{U}(y,:),0)) \\ \min(\max(\mathbf{U}(x,:),0),\max(\mathbf{U}(y,:),1)) \end{array}\right] \end{array}$$

In the positive quadrant (I), the x value is replaced with the corresponding y value because

$$\min(\max(\mathbf{U}(x,\mathrm{I}),1),\max(\mathbf{U}(y,\mathrm{I}),0)) = \min(1,\mathbf{U}(y,\mathrm{I})) = \mathbf{U}(y,\mathrm{I})$$

Likewise, the y value is replaced with the corresponding x value

$$\min(\max(\mathbf{U}(x,\mathrm{I}),0),\max(\mathbf{U}(y,\mathrm{I}),1)) = \min(\mathbf{U}(x,\mathrm{I}),1) = \mathbf{U}(x,\mathrm{I})$$

resulting in

$$\mathbb{I}_{+.\times} \text{ min.max } \mathbf{U}(:,\mathrm{I}) = \begin{array}{c} \\ x \\ y \end{array}\begin{array}{c} \mathrm{I} \\ \left[\begin{array}{c} \mathbf{U}(y,\mathrm{I}) \\ \mathbf{U}(x,\mathrm{I}) \end{array}\right] \end{array}$$

The quadrant I unit circle under the min.max transformation appears the same, but the points are swapped about the line $y = x$

In the upper left quadrant (II), the x value is swapped with its y values (as in quadrant I)

$$\min(\max(\mathbf{U}(x,\mathrm{II}),1),\max(\mathbf{U}(y,\mathrm{II}),0)) = \min(1,\mathbf{U}(y,\mathrm{II})) = \mathbf{U}(y,\mathrm{II})$$

However, all the y values become 0

$$\min(\max(\mathbf{U}(x,\mathrm{II}),0),\max(\mathbf{U}(y,\mathrm{II}),1)) = \min(0,1) = 0$$

resulting in

$$\mathbb{I}_{+.\times} \text{ min.max } \mathbf{U}(:,\mathrm{II}) = s \begin{array}{c} \\ x \\ y \end{array}\begin{array}{c} \mathrm{II} \\ \left[\begin{array}{c} \mathbf{U}(y,\mathrm{II}) \\ 0 \end{array}\right] \end{array}$$

The quadrant II min.max transformation shifts and projects the unit circle onto the line $y = 0$ as shown by

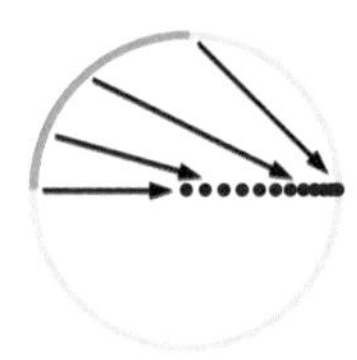

In the negative quadrant (III), x values are transformed to 0 since

$$\min(\max(\mathbf{U}(x,\text{III}),1),\max(\mathbf{U}(y,\text{III}),0)) = \min(0,1) = 0$$

Likewise, the y values also become 0

$$\min(\max(\mathbf{U}(x,\text{III}),0),\max(\mathbf{U}(y,\text{III}),1)) = \min(0,1) = 0$$

resulting in

$$\mathbb{I}_{+.\times} \text{ min.max } \mathbf{U}(:,\text{III}) = \begin{array}{c} \\ x \\ y \end{array} \begin{array}{c} \text{III} \\ \begin{bmatrix} 0 \\ 0 \end{bmatrix} \end{array}$$

The quadrant III min.max transformation thus has the property of mapping all points to (0,0), which can be illustrated as

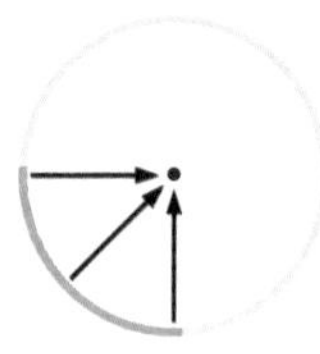

In the lower right quadrant (IV), the behavior is similar to that of quadrant II. All the x values become 0 because

$$\min(\max(\mathbf{U}(x,\text{IV}),1),\max(\mathbf{U}(y,\text{IV}),0)) = \min(1,0) = 0$$

Likewise, all the y values become their x values because

$$\min(\max(\mathbf{U}(x,\text{IV}),0),\max(\mathbf{U}(y,\text{IV}),1)) = \min(\mathbf{U}(x,\text{IV}),1) = \mathbf{U}(x,\text{IV})$$

resulting in

$$\mathbb{I}_{+.\times} \text{ min.max } \mathbf{U}(:,\text{IV}) = \begin{array}{c} \\ x \\ y \end{array} \begin{array}{c} \text{IV} \\ \begin{bmatrix} 0 \\ \mathbf{U}(x,\text{IV}) \end{bmatrix} \end{array}$$

The min.max transformation in quadrant IV appears as a translated projection of the unit circle onto the line $x = 0$

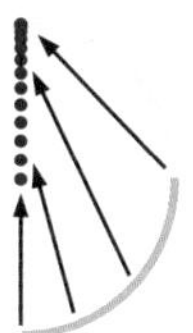

Multiplication Over max.$\times$

Multiplying the unit circle **U** by $\mathbb{I}_{+.\times}$ in the max.$\times$ semiring is shown in Figure 12.1 (middle, right) and performs the following transformation

$$\begin{array}{c} \\ x \\ y \end{array}\begin{array}{c} \begin{array}{cc} x & y \end{array} \\ \begin{bmatrix} 1 & 0 \\ 0 & 1 \end{bmatrix} \end{array} \text{max.}\times\ \mathbf{U} = \begin{array}{c} \\ x \\ y \end{array}\begin{array}{c} 1 \quad \cdots \quad n \\ \begin{bmatrix} \max(\mathbf{U}(x,:)\times 1, \mathbf{U}(y,:)\times 0) \\ \max(\mathbf{U}(x,:)\times 0, \mathbf{U}(y,:)\times 1) \end{bmatrix} \end{array}$$

$$= \begin{array}{c} \\ x \\ y \end{array}\begin{array}{c} 1 \quad \cdots \quad n \\ \begin{bmatrix} \max(\mathbf{U}(x,:),0) \\ \max(0,\mathbf{U}(y,:)) \end{bmatrix} \end{array}$$

The max.$\times$ is only a semiring in the positive quadrant. In the other quadrants, max.$\times$ does not satisfy all the conditions required of a semiring. For completeness, the result of applying max.$\times$ across all four quadrants is shown Figure 12.1 (middle, right) and is described as follows.

In the positive quadrant (I), the x values remain the same because

$$\max(\mathbf{U}(x,\text{I}),0) = \mathbf{U}(x,\text{I})$$

Likewise, the y values are also unchanged

$$\max(0,\mathbf{U}(y,\text{I})) = \mathbf{U}(y,\text{I})$$

resulting in

$$\mathbb{I}_{+.\times}\ \text{max.}\times\ \mathbf{U}(:,\text{I}) = \begin{array}{c} \\ x \\ y \end{array}\begin{array}{c} \text{I} \\ \begin{bmatrix} \mathbf{U}(x,\text{I}) \\ \mathbf{U}(y,\text{I}) \end{bmatrix} \end{array}$$

The max.$\times$ transformation in quadrant I is thus an identity transformation and the unit circle stays in place

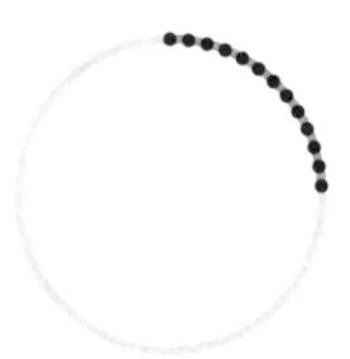

In the (non-semiring) upper left quadrant (II), all the x values are transformed to 0 because

$$\max(\mathbf{U}(x, \mathrm{II}), 0) = 0$$

In contrast, the y values are all unchanged

$$\max(0, \mathbf{U}(y, \mathrm{II})) = \mathbf{U}(y, \mathrm{II})$$

resulting in

$$\mathbb{I}_{+.\times} \ \max.\times \ \mathbf{U}(:, \mathrm{II}) = \begin{array}{c} \\ x \\ y \end{array} \begin{array}{c} \mathrm{II} \\ \left[\begin{array}{c} 0 \\ \mathbf{U}(y, \mathrm{II}) \end{array} \right] \end{array}$$

The quadrant II max.× transformation can be drawn as a direct projection of the unit circle onto the line $x = 0$

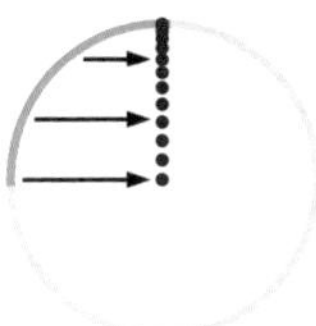

In the (non-semiring) negative quadrant (III), the x values are all set to 0 as a result of

$$\max(\mathbf{U}(x, \mathrm{III}), 0) = 0$$

The y values are also set to 0 since

$$\max(0, \mathbf{U}(y, \mathrm{III})) = 0$$

resulting in

$$\mathbb{I}_{+.\times} \ \max.\times \ \mathbf{U}(:, \mathrm{III}) = \begin{array}{c} \\ x \\ y \end{array} \begin{array}{c} \mathrm{III} \\ \left[\begin{array}{c} 0 \\ 0 \end{array} \right] \end{array}$$

Thus, the quadrant III max.× is simply shown as a mapping of all points on the unit circle going to the point $(0, 0)$

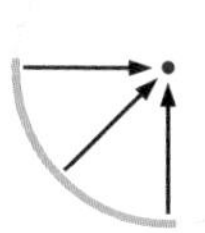

The (non-semiring) lower right quadrant (IV) is similar to quadrant II. All the x values remain unchanged.

$$\max(\mathbf{U}(x,\text{IV}),0) = \mathbf{U}(x,\text{IV})$$

In contrast, all the y values are changed to 0 as a result of

$$\max(0,\mathbf{U}(y,\text{IV})) = 0$$

and thus

$$\mathbb{I}_{+.\times} \text{ max.}\times \ \mathbf{U}(:,\text{IV}) = \begin{array}{c} \\ x \\ y \end{array} \begin{array}{c} \text{IV} \\ \begin{bmatrix} \mathbf{U}(x,\text{IV}) \\ 0 \end{bmatrix} \end{array}$$

The quadrant IV max.× transformation is a projection of the unit circle directly onto the line $y = 0$

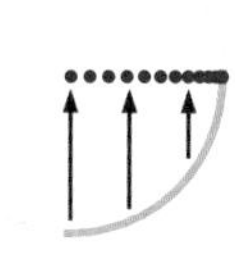

Multiplication Over min.×

Multiplying the unit circle $\mathbf{U}$ by $\mathbb{I}_{+.\times}$ in the min.× semiring is shown in Figure 12.1 (bottom, right) and performs the following transformation

$$\begin{array}{c} \\ x \\ y \end{array} \begin{array}{c} x \quad y \\ \begin{bmatrix} 1 & 0 \\ 0 & 1 \end{bmatrix} \end{array} \text{min.}\times \ \mathbf{U} = \begin{array}{c} \\ x \\ y \end{array} \begin{array}{c} 1 \quad \cdots \quad n \\ \begin{bmatrix} \min(\mathbf{U}(x,:)\times 1, \mathbf{U}(y,:)\times 0) \\ \min(\mathbf{U}(x,:)\times 0, \mathbf{U}(y,:)\times 1) \end{bmatrix} \end{array}$$

$$= \begin{array}{c} \\ x \\ y \end{array} \begin{array}{c} 1 \ \cdots \ n \\ \begin{bmatrix} \min(\mathbf{U}(x,:),0) \\ \min(0,\mathbf{U}(y,:)) \end{bmatrix} \end{array}$$

The min.× is only a semiring in the positive quadrant. In the other quadrants, min.× does not meet every condition required of a semiring. Nevertheless, the application of min.× on all four quadrants is given in Figure 12.1 (bottom, right) and analyzed as follows.

In the positive quadrant (I), the x values all project to 0 because

$$\min(\mathbf{U}(x,\mathrm{I}),0) = 0$$

Similarly, all the y values also transform to 0

$$\min(0,\mathbf{U}(y,\mathrm{I})) = 0$$

resulting in

$$\mathbb{I}_{+.\times} \text{ max.}\times \ \mathbf{U}(:,\mathrm{I}) = \begin{array}{c} \\ x \\ y \end{array} \begin{array}{c} \mathrm{I} \\ \left[\begin{array}{c} 0 \\ 0 \end{array}\right] \end{array}$$

In quadrant I, the min.× transformation takes all points on the unit circle and maps them directly onto the point (0,0) as shown

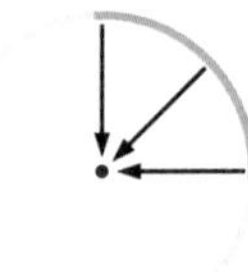

In the (non-semiring) upper left quadrant (II), the x value is unchanged

$$\min(\mathbf{U}(x,\mathrm{II}),0) = \mathbf{U}(x,\mathrm{II})$$

However, all the y values are changed to 0

$$\min(0,\mathbf{U}(y,\mathrm{II})) = 0$$

resulting in

$$\mathbb{I}_{+.\times} \text{ min.}\times \ \mathbf{U}(:,\mathrm{II}) = \begin{array}{c} \\ x \\ y \end{array} \begin{array}{c} \mathrm{II} \\ \left[\begin{array}{c} \mathbf{U}(x,\mathrm{II}) \\ 0 \end{array}\right] \end{array}$$

The quadrant II min.× transformation appears as the unit circle projected onto the line given by the equation $y = 0$

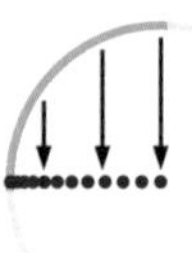

In the (non-semiring) negative quadrant (III), the x values do not change because

$$\min(\mathbf{U}(x,\text{III}),0) = \mathbf{U}(x,\text{III})$$

In addition, the y values also remain the same since

$$\min(0,\mathbf{U}(y,\text{III})) = \mathbf{U}(y,\text{III})$$

resulting in

$$\mathbb{I}_{+.\times}\ \text{min.}\times\ \mathbf{U}(:,\text{III}) = \begin{array}{c} \\ x \\ y \end{array} \begin{array}{c} \text{III} \\ \left[\begin{array}{c} \mathbf{U}(x,\text{III}) \\ \mathbf{U}(y,\text{III} \end{array}\right] \end{array}$$

Thus, the quadrant III min.× transformation is an identity transformation that leaves the points on the unit circle in place

The (non-semiring) lower right quadrant (IV) is similar to quadrant II. The x values all compute to 0 because

$$\min(\mathbf{U}(x,\text{IV}),0) = 0$$

On the other hand, the y values remain the same

$$\min(0,\mathbf{U}(y,\text{IV})) = \mathbf{U}(y,\text{IV})$$

resulting in

$$\mathbb{I}_{+.\times}\ \text{min.}\times\ \mathbf{U}(:,\text{IV}) = \begin{array}{c} \\ x \\ y \end{array} \begin{array}{c} \text{IV} \\ \left[\begin{array}{c} 0 \\ \mathbf{U}(y,\text{IV}) \end{array}\right] \end{array}$$

So, the quadrant IV min.× transformation is shown as the unit circle projected onto the line $x = 0$

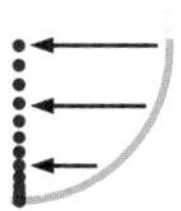

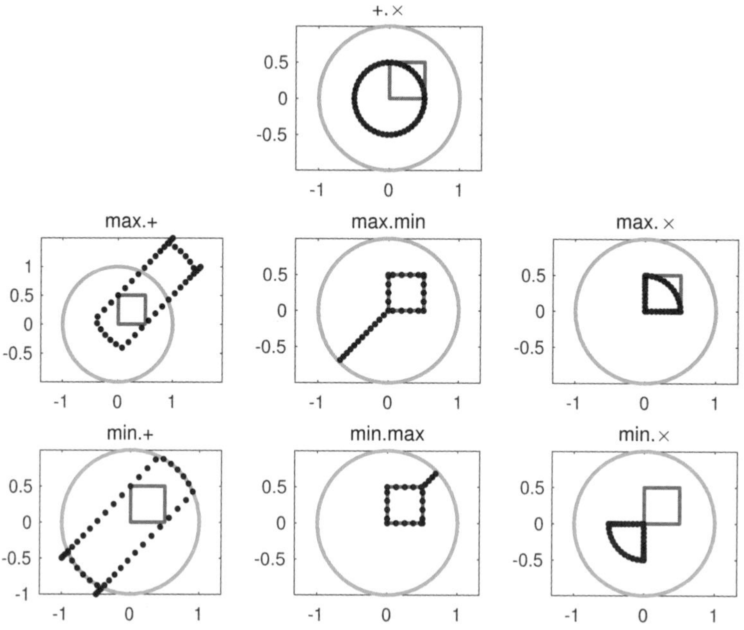

Figure 12.2
Transformation of points on the unit circle (green circle) via array multiplication by a 2×2 associative array (blue square), resulting in a new set of points (black dots). The title of each plot lists the ⊕ and ⊗ operators used in the array multiplication. The array values $\begin{bmatrix} 0.5 & 0 \\ 0 & 0.5 \end{bmatrix}$ correspond to a contraction transformation in the +.× semiring.

12.3 Contraction

The identity transformation of the unit circle is the simplest to visualize over different semirings. Perhaps the next simplest transformation is the contraction or expansion of the unit circle by a constant factor c. Multiplying the identity relation $\mathbb{I}_{+.\times}$ by a scalar c results in the diagonal array

$$c\mathbb{I}_{+.\times} = \begin{array}{c} \\ x \\ y \end{array} \begin{array}{c} \begin{array}{cc} x & y \end{array} \\ \begin{bmatrix} c & 0 \\ 0 & c \end{bmatrix} \end{array}$$

Array multiplication of the above array by the unit circle in the +.× semiring will contract or expand the unit circle. If $c > 1$, the unit circle will expand and if $0 < c < 1$ the unit circle will contract.

The specific case of $c = 0.5$ is shown in Figure 12.2 and corresponds to

$$\begin{array}{c} \\ x \\ y \end{array}\begin{array}{c} \begin{array}{cc} x & y \end{array} \\ \begin{bmatrix} 0.5 & 0 \\ 0 & 0.5 \end{bmatrix} \end{array} \oplus.\otimes \mathbf{U} = \begin{array}{c} \\ x \\ y \end{array}\begin{array}{c} \begin{array}{ccc} 1 & \cdots & n \end{array} \\ \begin{bmatrix} (\mathbf{U}(x,:)\otimes 0.5)\oplus(\mathbf{U}(y,:)\otimes\ 0\ ) \\ (\mathbf{U}(x,:)\otimes\ 0\ )\oplus(\mathbf{U}(y,:)\otimes 0.5) \end{bmatrix} \end{array}$$

The overall impact of contraction on the unit circle in the different semirings is to shrink the unit circle (with the exception of the max.+ semiring). Thus, the concept of contraction carries some meaning in these different semirings. In the max.+ and min.+ semirings, the contraction is similar to the identity transformation but is further truncated so that it touches the corners of the square corresponding to the contraction. In the max.min and min.max semirings, the contraction transformation pulls in the identity transformation further so that for much of the unit circle it conforms to the square corresponding to the contraction. Finally, in the max.× and min.× semirings, the identity transformation is further truncated so that it touches the corners of the square corresponding to the contraction.

From a graph perspective, contraction means that traversing the graph with the above array will result in a graph with edge values that get smaller with each transformation. Such a graph traversal is useful in circumstances when it is desired that the edge values decrease over the course of the traversal. A detailed analysis—similar to that of the previous section—of contraction on each quadrant for each semiring would reveal the exact effect of contraction on the unit circle. Qualitatively, the impact of contraction is clarified by comparing with the corresponding identity transformation for each semiring.

In the standard +.× semiring, the identity and the contraction appear as follows

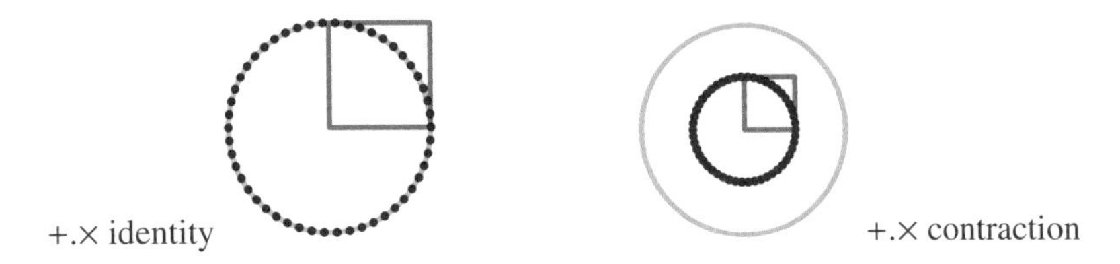

As expected, the contraction array shrinks the unit circle by exactly half when compared to the identity transformation. As in the identity transformation, the resulting unit circle also aligns exactly with the axis of the transformation, as shown by the blue boxes. In this semiring, the contraction behavior works on all points in the same manner and pulls them all equally closer to the origin. The graph equivalent of this operation would be to decrease all edge values by half.

In the max.+ semiring, the identity and the corresponding contraction can be depicted as

The max.+ semiring is distinct amongst the other semirings in that much of the unit circle increases under the contraction transformation. That said, when compared to the identity transformation, the overall area of the unit circle shrinks. Furthermore, like the identity transformation, the diagonals align with the corners of the axis of the transformation shown in the blue boxes. In the max.+ semiring, contraction effectively stretches the unit circle by expanding the unit circle along one diagonal and compressing the unit circle in the other direction. From a graph traversal perspective, this transformation would allow some edge values to grow and some edge values to shrink during the course of the traversal.

For the min.+ semiring, the identity transformation and the corresponding contraction are illustrated by

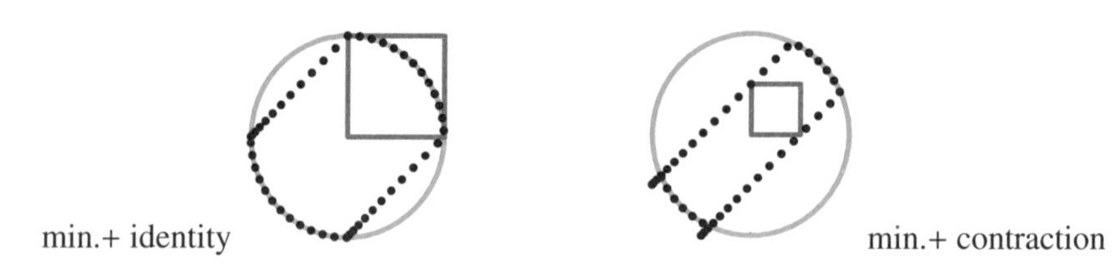

Contraction in the min.+ more closely resembles +.× in that nearly all points get closer to the origin and the overall area of the unit circle shrinks. In addition, the identity and contraction in min.+ look very similar, with the exception of a small number of points that are pushed outside the unit circle. Graph traversal via contraction in the min.+ semiring has the effect of reducing the values of many edges, preserving the values of many edges, and selecting a few edge values to be transformed outside the unit circle.

In the previous semirings, the resulting contraction of the unit circle is simply a distortion of the unit circle under the influence of the contraction transformation. The max.min identity and contraction illustrate a different behavior

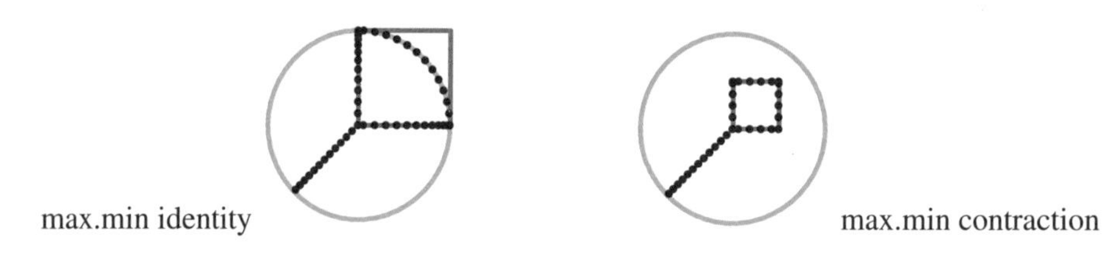

The contraction applied in the max.min semiring wraps much of the positive unit circle around the axis of the contraction (depicted in blue). The remainder of the unit circle behaves much the same as the identity transformation. In a graph context, this contraction allows the resulting edge values to be set by the contraction operation.

Contraction under the min.max semiring is similar to that of the min.max semiring. The identity and the corresponding contraction differ significantly as seen by

The min.max semiring, like max.min semiring, also wraps much of the unit circle around the axis of the contraction (depicted in blue), but in a somewhat opposite fashion. In a graph context, this contraction allows the resulting edge values of the negative parts of the unit circle to be determined by the contraction operation and the positive parts to be projected onto to the line $y = x$.

Both the max.× and min.× pairs of operations are semirings only in the positive (I) quadrant. These two pairs of operations have contraction behavior that is similar to their identity behavior. In both cases, the unit circle shrinks onto the axis (shown in blue) corresponding to the contraction as shown

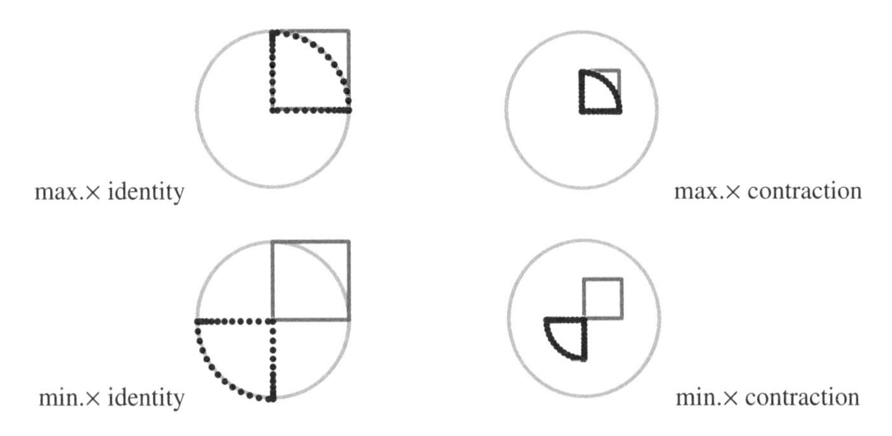

12.4 Stretching

The contraction and expansion transformations of the unit circle can be combined to stretch the unit circle by contracting one direction while expanding another direction. An example

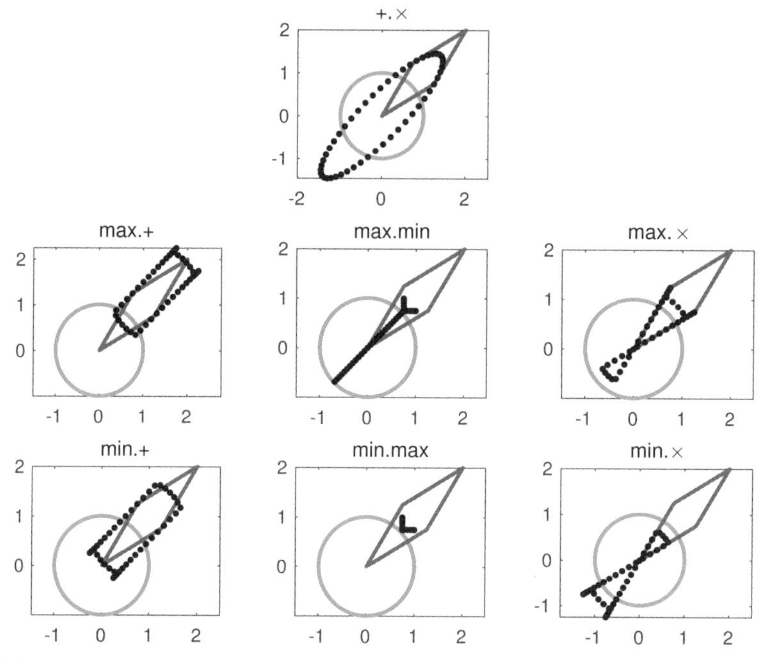

Figure 12.3

Transformation of points on the unit circle (green circle) via array multiplication by a 2×2 associative array (blue parallelogram) resulting in a new set of points (black dots). The title of each plot lists the ⊕ and ⊗ operators used in the array multiplication. The array values $\begin{bmatrix} 1.25 & 0.75 \\ 0.75 & 1.25 \end{bmatrix}$ correspond to a stretching transformation in the +.× semiring.

of stretching whereby the unit circle is expanded by 1.25 in one direction and contracted by 0.75 in another direction is shown in Figure 12.3 and corresponds to the array transformation

$$\begin{matrix} & x & y \\ x & 1.25 & 0.75 \\ y & 0.75 & 1.25 \end{matrix} \oplus.\otimes \mathbf{U} = \begin{matrix} & 1 \quad \cdots \quad n \\ x & (\mathbf{U}(x,:)\otimes 1.25)\oplus(\mathbf{U}(y,:)\otimes 0.75) \\ y & (\mathbf{U}(x,:)\otimes 0.75)\oplus(\mathbf{U}(y,:)\otimes 1.25) \end{matrix}$$

The concept of stretching has similar behavior in many of these semirings. Stretching the unit circle with the different semirings makes the unit circle longer and thinner (with the exception of the min.max semiring). In the max.+ and min.+ semirings, stretching is like the identity transformation but is more elongated. In the max.min and min.max semirings, stretching pulls in the identity transformation and pinches it off from the unit circle. Lastly, in the max.× and min.× semirings, the stretching elongates and twists the unit circle.

From the perspective of a graph operation, stretching implies that traversing the graph with the above array results in a graph with edge values that are smaller for some edge values and larger for other edge values. Such a graph traversal is useful in circumstances when it is desired that some edge values are enhanced and other edge values are suppressed. As in the previous section, the qualitative impact of stretching can be understood by comparing stretching with the equivalent identity transformation for each semiring.

In the standard +.× semiring, the identity and the stretching transformation look like

Stretching elongates the unit circle along the line $y = x$ and compresses the unit circle along the line $y = -x$. The stretched unit circle passes through and is tangent to the blue parallelogram defined by the stretching transformation. Applying the stretching transformation to a graph causes edge values along the line $y = x$ to be increased and edge values along $y = -x$ to be decreased.

The max.+ and min.+ semirings exhibit similar stretching behavior when compared to their corresponding identity transformations

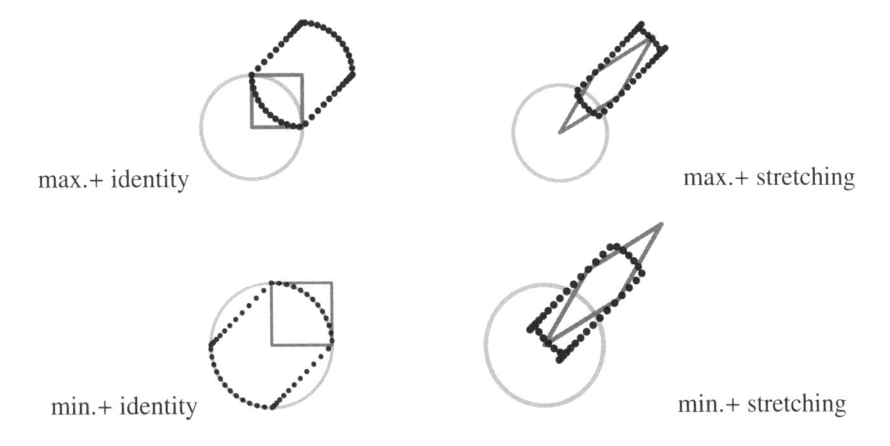

In both cases, the resulting unit circles are elongated and narrowed in a similar manner and have symmetry around the anti-diagonal of their tranformations shown by the blue parallelograms. In graph traversal, both would result in similar (but opposite) enhancements of edge values.

Stretching in the max.min semiring extends the concept of stretching by projecting most of the unit circle onto the line $y = x$ as compared with the identity transformation

In this instance, the stretching array acts as a filter on graph edge values by projecting nearly all of them onto the diagonal line, with the exception of a few values that are set by the coordinates of the stretching parallelogram.

Stretching in the min.max semiring is significantly different from the identity and shrinks the unit circle onto a narrow elbow of points defined by the axis of the transformation depicted by the blue parallelgram

This kind of stretching transformation replaces a variety of graph edge values with a narrow range of graph edge values.

Stretching in the max.× and min.× operation pairs are only semirings in the positive quadrant (I). They are similar to each other but different from their respective identify transformations. Both project and twist the unit circle and are symmetric with each other (similar to the max.+ and min.+ semirings)

12.5 Rotation

The above transformations give a hint at how easy it is to create complex transformations with very simple operations. The rotation transformation takes this complexity to another level to produce shapes that are typically not found in standard computations. Rotation of the unit circle by $\pi/4$ is one of the simplest transformations in the +.× semiring, but it exhibits a wide array of behaviors in other semirings. This rotation is shown in Figure 12.4 and corresponds to the array transformation

$$\begin{array}{c} \\ x \\ y \end{array} \begin{array}{c} \begin{array}{cc} x & y \end{array} \\ \begin{bmatrix} \cos(\pi/4) & -\sin(\pi/4) \\ \sin(\pi/4) & \cos(\pi/4) \end{bmatrix} \end{array} \oplus.\otimes \mathbf{U} = \begin{array}{c} \\ x \\ y \end{array} \begin{array}{c} \begin{array}{ccc} 1 & \cdots & n \end{array} \\ \begin{bmatrix} (\mathbf{U}(x,:) \otimes \cos(\pi/4)) \oplus (\mathbf{U}(y,:) \otimes -\sin(\pi/4)) \\ (\mathbf{U}(x,:) \otimes \sin(\pi/4)) \oplus (\mathbf{U}(y,:) \otimes \cos(\pi/4)) \end{bmatrix} \end{array}$$

The concept of rotation has a fairly different meaning in many of these semirings. The max.+, min.+, max.×, and min.× operation pairs all transform the unit circle into a half circle with a handle. The max.min semiring projects the unit circle onto a triangle, and the max.min projects the unit circle onto a line.

The rotation transformation of the unit circle in the max.+ and min.+ semirings produces similar geometric shapes that are very distinct from their identity transformation

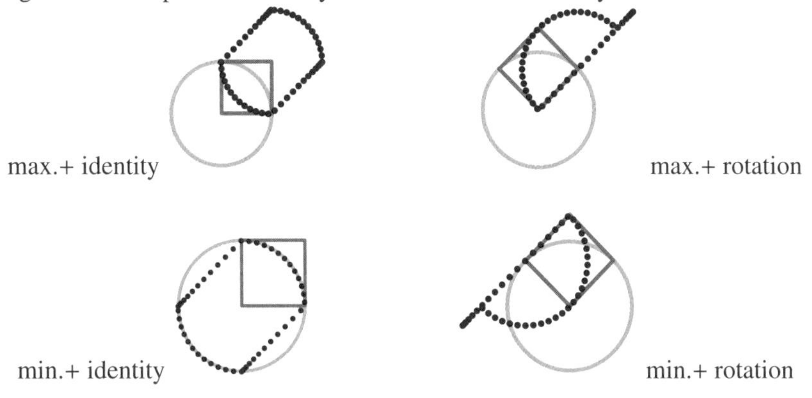

The min.+ rotation produces a half circle with handle shape; this is a $\pi/2$ rotation of the

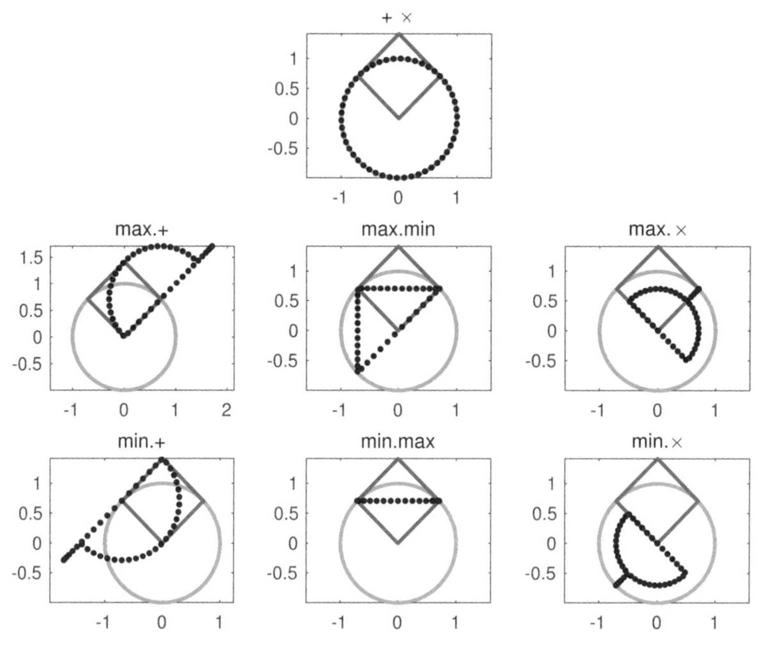

Figure 12.4
Transformation of points on the unit circle (green circle) via array multiplication by a 2×2 associative array (blue parallelogram), resulting in a new set of points (black dots). The title of each plot lists the ⊕ and ⊗ operator used in the array multiplication. The array values $\begin{bmatrix} \cos(\pi/4) & -\sin(\pi/4) \\ \sin(\pi/4) & \cos(\pi/4) \end{bmatrix}$ correspond to a rotation transformation in the +.× semiring.

min.+ rotation. The main similarity between these rotation transformations and their identities is that both have a part of the unit circle that connects the diagonal corners of the blue square.

In the max.min and min.max semirings, the identity and the rotation transformations have little similarity

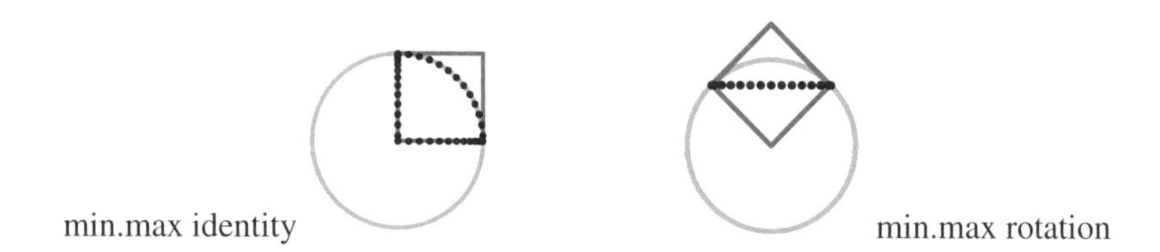

The max.min rotation transforms the unit circle into a triangle connecting three points on the unit circle. The min.max rotation projects the points onto a straight line. One similarity that all the identities and rotations share in the five +.×, max.+, min.+, max.min, and min.max is that they all connect diagonal points of their respective squares shown in blue.

The max.× and min.× operation pairs can only be semirings in the positive quadrant (I). They have little similarity to their respective identity transformations in that each projects the unit circle onto a half circle with handle shape. As with max.+ and min.+, the half circles with handle shape are $\pi/2$ rotations of each other

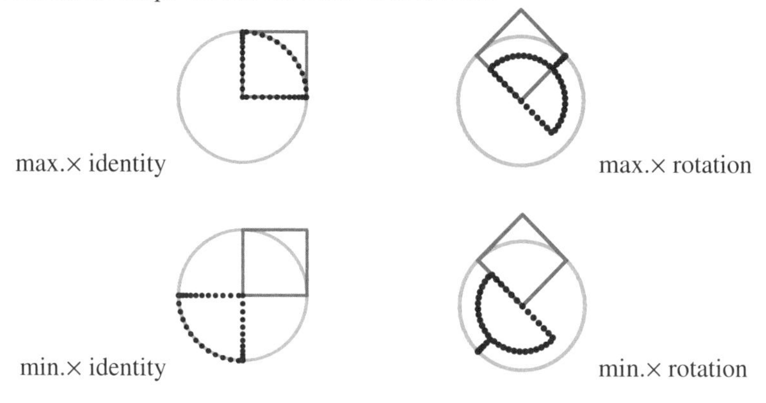

12.6 Conclusions, Exercises, and References

The transformation of the values of an array is a key capability of array multiplication. Using different combinations of ⊕ and ⊗ operations enables a wide variety of transformations. The intuition of these transformations can be visualized by looking at various 2×2 arrays and the ways they change the unit circle. Identity, contraction, dilation, stretching, and rotation are common transformations that can be composed to build more complicated transformations. Perhaps the most impressive aspect of these diverse transformations is that they maintain the useful properties of semirings such as associativity, distributivity, and identities. The differences among semirings such as +.×, max.+, min.+, max.min, min.max, max.×, and min.× when performing array multiplication motivate the need to revisit other properties of matrices, such as eigenvalues and eigenvectors, in the context of these semirings.

Exercises

Exercise 12.1 — Sketch the result of multiplying the $\mathbb{I}_{\max.+}$ identity array given in Section 12.2 by the unit circle over the following pairs of operations

(a) $+.\times$

(b) max.+

(c) min.+

(d) max.min

(e) min.max

(f) max.$\times$

(g) min.$\times$

Exercise 12.2 — Sketch the result of multiplying the $\mathbb{I}_{\min.+}$ identity array given in Section 12.2 by the unit circle over the following pairs of operations

(a) $+.\times$

(b) max.+

(c) min.+

(d) max.min

(e) min.max

(f) max.$\times$

(g) min.$\times$

Exercise 12.3 — Sketch the result of multiplying the $\mathbb{I}_{\max.\min}$ identity array given in Section 12.2 by the unit circle over the following pairs of operations

(a) $+.\times$

(b) max.+

(c) min.+

(d) max.min

(e) min.max

(f) max.$\times$

(g) min.$\times$

Exercise 12.4 — Sketch the result of multiplying the $\mathbb{I}_{\text{min.max}}$ identity array given in Section 12.2 by the unit circle over the following pairs of operations

(a) +.×

(b) max.+

(c) min.+

(d) max.min

(e) min.max

(f) max.×

(g) min.×

Exercise 12.5 — Sketch the result of multiplying the $\mathbb{I}_{\text{max.}\times}$ identity array given in Section 12.2 by the unit circle over the following pairs of operations

(a) +.×

(b) max.+

(c) min.+

(d) max.min

(e) min.max

(f) max.×

(g) min.×

Exercise 12.6 — Sketch the result of multiplying the $\mathbb{I}_{\text{min.}\times}$ identity array given in Section 12.2 by the unit circle over the following pairs of operations

(a) +.×

(b) max.+

(c) min.+

(d) max.min

(e) min.max

(f) max.×

(g) min.×

References

[1] R. A. Cuninghame-Green, *Minimax Algebra*, vol. 166. New York: Springer Science & Business Media, 2012.

[2] M. Akian, G. Cohen, S. Gaubert, R. Nikoukhah, and J. P. Quadrat, "Linear systems in (max,+) algebra," in *Proceedings of the 29th IEEE Conference on Decision and Control*, pp. 151–156, IEEE, 1990.

[3] P. Butkovič, "Strong regularity of matrices–A survey of results," *Discrete Applied Mathematics*, vol. 48, no. 1, pp. 45–68, 1994.

[4] K. Cechlárová and J. Plávka, "Linear independence in bottleneck algebras," *Fuzzy Sets and Systems*, vol. 77, no. 3, pp. 337–348, 1996.

[5] S. Gaubert and M. Plus, "Methods and applications of (max,+) linear algebra," in *Annual Symposium on Theoretical Aspects of Computer Science (STACS)*, pp. 261–282, Berlin, Heidelberg: Springer, 1997.

[6] M. Gondran and M. Minoux, "Dioïds and semirings: Links to fuzzy sets and other applications," *Fuzzy Sets and Systems*, vol. 158, no. 12, pp. 1273–1294, 2007.

[7] W. B. Person and B. Crawford Jr., "A geometric visualization of normal-coordinate transformations. Application to the calculation of bond-moment parameters and force constants," *The Journal of Chemical Physics*, vol. 26, no. 5, pp. 1295–1301, 1957.

[8] T. Kanade and J. R. Kender, *Mapping Image Properties into Shape Constraints: Skewed Symmetry, Affine-Transformable Patterns, and the Shape-from-Texture Paradigm.* Carnegie-Mellon University, Department of Computer Science, 1980.

[9] J. F. Blinn, "How to solve a cubic equation, part 1: The shape of the discriminant," *IEEE Computer Graphics and Applications*, vol. 26, no. 3, pp. 84–93, 2006.

[10] "Mathable: Matrices and geometry." http://www.mathable.io/courseware/matrices-geometry. Accessed: 2017-04-08.

13 Maps and Bases

Summary

The diverse data transformations that are possible with associative arrays can be composed into powerful analytics by using the linear properties of associativity, commutativity, and distributivity inherited from semirings. Proving these linear properties for associative arrays requires building up the corresponding mathematical objects, beginning with semimodules and linear maps, from which the foundational concept of a base can be proven. The existence of bases sets the stage for proving the additional properties of associative array algebra. This chapter defines the mathematical objects used in the remaining chapters to prove the linear properties of associative arrays.

13.1 Semimodules

The calculations of the previous chapter show the diversity of transformations that are possible with simple 2×2 array multiplication using different pairs of operations. Enumerating all the transformations exhibited by a given associative array is an extensive undertaking. The matrix multiplication properties of max.min and min.max algebras (sometimes referred to as fuzzy or bottleneck algebras) have been explored by many researchers. These explorations include linear independence [1–3], matrix periodicity [4], image sets [5], and a variety of applications [6–8]. Likewise, the properties of max.+ and min.+ algebras (sometimes referred to as tropical algebras) have also been explored (see survey in [9]). Examples of these explorations include systems of linear systems [10], spectral inequalities [11], categorical equivalences [12], and a wide range of applications [13].

The role of associative array multiplication in associative array algebra can be understood with concepts similar to matrix multiplication in linear algebra. The first step toward achieving this is to extend semirings into objects similar to vector spaces. Throughout this construction process, examples will be given using well-known sets of values

$\mathbb{N}$ the natural numbers $\{0, 1, 2, \ldots\}$

$\mathbb{Z}$ the integers $\{\ldots, \text{-}1, 0, 1, \ldots\}$

$\mathbb{Q}$ the rational numbers $\{m/n \mid n \neq 0, m, n \in \mathbb{N}\}$

$\mathbb{R}$ the real numbers $\{x \mid \text{-}\infty < x < \infty\}$

$\mathbb{C}$ the complex numbers $\{x + y\sqrt{\text{-}1} \mid x, y \in \mathbb{R}\}$

The concept of a semiring over a set of values V is extended to a semimodule by adding a scalar multiplication operation, which naturally leads to semimodules that are $m \times n$ arrays of elements in V that are denoted $V^{m \times n}$. Associative array multiplication transforms one semimodule to another semimodule. Such a transformation is a linear mapping between semimodules. Linear maps over semimodules thus provide the key mathematical foundation for understanding associative array multiplication. The definition of a semimodule over a semiring is as follows.

Definition 13.1

Semimodule

[14, p. 149] A *semimodule* M over a semiring V, or a V-*semimodule*, is a quadruple $(M, \oplus, \otimes, \mathbf{0})$ where M is a set, $\oplus$ a binary *vector addition* operation

$$\oplus : M \times M \to M$$

the binary *scalar multiplication* operation $\otimes$ is

$$\otimes : V \times M \to M$$

and $\mathbf{0}$ is an element of M. For scalar $c \in V$ and semimodule set member $\mathbf{v} \in M$, scalar multiplication $c \otimes \mathbf{v}$ is typically denoted by $c\mathbf{v}$ unless $\otimes$ must be explicitly named. To be a V-*semimodule* then for any scalars

$$a, b, c \in V$$

and semimodule set members

$$\mathbf{u}, \mathbf{v}, \mathbf{w} \in M$$

the quadruple must satisfy the following axioms

1. $(M, \oplus, \mathbf{0})$ is a commutative monoid
2. 1 is the scalar identity $1\mathbf{v} = \mathbf{v}$
3. Scalar multiplication distributes over monoid addition

$$c(\mathbf{u} \oplus \mathbf{v}) = (c\mathbf{u}) \oplus (c\mathbf{v})$$

4. Scalar multiplication distributes over semiring addition

$$(a \oplus b)\mathbf{v} = (a\mathbf{v}) \oplus (b\mathbf{v})$$

5. Scalar multiplication is compatible with semiring multiplication

$$a(b\mathbf{v}) = (a \otimes b)\mathbf{v}$$

A semimodule can easily be extended to a module or to the commonly used linear algebra concept of a vector space. If V is a ring, then call M a *module* over V or a V*-module*, and if V is a field, then call M a *vector space over* V or a V*-vector space*.

Example 13.1

A semiring $(V,\oplus,\otimes,0,1)$ is itself a V-semimodule $(M,\oplus,\otimes,\mathbf{0})$, where vector addition is the semiring addition and scalar multiplication is the semiring multiplication.

Example 13.2

If V is a semiring, then V^n is a semimodule over V where $\oplus$ is defined by component-wise addition in V

$$\mathbf{w} = \mathbf{u} \oplus \mathbf{v} \quad \text{where} \quad \mathbf{w}(k) = \mathbf{v}(k) \oplus \mathbf{w}(k)$$

scalar multiplication is component-wise multiplication in V

$$\mathbf{w} = c\mathbf{v} \quad \text{where} \quad \mathbf{w}(k) = c\mathbf{v}(k)$$

and the identity is the tuple $\mathbb{0} = (0,\ldots,0)$.

Example 13.3

If V is a semiring and K_1,K_2 are key sets, then the set $\mathbb{A} = \mathbb{A}(K_1,K_2;V)$ of all two-dimensional associative arrays $\mathbf{A} : K_1 \times K_2 \to V$ is a semimodule over V. Vector addition is element-wise addition, and scalar multiplication is array scalar multiplication.
More discussion about this structure and further structure on $\mathbb{A}$ is explored in Section 13.6.

The previous two examples can be generalized considerably: Suppose K is a product $K_1 \times \cdots \times K_d$ of key sets. Then the set $V^{\boxplus K}$ of all functions $K \to V$ with finite support, called the *direct sum of* K*-copies of* V, forms a semimodule over V in a similar way to V^n and $\mathbb{A}(K_1,K_2;V)$.

Recall that $\mathbf{v} : K \to V$ has finite support when $\mathbf{v}(k) \neq 0$ for a finite number of values of $k \in K$. With $K = K_1 \times \cdots \times K_d$, these are precisely the d-dimensional associative arrays over V.

The definitions of $\oplus$ and scalar multiplication are exactly as in Example 13.2. In other words, given $\mathbf{u}, \mathbf{v}, \mathbf{w} \in V^{\boxplus K}$, define

$$\mathbf{w} = \mathbf{u} \oplus \mathbf{v} \quad \text{where} \quad \mathbf{w}(k) = \mathbf{u}(k) \oplus \mathbf{v}(k)$$
$$\mathbf{w} = c\mathbf{v} \quad \text{where} \quad \mathbf{w}(k) = c\mathbf{v}(k)$$

and the identity is the constant function $\mathbb{0}(k) = 0$.

This construction works when K is *any* set, not just one of the form $K = K_1 \times \cdots \times K_d$. There are examples of semimodules that do not arise as $V^{\boxplus K}$ for some set K.

Example 13.4

The same definitions of addition and scalar multiplication used in the construction of $V^{\boxplus K}$ can be extended to the set V^K of *all* functions from K into V, making this set into a semimodule over V.

Example 13.5

The set $\mathbb{Z}/n\mathbb{Z} = \{0, \ldots, n-1\}$ is a module over $\mathbb{Z}$ where vector addition is addition modulo n, the vector addition identity is 0, and the scalar multiple $m \otimes v$ is defined to be the unique element of $\mathbb{Z}/n\mathbb{Z}$ congruent to mv modulo n.

Example 13.6

$\mathbb{R}$ is a $\mathbb{Q}$-vector space, with vector addition given by the standard addition on $\mathbb{R}$, the vector addition identity is 0, and scalar multiplication is the standard multiplication on $\mathbb{R}$.

Example 13.7

$\mathbb{Q}$ is a $\mathbb{Z}$-module, with vector addition given by the standard addition on $\mathbb{Q}$, the vector identity 0, and scalar multiplication is the standard multiplication on $\mathbb{Q}$.

Example 13.8

A commutative monoid $(M, \oplus, \mathbf{0})$ can be regarded as an $\mathbb{N}$-semimodule by taking vector addition to be the monoid operation $\oplus$, vector addition identity as $\mathbf{0}$, and scalar multiplication defined by setting

$$n\mathbf{v} = \underbrace{\mathbf{v} \oplus \cdots \oplus \mathbf{v}}_{n \text{ times}}$$

for n a natural number and $\mathbf{v} \in M$.

Example 13.9

Suppose $(M, \oplus, \mathbf{0})$ is a commutative group, so that for every $\mathbf{v} \in M$ there is an inverse $-\mathbf{v}$. Then Example 13.8 shows how to consider it as a $\mathbb{N}$-semimodule. It can be further considered a $\mathbb{Z}$-module by setting

$$(-n)\mathbf{v} = -(n\mathbf{v})$$

for n a non-negative integer and $\mathbf{v} \in M$.

13.2 Linear Maps

Array multiplication transforms one array into another. In more general terms, the transformation from one semimodule to another is a map. Of particular interest is the concept of a linear map that distributes over addition and is homogeneous with respect to scalar multiplication.

Definition 13.2

Linear Map

A *linear map* between two V-semimodules M and M' is a map

$$f : M \to M'$$

satisfying

$$f(c\mathbf{u} \oplus \mathbf{v}) = cf(\mathbf{u}) \oplus f(\mathbf{v})$$

or equivalently

$$f(\mathbf{u} \oplus \mathbf{v}) = f(\mathbf{u}) \oplus f(\mathbf{v}) \quad \text{and} \quad f(c\mathbf{u}) = cf(\mathbf{u})$$

for every scalar $c \in V$ and $\mathbf{u}, \mathbf{v} \in M$.

A bijective linear map is a *linear isomorphism*, or simply an *isomorphism*.

Example 13.10

Consider V^n and $V^{1 \times n}$; there is a linear isomorphism $f : V^n \to V^{1 \times n}$ sending $\mathbf{v}$ to

$$\begin{array}{cc} & \begin{array}{ccc} 1 & \cdots & n \end{array} \\ 1 & \begin{bmatrix} \mathbf{v}(1) & \cdots & \mathbf{v}(n) \end{bmatrix} \end{array}$$

and the two V-semimodules are identified with one another.

Example 13.11

The maps

$$f, f' : V^{\mathbb{N}} \to V^{\mathbb{N}}$$

defined by

$$f(v_0, v_1, \ldots) = (v_1, v_2, \ldots)$$

and

$$f'(v_0, v_1, \ldots) = (0, v_0, v_1, \ldots)$$

are linear maps.

Matrices are perhaps the most significant linear mapping for understanding array multiplication. Given an $m \times n$ matrix $\mathbf{A}$ with entries in V, such a matrix defines a linear map

$$f : V^n \to V^m$$

by sending $\mathbf{v} \in V^n$ to $\mathbf{A}\mathbf{v} \in V^m$.

In fact, all linear maps from V^n into V^m are of this form. Let $f : V^n \to V^m$ be a linear map, and denote by $\mathbf{e}_k$ the m-element vector defined by $\mathbf{e}_k(j) = 1$ if $k = j$ and 0 otherwise. The $m \times n$ matrix $\mathbf{A}$ can be formed by

$$\mathbf{A} = \begin{array}{c} \\ \end{array}\begin{matrix} 1 & 2 & \cdots & n \\ \end{matrix}$$
$$\mathbf{A} = \left[\begin{matrix} f(\mathbf{e}_1) & f(\mathbf{e}_2) & \cdots & f(\mathbf{e}_n) \end{matrix} \right]$$

Then given any n-element vector

$$\mathbf{v} = (\mathbf{v}(1), \ldots, \mathbf{v}(n)) \in V^n$$

it must be that the linear map

$$f(\mathbf{v}) = \mathbf{A}\mathbf{v}$$

for

$$\mathbf{v} = \mathbf{v}(1)\mathbf{e}_1 \oplus \cdots \oplus \mathbf{v}(n)\mathbf{e}_n$$

and so by linearity

$$f(\mathbf{v}) = \mathbf{v}(1) f(\mathbf{e}_1) \oplus \cdots \oplus \mathbf{v}(n) f(\mathbf{e}_n) = \mathbf{A}\mathbf{v}$$

More generally, this same idea shows that a linear map from V^n is uniquely determined by its values on the vectors $\mathbf{e}_i$. This is also true for maps from $V^{\boxplus K}$, where $\mathbf{e}_k(j) = 1$ if $k = j$

and 0 otherwise for each $k \in K$. In fact, assigning *any* values to the vectors $\mathbf{e}_i$ uniquely determines a linear map.

Proposition 13.1

Uniqueness and Existence of Extension of Map from Standard Basis

Let $K = K_1 \times \cdots \times K_d$, V be a semiring, and M be any V-semimodule. Then given any function

$$f : \{\mathbf{e}_k \mid k \in K\} \to M$$

there exists a unique linear map

$$f' : V^{\boxplus K} \to M$$

such that $f'(\mathbf{e}_k) = f(\mathbf{e}_k)$ for each $k \in K$.

Proof. See Exercise 13.4. □

13.3 Linear Independence and Bases

As seen in Proposition 13.1, linear maps are particularly nice for dealing with semimodules of the form $V^{\boxplus K}$ for some K. So it is worth questioning what the essential properties of $V^{\boxplus K}$ are that make Proposition 13.1 possible, such as when the size of K is unique. For example, can V^2 and V^1 be linearly isomorphic?

Definition 13.3

Linear Independence and Dependence

Let M be a V-semimodule and U a subset of M.
U is *linearly independent* if for all

$$\mathbf{v}_1, \ldots, \mathbf{v}_n \in U \quad \text{and} \quad u_1, \ldots, u_n, w_1, \ldots, w_n \in V$$

then

$$\bigoplus_{k=1}^{n} u_k \mathbf{v}_k = \bigoplus_{k=1}^{n} w_k \mathbf{v}_k$$

implies $u_k = w_k$ for each k. If a subset U is not linearly independent, then it is *linearly dependent*.

Definition 13.4

Span

Span(U) is the set of all elements of the form

$$\bigoplus_{k=1}^{n} u_k \mathbf{v}_k$$

where

$$\mathbf{v}_1, \ldots, \mathbf{v}_n \in U \quad \text{and} \quad u_1, \ldots, u_n \in V$$

is called the *(linear) span of* U.

Definition 13.5

Generating Set, Basis

U is a *generating set* of M if

$$\text{Span}(U) = M$$

A subset $B \subset M$ is a *basis* if B is linearly independent and a generating set.

Proposition 13.2

Equivalent Definitions of Linear Dependence

Suppose M is a V-semimodule and $U \subset M$. Consider the conditions

(i) U is linearly dependent

(ii) there exist $\mathbf{v}_1, \ldots, \mathbf{v}_n \in U$ and $u_1, \ldots, u_n \in V$ not all zero such that

$$\bigoplus_{k=1}^{n} u_k \mathbf{v} = \mathbf{0}$$

(iii) there exists $\mathbf{v} \in U$ such that

$$\mathbf{v} \in \text{Span}(U \setminus \{\mathbf{v}\})$$

If M is a module, then (i) and (ii) are equivalent. If M is a vector space, then all of the above are equivalent.

Proof. See Exercise 13.6. □

Example 13.12

For $V^{\boxplus K}$, the set $\{\mathbf{e}_k \mid k \in K\}$ forms a basis.

Example 13.13

For $V^{n\times m}$, the set of the matrices $\mathbf{e}_{i,j}$ with ℓ,k-th entry equal to 1 if $i=\ell$ and $j=k$, and 0 otherwise, forms a basis.

Example 13.14

The set $\{(1,\text{-}1),(0,1)\}$ is a basis for $\mathbb{R}^2$.

Example 13.15

Consider $\mathbb{Z}$ as a module over itself. Then $\{2,3\}$ is linearly dependent since $3\times 2 + \text{-}2\times 3 = 0$.
Note that neither 2 nor 3 can be written as a multiple of the other, so it is not linearly dependent in the other sense [15], showing that the equivalence of Proposition 13.2(ii) and (iii) need not hold in an arbitrary module.

The above example shows that the definition of linear dependence in a semimodule is not out of mere convenience, but out of necessity to properly capture the notion. The equivalence in the case of vector spaces is an important fact used to show that every vector space has a basis. See [16] for further discussion.

The existence of a basis can be interpreted as the existence of a linear isomorphism with a V-semimodule of the form $V^{\boxplus K}$.

Proposition 13.3

Existence of Basis Equivalent to Freeness

Suppose M is a V-semimodule and that M has a basis. Then for some set K, there exists a linear isomorphism

$$\varphi : V^{\boxplus K} \to M$$

Conversely, if there exists a linear isomorphism

$$\varphi : V^{\boxplus K} \to M$$

then M has a basis.

Proof. See Exercise 13.9. □

The above proposition allows Proposition 13.1 to be extended to semimodules which have bases.

Corollary 13.4

Uniqueness and Existence of Linear Extension of Map of Basis

Suppose M is a semimodule and B is a basis of M. Then any function

$$f : B \to M'$$

uniquely determines a linear map

$$f' : M \to M'$$

Proof. See Exercise 13.10. □

13.4 Existence of Bases

The first question to ask regarding semimodules and bases is whether a basis need exist for each semimodule. Ultimately, the answer is negative, though in the case of vector spaces it can be proven that bases do exist.

Proposition 13.5

Vector Spaces Have Bases

Let M be a $\mathbb{F}$-vector space. Then there exists a basis.

Proof. See [17, p. 518] for the details. The proof makes use of Zorn's Lemma, which states

> "If $(P, \leq)$ is a partially ordered set in which every linearly ordered subset $Q \subset P$ has an upper bound, then there exists a maximal element of P."

which is equivalent to the Axiom of Choice.

Proposition 13.2 implies that in a vector space, a subset U is linearly independent if there does not exist $\mathbf{v} \in U$ such that $\mathbf{v} \in \text{Span}(U \setminus \{\mathbf{v}\})$. As a result, a basis of a vector space is the same as a maximal linearly ordered set.

Let P be the set of linearly independent subsets of M and $\leq$ be $\subset$. If $Q \subset P$ is linearly ordered by $\subset$, then $\bigcup Q$ is a linearly independent subset of M, showing that Q has an upper bound. Thus, Zorn's Lemma implies that there is a maximal element of P, which will be a basis of M. □

After moving beyond the case of vector spaces, however, there are no longer any guarantees that a basis exists.

Example 13.16

$\mathbb{Q}$ forms a $\mathbb{Z}$-module, but has no basis. To see this, note that given two nonzero rational numbers $r = \frac{n}{m}$ and $s = \frac{p}{q}$, there exist nonzero integers x, y such that $xr + ys = 0$, including $x = mp$ and $y = -qn$ as

$$mp\frac{n}{m} + -qn\frac{p}{q} = np - np = 0$$

Thus, the only linearly independent sets are those containing only a single element, and the span of $\{r\}$ cannot be equal to all $\mathbb{Q}$ (for example, it does not contain $r/2$).

The above example illustrates that the existence of a basis is a nontrivial property. Here, the primary interest lies in semimodules for which a basis exists.

Definition 13.6

Free Semimodule

A semimodule for which a basis exists is known as a *free semimodule*.

13.5 Size of Bases

The second question to ask concerns the size of bases, and in particular whether a unique size exists. With this discussion of size, precision is needed about when set U has greater size than set V or when set U and set V have the same size.

Definition 13.7

Surjections, Injections, and Bijections

A function

$$f : U \to V$$

is

- an *injection* if $f(u) = f(u')$ implies $u = u'$,
- a *surjection* if for every $v \in V$ there is $u \in U$ such that $f(u) = v$, and
- a *bijection* if it is both an injection and a surjection, or equivalently f has an inverse.

Definition 13.8

Cardinality

Given two sets U, V, U has *smaller size* or *smaller cardinality*, written

$$|U| < |V|$$

if there does not exist a surjection $f : U \to V$. If there is a surjection $f : U \to V$, or equivalently an injection $g : V \to U$, then this is written

$$|U| \geq |V|$$

If there is a bijection $f : U \to V$, then this is written

$$|U| = |V|$$

and U and V are said to have *equal size* or *equal cardinality*.

To understand why these definitions are what they are, it is best to examine the definitions in the context of finite sets: counting the elements of a finite set is ultimately the same thing as establishing a bijection with a set of the form $\{1, \ldots, n\}$ for some natural number n.

Thus, the question of whether the bases of a semimodule (given any exist to begin with) have a unique size is meant to ask if for every two bases, there exists a bijection between them as sets (so no algebraic structure is being considered here). It ends up that the question only becomes ambiguous when there are finite bases.

Theorem 13.6

Uniqueness of Size of Infinite Bases

Let M be a free V-semimodule, and let B be an infinite basis of M. Then no set U that has size smaller than B is a basis of M.

Proof. Because B is a basis, each element of U is a (finite) linear combination of elements of B. Let B' be the subset of B that consists of all those elements of B that show up in a linear combination representation of an element of U.

Now, if U is infinite, then it ends up that B' has the same size as U so that $|U| = |B'|$. If U is finite, B' may still be larger than U, but still finite, and so still smaller in size than B. In either case, $|B'| < |B|$. This condition guarantees that $\text{Span}(B')$ is a proper subset of $\text{Span}(B) = M$, and in particular it means that B' cannot generate all of M. Finally, every finite linear combination of elements in U, by rewriting them as linear combinations of elements in B', can be rewritten as a linear combination of elements in B'. Thus, $\text{Span}(U) \subset \text{Span}(B')$, and so $\text{Span}(U)$ is unequal to M as well, showing that U cannot be a basis. □

Corollary 13.7

Bases Are Either All Finite or All Infinite

Let M be a V-semimodule. If there is a finite basis of M, then every basis is finite.

Proof. See Exercise 13.12. □

Say that a semimodule M is *finitely generated* if there exists a finite basis of M. Corollary 13.7 then says that the only remaining ambiguity concerning the size of such finitely generated free semimodules is whether it is possible that V^n is linearly isomorphic to V^m when $n \neq m$. The following example shows that the ambiguity can exist.

Example 13.17

Let V be any ring, and define V' to be the set of all V-matrices whose entries are labeled by $\mathbb{N} \times \mathbb{N}$ (so it is an infinite matrix) and whose columns each contain only finitely many nonzero entries. For example, one such element is

$$\begin{array}{c} \\ 0 \\ 1 \\ 2 \\ 3 \\ \vdots \\ n \\ \vdots \end{array} \begin{array}{c} \begin{array}{cccccc} 0 & 1 & 2 & \cdots & n-1 & \cdots \end{array} \\ \begin{bmatrix} 1 & 2 & 3 & \cdots & n & \cdots \\ 0 & 2 & 3 & \cdots & n & \cdots \\ 0 & 0 & 3 & \cdots & n & \cdots \\ 0 & 0 & 0 & \cdots & n & \cdots \\ \vdots & \vdots & \vdots & \ddots & \vdots & \ddots \\ 0 & 0 & 0 & \cdots & 0 & \cdots \\ \vdots & \vdots & \vdots & \vdots & \vdots & \vdots \end{bmatrix} \end{array}$$

With this condition on the number of nonzero entries in each column, the product of such matrices is well-defined, producing a ring. Then there is a left V-module isomorphism $f : V' \to V'^2$ given by

$$f : \mathbf{A} \mapsto (\text{odd columns of } \mathbf{A}, \text{even columns of } \mathbf{A})$$

so that the V'-module V' is isomorphic to the V'-module V'^2. V', as a V'-module, has a basis given by the identity $\mathbb{I}$, whereas V'^2, as a V'-module, has a basis given by

$$\{(\mathbb{I}, \mathbb{O}), (\mathbb{O}, \mathbb{I})\}$$

But then $\{f(\mathbb{I})\}$ is also a basis for V'^2, and so the sizes of bases of V'^2 are unequal.

Commutativity of the multiplication operation is enough to remove this ambiguity. The key property from which commutativity is used is in the following result.

Lemma 13.8

Square Matrix Left Invertible if and only if Right Invertible

[18] Let $\mathbf{A}$ and $\mathbf{B}$ be two square $n \times n$ matrices defined over a commutative semiring. Then

$$\mathbf{AB} = \mathbb{I} \quad \text{if and only if} \quad \mathbf{BA} = \mathbb{I}$$

With this fact in mind, the traditional proof for proving the result for commutative modules goes through.

Proposition 13.9

Commutative Semirings Have Unique Size of Bases

[17, p. 416] Let V be a commutative semiring and let M be a V-semimodule. If B and B' are two finite bases for M, then they have equal size.

Proof. Let

$$B = \{\mathbf{v}_1, \ldots, \mathbf{v}_n\} \quad \text{and} \quad B' = \{\mathbf{w}_1, \ldots, \mathbf{w}_m\}$$

Then for each $k \in \{1, \ldots, m\}$, there are

$$\mathbf{A}(k, 1), \ldots, \mathbf{A}(k, n)$$

such that

$$\mathbf{w}_k = \bigoplus_{j=1}^{n} \mathbf{A}(k, j)\mathbf{v}_j$$

The above sum defines a matrix $\mathbf{A}$, which sends

$$(\mathbf{v}_1, \ldots, \mathbf{v}_n)^\intercal \mapsto (\mathbf{w}_1, \ldots, \mathbf{w}_m)^\intercal$$

Likewise, there is the matrix $\mathbf{B}$ which sends

$$(\mathbf{w}_1, \ldots, \mathbf{w}_m)^\intercal \mapsto (\mathbf{v}_1, \ldots, \mathbf{v}_n)^\intercal$$

and so $\mathbf{BA}$ sends $(\mathbf{v}_1, \ldots, \mathbf{v}_n)^\intercal$ to itself. Since $\{\mathbf{v}_1, \ldots, \mathbf{v}_n\}$ is a basis and multiplication by $\mathbf{I}$ gives the same action on that basis, by Corollary 13.4 it follows that

$$\mathbf{BA} = \mathbb{I}_n$$

and likewise

$$\mathbf{AB} = \mathbb{I}_m$$

But if $\mathbf{A}$ is any invertible matrix over V, then it must be square. Let $\mathbf{B}$ be the inverse of $\mathbf{A}$, and suppose $\mathbf{A}$ has size $m \times n$ and $\mathbf{B}$ has size $n \times m$. Suppose without loss of generality that $m \geq n$. Then if $n \neq m$, add columns of zeroes to $\mathbf{A}$ on the right-hand side and rows of zeroes to $\mathbf{B}$ at the bottom, giving new matrices $\mathbf{A}'$ and $\mathbf{B}'$, which are square and also satisfy

$$\mathbf{A}'\mathbf{B}' = \mathbb{I}_m$$

By Lemma 13.9, applied over a commutative semiring, it is the case that

$$\mathbf{A}'\mathbf{B}' = \mathbb{I}_m \quad \text{if and only if} \quad \mathbf{B}'\mathbf{A}' = \mathbb{I}_m$$

But it can be directly checked that the m,m-th entry of $\mathbf{B}'\mathbf{A}'$ is 0, giving a contradiction. As such, $n = m$. □

When the size of a basis is unique in a finitely generated semimodule, call that size the *rank* of the semimodule. In vector spaces, the term *dimension* is instead used.

13.6 Semialgebras and the Algebra of Arrays

Some semimodules admit a vector multiplication, which is compatible with the vector addition and scalar multiplication operators. The addition of a vector multiplication gives rise to the notion of a semialgebra.

Definition 13.9

Semialgebra

[14, p. 53] A *semialgebra* M over a semiring V is a quintuple

$$(M, \oplus, \otimes, \mathbf{0}, *)$$

where

$$(M, \oplus, \otimes, \mathbf{0})$$

is a semimodule over V and

$$* : M \times M \to M$$

is *bilinear*, so for any $\mathbf{v}, \mathbf{v}', \mathbf{v}'' \in M$ and $u, w \in V$

$$(u\mathbf{v} \oplus w\mathbf{v}') * \mathbf{v}'' = u(\mathbf{v} * \mathbf{v}'') \oplus w(\mathbf{v}' * \mathbf{v}'')$$
$$\mathbf{v}'' * (u\mathbf{v} \oplus w\mathbf{v}') = u(\mathbf{v}'' * \mathbf{v}) \oplus w(\mathbf{v}'' * \mathbf{v}')$$

Definition 13.10

A semialgebra

$$(M, \oplus, \otimes, \mathbf{0}, *)$$

is an *associative semialgebra* or a *commutative semialgebra* if $*$ is associative or commutative, respectively.
If there exists an identity element for $*$ in M, then M is a *unital semialgebra*. Note that a unital associative semialgebra is also a semiring in its own right.

Recall that

$$\mathbb{A} = \mathbb{A}(K_1, K_2; V)$$

is the set of all associative arrays

$$\mathbf{A} : K_1 \times K_2 \to V$$

over a fixed semiring V and K_1, K_2 fixed finite sets. Furthermore, on $\mathbb{A}$, element-wise addition, element-wise multiplication, and scalar multiplication (by elements of V) are defined. When $K_1 = K_2 = K$, array multiplication is also defined on $\mathbb{A}$.

Proposition 13.10

$\mathbb{A}$ is a Unital Semialgebra under Element-Wise Multiplication

Suppose V is a semiring and K a set. Then the quintuple

$$(\mathbb{A}(K, K; V), \oplus, \otimes, \mathbb{0}, \otimes, \mathbb{1})$$

is an associative unital semialgebra where $\oplus$ is element-wise addition, the first use of $\otimes$ is scalar multiplication, $\mathbb{0}$ is the zero array, the second use of $\otimes$ is element-wise multiplication, and $\mathbb{1}$ is the associative array which is 1 everywhere.

Proposition 13.11

$\mathbb{A}$ is a Unital Semialgebra under Array Multiplication

Suppose V is a semiring and K_1, K_2 sets. Then the quintuple

$$(\mathbb{A}(K_1, K_2; V), \oplus, \otimes, \mathbb{0}, \oplus.\otimes, \mathbb{I})$$

is an associative unital semialgebra where $\oplus$ is element-wise addition, $\otimes$ is scalar multiplication, $\mathbb{0}$ is the zero array, $\oplus.\otimes$ is array multiplication, and $\mathbb{I}$ is the identity array.

The proofs of these propositions only involve confirming the semialgebra axioms.

Example 13.18

Consider the case where $K = \{1,\ldots,n\}$, V is any commutative semiring, and $d = 1$. In this case, a square d-dimensional associative array corresponds to an n-tuple of elements in V, which are regarded as column vectors. Then the semialgebra structure defined in Proposition 13.10 $\mathbb{A}(K,K;V)$ is an associative semialgebra. Moreover, an identity

$$\mathbb{1} = (1,\ldots,1)$$

exists, and $\mathbb{A}(K,K;V)$ is isomorphic to the product semiring

$$V^n = \prod_{k=1}^{n} V$$

It should be noted that having associative arrays over infinite key sets naturally requires the relaxation of the finite support condition to allow, for example, identities for element-wise multiplication and array multiplication. However, the following results can be recovered.

Proposition 13.12

Every Array is Contained in a Finite-Dimensional Sub-Semimodule of $\mathbb{A}$

Let a set of associative arrays be given by

$$\{\mathbf{A}_1,\ldots,\mathbf{A}_n\} \subset \mathbb{A}$$

Then there exists a sub-semialgebra M of $\mathbb{A}$ containing $\{\mathbf{A}_1,\ldots,\mathbf{A}_n\}$ that is the span of some finite collection of unit arrays

$$\{\mathbf{e}_{(k_1,k_2)} \in \mathbb{A} \mid (k_1,k_2) \in I \subset K_1 \times K_2\}$$

where I is finite. That is, every element of M can be written in the form

$$\bigoplus_{(k_1,k_2)\in I} v_{(k_1,k_2)}\mathbf{e}_{(k_1,k_2)}$$

where $v_{(k_1,k_2)} \in V$ for each $(k_1,k_2) \in I$.

Proof. Consider the finite set

$$I = \bigcup_{k=1}^{n} \mathrm{supp}(\mathbf{A}_k)$$

and note that the sub-semimodule

$$M = \mathrm{Span}\{\mathbf{e}_{(k_1,k_2)} \mid (k_1,k_2) \in I\}$$

of $\mathbb{A}$ contains each $\mathbf{A}_k$, as

$$\mathbf{A}_k = \bigoplus_{(k_1,k_2)\in\mathrm{supp}(\mathbf{A}_k)} \mathbf{A}_k(k_1,k_2)\mathbf{e}_{(k_1,k_2)}$$

It only remains to see that M is closed under element-wise multiplication. This follows from the fact that

$$\mathrm{supp}(\mathbf{A}\otimes\mathbf{B}) \subset \mathrm{supp}(\mathbf{A}) \cap \mathrm{supp}(\mathbf{B})$$

Thus, M is a semialgebra. □

Corollary 13.13

$\mathrm{Span}\{\mathbf{e}_{(k_1,k_2)} \mid (k_1,k_2) \in I\}$ is a Unital Semialgebra When I is Finite

M is, in fact, a unital semialgebra, with multiplicative unit

$$\bigoplus_{\mathbf{k}\in I} \mathbf{e}_\mathbf{k}$$

These above theorems and corollaries signify that to understand the properties of the space of associative arrays spanned by a finite number of associative arrays, it suffices to use the theory of free semimodules over the underlying semiring. This result is comparable to the theory of vector spaces, but without the requirement that the values be in a field. In other words, "vector spaces" where the values do not require additive and multiplicative inverses.

13.7 Conclusions, Exercises, and References

Building up the algebra of associative arrays begins with the construction of an equivalent semimodule construct. Using direct sums of semimodules, it is possible define linear maps that can be used to build bases. The provable existence of these bases provides the foundation of the formal algebra of associative arrays from which many additional useful matrix-like properties can be derived.

Exercises

Exercise 13.1 — How does the set $V^{n\times m}$ of all $n\times m$ matrices with entries in the semiring V arise as an example of $V^{\boxplus K}$? In other words, what should K be?

Exercise 13.2 — Show that the set V^K of *all* functions $K \to V$ is strictly larger than $V^{\boxplus K}$ when K is infinite. What about when K is finite?

Exercise 13.3 — What are the benefits to considering $V^{\boxplus K}$ instead of the (possibly larger) semimodule V^K?

Exercise 13.4 — Prove Proposition 13.1.

Exercise 13.5 — Explicity define the unique linear map $f' : \mathbb{R}^{2\times 1} \to \mathbb{R}$ extending $f : \{\mathbf{e}_1, \mathbf{e}_2\} \to \mathbb{R}$ in Proposition 13.1 where f is defined by $f(\mathbf{e}_1) = 1$ and $f(\mathbf{e}_2) = -2$. In other words, give a matrix which defines this linear map.

Exercise 13.6 — Prove Proposition 13.2.

Exercise 13.7 — Show that if V is a ring and M is a module over V, then $U \subset M$ is linearly independent if and only if for every $\{\mathbf{v}_1, \ldots, \mathbf{v}_n\} \subset U$ the equality $\bigoplus_{k=1}^{n} u_k \mathbf{v}_k = \mathbf{0}$ implies $u_k = 0$ for each k.

Exercise 13.8 — Show that the two vectors $\begin{array}{c} \\ 1 \\ 2 \end{array}\!\!\begin{array}{c} 1 \\ \left[\begin{array}{c} 1 \\ -1 \end{array}\right] \end{array}, \begin{array}{c} \\ 1 \\ 2 \end{array}\!\!\begin{array}{c} 1 \\ \left[\begin{array}{c} 2 \\ 0 \end{array}\right] \end{array}$ in $\mathbb{R}^{2\times 1}$ (being considered as a vector space over $\mathbb{R}$) are linearly independent.

Exercise 13.9 — Prove Proposition 13.3.

Exercise 13.10 — Prove Corollary 13.4.

Exercise 13.11 — Is $\mathbb{Z}/n\mathbb{Z}$ a free semimodule over $\mathbb{Z}$? What about $\mathbb{R}$ as a semimodule over $\mathbb{Q}$?

Exercise 13.12 — Prove Corollary 13.7.

References

[1] P. Butkovič, K. Cechlárová, and P. Szabó, "Strong linear independence in bottleneck algebra," *Linear Algebra and its Applications*, vol. 94, pp. 133–155, 1987.

[2] J. Plavka, "Linear independences in bottleneck algebra and their coherences with matroids," *Acta Math. Univ. Comenianae*, vol. 64, no. 2, pp. 265–271, 1995.

[3] K. Cechlárová and J. Plávka, "Linear independence in bottleneck algebras," *Fuzzy Sets and Systems*, vol. 77, no. 3, pp. 337–348, 1996.

[4] M. Gavalec, "Computing matrix period in max-min algebra," *Discrete Applied Mathematics*, vol. 75, no. 1, pp. 63–70, 1997.

[5] M. Gavalec and J. Plavka, "Simple image set of linear mappings in a max–min algebra," *Discrete Applied Mathematics*, vol. 155, no. 5, pp. 611–622, 2007.

[6] R. A. Cuninghame-Green, *Minimax Algebra*, vol. 166. New York: Springer Science & Business Media, 2012.

[7] J. Gunawardena, "Min-max functions," *Discrete Event Dynamic Systems*, vol. 4, no. 4, pp. 377–407, 1994.

[8] R. A. Cuninghame-Green, "Minimax algebra and applications," *Advances in Imaging and Electron Physics*, vol. 90, pp. 1–121, 1994.

[9] S. Gaubert and M. Plus, "Methods and applications of (max,+) linear algebra," in *Annual Symposium on Theoretical Aspects of Computer Science (STACS)*, pp. 261–282, Berlin, Heidelberg: Springer, 1997.

[10] M. Akian, G. Cohen, S. Gaubert, R. Nikoukhah, and J. P. Quadrat, "Linear systems in (max,+) algebra," in *Proceedings of the 29th IEEE Conference on Decision and Control*, pp. 151–156, IEEE, 1990.

[11] R. Bapat, D. P. Stanford, and P. Van den Driessche, "Pattern properties and spectral inequalities in max algebra," *SIAM Journal on Matrix Analysis and Applications*, vol. 16, no. 3, pp. 964–976, 1995.

[12] A. Di Nola and C. Russo, "Semiring and semimodule issues in MV-algebras," *Communications in Algebra*, vol. 41, no. 3, pp. 1017–1048, 2013.

[13] S. Gaubert, "Performance evaluation of (max,+) automata," *IEEE Transactions on Automatic Control*, vol. 40, no. 12, pp. 2014–2025, 1995.

[14] J. S. Golan, *Semirings and Their Applications*. New York: Springer Science & Business Media, 2013.

[15] S. Roman, *Advanced Linear Algebra*, vol. 3. New York: Springer-Verlag, 2005.

[16] Y.-J. Tan, "Bases in semimodules over commutative semirings," *Linear Algebra and Its Applications*, vol. 443, pp. 139–152, 2014.

[17] M. Artin, *Algebra*. New York: Pearson, 2010.

[18] C. Reutenauer and H. Straubing, "Inversion of matrices over a commutative semiring," *Journal of Algebra*, vol. 88, no. 2, pp. 350–360, 1984.

14 Linearity of Associative Arrays

Summary

The linear systems problem, which motivates much of elementary matrix mathematics, presents unique challenges in the setting of general semimodules. Thus, the focus of this chapter is on families of semimodules that include the semirings of interest described in previous chapters. Theorems dealing with the existence and uniqueness of solutions to systems over tropical, max-plus, and Boolean algebras are given. Among those results is a structure theorem for solutions to linear systems over *supremum-blank algebras*, which are defined and include max-plus, tropical algebras, and power set algebras.

14.1 The Null Space of Linear Maps

The linear systems properties of max.min and min.max algebras, also referred to as fuzzy or bottleneck algebras, have been extensively investigated. These investigations have looked at the solutions of equations [1], solvability conditions [2, 3], uniqueness of solutions (strong regularity) [4–6], uniqueness of solutions in discrete sets (discrete strong regularity) [7, 8], solution complexity [9], and regularity complexity [10].

While the null space plays a large role in the theory of linear systems over fields, it can be demonstrated here that, in the context of the semirings of interest, it is decoupled from the existence, uniqueness, or computability of solutions to linear systems. To begin the analysis requires placing additional constraints on the set of values. Specifically, it is convenient to examine sets of values that do not sum to zero, which are referred to as zero-sum-free semirings.

Definition 14.1

Zero-Sum-Free Semiring

[11, p. 1] A semiring over the set V is *zero-sum-free* if $u \oplus v = 0$ implies $u = v = 0$.

Example 14.1

Many of the semirings of interest, namely tropical algebras, power set algebras, and max-plus algebras, are all zero-sum-free.

Example 14.2

The semiring $(\mathbb{N},+,\times,0,1)$ is zero-sum-free.

Definition 14.2

Null space

Given a semiring V and a linear map

$$\mathbf{A}: M \to N$$

between semimodules M,N over V, the *null space* of $\mathbf{A}$ is the set

$$\{\mathbf{v} \in M \mid \mathbf{A}(\mathbf{v}) = \mathbf{0}\}$$

For a zero-sum-free semiring V, the following theorem provides a useful statement about the null space of any linear operator over V.

Theorem 14.1

Null Space Equivalence for Zero-Sum-Free Commutative Semirings

Let V be a zero-sum-free commutative semiring and let $\mathbf{A}$ be an $n \times m$ matrix. Then

$$\mathbf{A}\mathbf{v} = \mathbf{0}$$

if and only if

$$\mathbf{v}(i) \otimes \mathbf{A}(i,j) = 0$$

for all $i \in \{1,\ldots,m\}$ and $j \in \{1,\ldots,n\}$.

Proof. Recall that by definition

$$(\mathbf{A}\mathbf{v})(i) = \bigoplus_{j=1}^{m} (\mathbf{A}(i,j) \otimes \mathbf{v}(j)) = 0$$

for each i. Thus, for any k,

$$(\mathbf{A}(k,j) \otimes \mathbf{v}(k)) \oplus \left(\bigoplus_{j=1}^{k-1} (\mathbf{A}(i,j) \otimes \mathbf{v}(j)) \oplus \bigoplus_{j=k+1}^{m} (\mathbf{A}(i,j) \otimes \mathbf{v}(j)) \right) = 0.$$

This implies that

$$\mathbf{A}(k,j) \otimes \mathbf{v}(k) = 0$$

by the zero-sum-free assumption on V. □

Corollary 14.2

Structure of Null Space for Zero-Sum-Free Commutative Semirings

Suppose $\mathbf{A}$ is an $n \times m$ matrix over a zero-sum-free commutative semiring V. Define $I_k \subset V$ by

$$I_j(\mathbf{A}) = \left\{ v \in V \;\middle|\; \left(\bigoplus_{i=1}^{m} \mathbf{A}(i, j) \right) \otimes v = 0 \right\}$$

Then the null space of $\mathbf{A}$ is

$$I_1(\mathbf{A}) \times I_2(\mathbf{A}) \times \cdots \times I_m(\mathbf{A})$$

Proof.

$$\mathbf{Av} = \mathbf{0}$$

if and only if

$$\mathbf{A}(i, j) \otimes \mathbf{v}(j) = 0$$

for every pair i, j. Because V is zero-sum-free, the above condition is true if and only if

$$0 = \bigoplus_{i=1}^{m} \mathbf{A}(i, j) \otimes \mathbf{v}(j) = \left(\bigoplus_{i=1}^{m} \mathbf{A}(i, j) \right) \otimes \mathbf{v}(j)$$

This is the same as saying that

$$\mathbf{v} \in I_1(\mathbf{A}) \times \cdots \times I_m(\mathbf{A})$$

Thus, the null space of $\mathbf{A}$ is equal to $I_1(\mathbf{A}) \times \cdots \times I_m(\mathbf{A})$. □

The latter corollary fully classifies the null space of an arbitrary linear map when V is zero-sum-free and provides an extra corollary when V has no zero divisors.

Corollary 14.3

Null Space in Zero-Sum-Free Commutative Semirings with No Zero Divisors

Suppose that V is a zero-sum-free commutative semiring with no zero divisors. Let $\mathbf{A}$ be an $n \times m$ matrix. Suppose that $\mathbf{A}$ has precisely $k \leq n$ rows with all 0 entries. Then

$$\{\mathbf{v} \in V^m \mid \mathbf{Av} = \mathbf{0}\}$$

is linearly isomorphic to V^k.

The above corollary is a helpful detail about the null space of maps in commutative semirings where the multiplicative monoid $\otimes$ has inverses and is in fact a group, such as in the max-plus algebra. In this case, there cannot exist any zero divisors, so Corollary 14.3 applies.

14.2 Supremum-Blank Algebras

Definition 14.3

Supremum-Blank Algebra

A *supremum-blank algebra* is a semiring

$$(V, \vee, \otimes, -\infty, 1)$$

such that

1. $V = [-\infty, \infty]$ is a complete lattice.
2. $u \vee v$ is the (binary) supremum of the lattice.
3. Suprema-preservation condition holds for any $v \in V$ and $U \subset V$

$$v \otimes \bigvee_{u \in U} u = \bigvee_{u \in U} (v \otimes u)$$

Note that in a supremum-blank algebra, sums of an arbitrary set of elements is possible. The name supremum-blank was selected because this definition places no limitations on the multiplication operation of the semirings in question, other than the suprema-preservation condition, so one can insert an arbitrary multiplication operation in the place of the word blank, provided it satisfies the suprema-preservation condition and makes $(V, \vee, \otimes, -\infty, 1)$ into a semiring.

Example 14.3

Each of the semirings of interest, including the tropical (max-min) algebra, the max-plus algebras with $\mathbb{R} \cup \{-\infty, \infty\}$ as the underlying set, and the power set algebras, are well-known examples of supremum-blank algebras; it is verified that the suprema-preservation condition holds for each of these semirings in the next subsection.

An important function in relation to a supremum-blank algebra $(V, \vee, \otimes, -\infty, 1)$ is the function

$$f_v(u) = v \otimes u$$

defined for each $v \in V$. The suprema-preservation condition implies that each function f_v is monotonic.

Lemma 14.4

Functions that Preserve Suprema or Infima are Monotonic

Let V be a complete lattice with a function

$$f : V \to V$$

which preserves suprema or preserves infima, meaning given a subset U of V that either has a supremum or an infimum, respectively, then

$$f\left(\bigvee_{v \in U} v\right) = \bigvee_{v \in U} f(v) \quad \text{or} \quad f\left(\bigwedge_{v \in U} v\right) = \bigwedge_{v \in U} f(v),$$

respectively. Then f is monotonic.

Proof. See Exercise 14.4. □

The above lemma will be indispensable in the coming results. Similarly, an *infimum-blank algebra* has the conditions

1. $V = [-\infty, \infty]$ is a complete lattice.
2. $a \wedge b$ is the (binary) infimum of the lattice.
3. Infima-preservation condition holds for any $v \in V$, the function

$$f_v : V \to V$$

defined by $u \mapsto v \otimes u$ preserves infima, in the sense that

$$v \otimes \bigwedge_{u \in U} u = \bigwedge_{u \in U} (v \otimes u)$$

for every subset $U \subset V$.

Solving linear systems in such an infimum-blank algebra is reduced to solving them in a supremum-blank algebra by taking the opposite lattice V^{op}, where the ordering on V^{op} is such that $a \leq b$ in V^{op} if and only if $a \geq b$ in V. Then infima become suprema, and V^{op} becomes a supremum-blank algebra. Because $(V^{\text{op}})^{\text{op}} = V$, solving linear systems in V^{op} reduces to solve linear systems in V.

To make use of the order theoretic properties of the supremum-blank algebras, an ordering must be put on V^n that works with the ordering on V and with the corresponding suprema and infima. This order is the *product order* on V^n where

$$\mathbf{u} = \begin{matrix} 1 \\ \vdots \\ n \end{matrix} \overset{1}{\begin{bmatrix} \mathbf{u}(1) \\ \vdots \\ \mathbf{u}(n) \end{bmatrix}} \leq \begin{matrix} 1 \\ \vdots \\ n \end{matrix} \overset{1}{\begin{bmatrix} \mathbf{v}(1) \\ \vdots \\ \mathbf{v}(n) \end{bmatrix}} = \mathbf{v} \quad \text{if and only if} \quad \mathbf{u}(i) \leq \mathbf{v}(i) \text{ for all } i \in \{1,\ldots,n\}$$

The above equation clearly captures the relationship between the product order and the order on V. The next proposition illuminates the connection between suprema and infima in V and in V^n with the product order.

Proposition 14.5

Product Order Inheritance of Order-Theoretic Properties

If V is a (bounded, distributive) lattice, then V^n is a (bounded, distributive) lattice in the product order. If V is closed under (non-empty) arbitrary suprema or infima, then V^n is closed under (non-empty) arbitrary suprema or infima, respectively. Moreover, $\bigvee_{\mathbf{v}\in U} \mathbf{v}$ exists if and only if $\bigvee_{\mathbf{v}\in U} \mathbf{v}(i)$ exists for each i, in which case

$$\left(\bigvee_{\mathbf{v}\in U} \mathbf{v}\right)(i) = \bigvee_{\mathbf{v}\in U} \mathbf{v}(i)$$

Likewise, $\bigwedge_{\mathbf{v}\in U} \mathbf{v}$ exists if and only if $\bigwedge_{\mathbf{v}\in U} \mathbf{v}(I)$ exists for each i, in which case

$$\left(\bigwedge_{\mathbf{v}\in U} \mathbf{v}\right)(i) = \bigwedge_{\mathbf{v}\in U} \mathbf{v}(i)$$

Proof. See Exercise 14.5. □

Now denote by

$$X(\mathbf{A}, \mathbf{w})$$

the *solution space* of $\mathbf{A}\mathbf{v} = \mathbf{w}$, the set of all solutions $\mathbf{v}$ to the equation

$$\mathbf{A}\mathbf{v} = \mathbf{w}$$

in other words

$$X(\mathbf{A}, \mathbf{w}) = \{\mathbf{v} \in V^m \mid \mathbf{A}\mathbf{v} = \mathbf{w}\}$$

where $\mathbf{A}$ is an $n \times m$ matrix and $\mathbf{w}$ is an element of V^n. The principal goal of the remainder of this section will be finding the structure of $X(\mathbf{A}, \mathbf{w})$, ultimately showing that it is closed under arbitrary non-empty suprema and convex, and hence a union of closed intervals in V^m with a common terminal point.

Proposition 14.6

Solution Space of Linear System is Closed under Arbitrary Non-Empty Suprema

Suppose V is a supremum-blank algebra, $\mathbf{A}$ an $n \times m$ matrix, and $\mathbf{w}$ an element of V^n, then

$$X(\mathbf{A}, \mathbf{w}) = \{\mathbf{v} \in V^m \mid \mathbf{A}\mathbf{v} = \mathbf{w}\}$$

is closed under arbitrary non-empty suprema.

Proof. Firstly, $X(\mathbf{A}, \mathbf{w})$ is a subset of V^m, so that $X(\mathbf{A}, \mathbf{w})$ is partially ordered. Additionally, because V is a complete lattice, by Proposition 14.5 it follows that V^n is also a complete lattice, and

$$\bigvee_{\mathbf{v} \in X(\mathbf{A}, \mathbf{w})} \mathbf{v}$$

is well-defined. It suffices to show that for every non-empty subset U of $X(\mathbf{A}, \mathbf{w})$

$$\bigvee_{\mathbf{v} \in U} \mathbf{v}$$

is a solution to

$$\mathbf{A}\mathbf{v} = \mathbf{w}$$

For each component

$$\left(\mathbf{A} \bigvee_{\mathbf{v} \in U} \mathbf{v}\right)(i) = \bigvee_{j=1}^{m} \left[\mathbf{A}(i,j) \otimes \bigvee_{\mathbf{v} \in U} \mathbf{v}(j)\right]$$

Because the functions f_v are suprema preserving and suprema commute with each other

$$\begin{aligned}
\left(\mathbf{A} \bigvee_{\mathbf{v} \in U} \mathbf{v}\right)(i) &= \bigvee_{j=1}^{m} \left[\mathbf{A}(i,j) \otimes \bigvee_{\mathbf{v} \in U} \mathbf{v}(j)\right] \\
&= \bigvee_{j=1}^{m} \left[\bigvee_{\mathbf{v} \in U} (\mathbf{A}(i,j) \otimes \mathbf{v})\right] \\
&= \bigvee_{\mathbf{v} \in U} \left[\bigvee_{j=1}^{m} (\mathbf{A}(i,j) \otimes \mathbf{v})\right] \\
&= \bigvee_{\mathbf{v} \in U} \mathbf{w} = \mathbf{w}
\end{aligned}$$

□

Since a supremum-blank algebra is a complete lattice satisfying the suprema-preservation condition, the above proposition clearly holds. The maximum solution to the equation $\mathbf{A}\mathbf{v} = \mathbf{w}$ is denoted $x(\mathbf{A}, \mathbf{w})$; such a maximum exists (*if* any solution exists) because $X(\mathbf{A}, \mathbf{w})$ is closed under arbitrary non-empty suprema and thus is bounded. There is a second order-theoretic property of $X(\mathbf{A}, \mathbf{w})$ of note.

Proposition 14.7

Solution Space of Linear System is Convex

Suppose $\mathbf{v}_1, \mathbf{v}_3$ are elements of $X(\mathbf{A}, \mathbf{w})$, and

$$\mathbf{v}_1 < \mathbf{v}_2 < \mathbf{v}_3$$

then $\mathbf{v}_2$ is an element of $X(\mathbf{A}, \mathbf{w})$.

Proof. The Suprema-Preservation Condition implies that each function f_v is monotonic. Consequently,

$$\mathbf{A}(i,j) \otimes \mathbf{v}_1(j) \leq \mathbf{A}(i,j) \otimes \mathbf{v}_2(j) \leq \mathbf{A}(i,j) \otimes \mathbf{v}_3(j)$$

for every i, j. Because suprema respect order, it follows that

$$\mathbf{w}(i) = \bigvee_{j=1}^{m} (\mathbf{A}(i,j) \otimes \mathbf{v}_1(j)) \leq \bigvee_{j=1}^{m} (\mathbf{A}(i,j) \otimes \mathbf{v}_2(j)) \leq \bigvee_{j=1}^{m} (\mathbf{A}(i,j) \otimes \mathbf{v}_3(j)) = \mathbf{w}(i)$$

for every $i \in \{1, \ldots, n\}$, so that

$$(\mathbf{A}\mathbf{v}_2)(i) = \mathbf{w}(i)$$

for every $i \in \{1, \ldots, n\}$, and thus

$$\mathbf{v}_2 \in X(\mathbf{A}, \mathbf{w})$$

□

Theorem 14.8

Structure Theorem for Supremum-Blank Algebras

Let V be a supremum-blank algebra. Then $X(\mathbf{A}, \mathbf{w})$ is the union of closed intervals all having the same terminal point so that there exists a subset U of V^m such that

$$X(\mathbf{A}, \mathbf{w}) = \bigcup_{\mathbf{v} \in U} [\mathbf{v}, \mathbf{x}]$$

where $\mathbf{x} = x(\mathbf{A}, \mathbf{w})$.

Proof. By Proposition 14.7, $X(\mathbf{A},\mathbf{w})$ is convex so if

$$\mathbf{v} \in X(\mathbf{A},\mathbf{w})$$

then the closed interval $[\mathbf{v},\mathbf{x}]$ is contained in $X(\mathbf{A},\mathbf{w})$. Hence

$$X(\mathbf{A},\mathbf{w}) = \bigcup_{\mathbf{v}\in X(\mathbf{A},\mathbf{w})} [\mathbf{v},\mathbf{x}]$$

□

The following theorem shows how the structures

$$X(\mathbf{A}(i,:),\mathbf{w}(i)) = \bigcup_{\mathbf{v}\in U_i} [\mathbf{v},\mathbf{x}_i]$$

of the solution spaces of the systems $\mathbf{A}(i,:)\mathbf{v} = \mathbf{w}(i)$ contribute to the structure

$$X(\mathbf{A},\mathbf{w}) = \bigcup_{\mathbf{v}\in U} [\mathbf{v},\mathbf{x}]$$

of the solution space of the system $\mathbf{A}\mathbf{v} = \mathbf{w}$.

Lemma 14.9

Switching Order of Union and Intersection

Let

$$\{U_{i,j} \mid i \in \{1,\ldots,m\}, j \in \{1,\ldots,n\}\}$$

be a collection of sets. Then

$$\bigcap_{j=1}^{n}\bigcup_{i=1}^{m} U_{i,j} = \bigcup_{1\le i_1,\ldots,i_j\ldots,i_n\le m}\ \bigcap_{j=1}^{n} U_{i_j,j}$$

Proof. See Exercise 14.7. □

Lemma 14.10

Intersection of Intervals is Interval

Let V be any lattice and I, J two intervals in V. Then $I \cap J$ is an interval. Moreover,

$$[v_1,w_1] \cap [v_2,w_2] = [v_1 \vee w_1, v_2 \wedge w_2]$$

Proof. See Exercise 14.8. □

Theorem 14.11

Intersection of Solution Spaces

Suppose V is a join-blank algebra, $\mathbf{A}$ is an $m \times n$ matrix V, and $\mathbf{w} \in V^m$. Write

$$X(\mathbf{A}(i,:),\mathbf{w}(i)) = \bigcup_{\mathbf{v} \in U_i} [\mathbf{v}, \mathbf{x}_i]$$

for each i. Then

$$X(\mathbf{A},\mathbf{w}) = \bigcup_{\mathbf{v}_1 \in U_1,\ldots,\mathbf{v}_m \in U_m} \left[\bigvee_{i=1}^{m} \mathbf{v}_i, \bigwedge_{i=1}^{m} \mathbf{x}_i \right]$$

Proof. See Exercise 14.9. □

Example 14.4

Consider the system

$$\begin{array}{c} \\ 1 \\ 2 \end{array} \begin{array}{c} \begin{array}{cc} 1 & 2 \end{array} \\ \begin{bmatrix} 0 & 0 \\ 0 & 1 \end{bmatrix} \end{array} \begin{array}{c} \\ 1 \\ 2 \end{array} \begin{array}{c} 1 \\ \begin{bmatrix} v_1 \\ v_2 \end{bmatrix} \end{array} = \begin{array}{c} \\ 1 \\ 2 \end{array} \begin{array}{c} 1 \\ \begin{bmatrix} 0 \\ 0 \end{bmatrix} \end{array}$$

over the max-min algebra. This system of equations leads to the conditions

$$\max(\min(v_1,0),\min(v_2,0)) = 0$$
$$\max(\min(v_1,0),\min(v_2,1)) = 0$$

The solution spaces of these individual equations are

$$([-\infty,\infty] \times [0,\infty]) \cup ([0,\infty] \times [-\infty,\infty])$$
$$([0,\infty] \times [-\infty,0]) \cup ([-\infty,\infty] \times [0,0])$$

respectively. The solution space $X(\mathbf{A},\mathbf{w})$ is the intersection of these two individual solution spaces

$$([-\infty,\infty] \times [0,0]) \cup ([0,\infty] \times [-\infty,0])$$

Using the interval notation from the product order on $V \times V$, the above union can be written instead as

$$X(\mathbf{A},\mathbf{w}) = \left[\begin{array}{c} \\ 1 \\ 2 \end{array} \begin{array}{c} 1 \\ \begin{bmatrix} -\infty \\ 0 \end{bmatrix} \end{array}, \begin{array}{c} \\ 1 \\ 2 \end{array} \begin{array}{c} 1 \\ \begin{bmatrix} \infty \\ 0 \end{bmatrix} \end{array} \right] \cup \left[\begin{array}{c} \\ 1 \\ 2 \end{array} \begin{array}{c} 1 \\ \begin{bmatrix} 0 \\ -\infty \end{bmatrix} \end{array}, \begin{array}{c} \\ 1 \\ 2 \end{array} \begin{array}{c} 1 \\ \begin{bmatrix} \infty \\ 0 \end{bmatrix} \end{array} \right]$$

Example 14.5

Consider the same system as in the prior example, but over the max-plus algebra. This system of equiations leads to the conditions

$$\max(v_1 + 0, v_2 + 0) = 0$$
$$\max(v_1 + 0, v_2 + 1) = 0$$

The solution spaces of these individual equations are

$$([0,0] \times [\text{-}\infty, 0]) \cup ([\text{-}\infty, 0] \times [0,0])$$
$$([0,0] \times [\text{-}\infty, \text{-}1]) \cup ([\text{-}\infty, 0] \times [\text{-}1, \text{-}1])$$

respectively. The solution space $X(\mathbf{A}, \mathbf{w})$ is the intersection of these two individual solution spaces

$$([0,0] \times [\text{-}\infty, \text{-}1]) \cup ([\text{-}\infty, 0] \times [\text{-}1, \text{-}1])$$

or

$$X(\mathbf{A}, \mathbf{w}) = \left[\begin{array}{c} \\ 1 \\ 2 \end{array} \begin{array}{c} 1 \\ \left[\begin{array}{c} 0 \\ \text{-}\infty \end{array} \right] \end{array} , \begin{array}{c} \\ 1 \\ 2 \end{array} \begin{array}{c} 1 \\ \left[\begin{array}{c} 0 \\ \text{-}1 \end{array} \right] \end{array} \right]$$

The above examples also indicate the slightly stronger result that the union can be made to be a finite union, at least when dealing with total orders. This isn't strictly correct, however, as the below examples show.

Example 14.6

Consider the system

$$\begin{array}{c} \\ 1 \end{array} \begin{array}{c} 1 \\ \left[\, \infty \, \right] \end{array} \begin{array}{c} \\ 1 \end{array} \begin{array}{c} 1 \\ \left[\, v \, \right] \end{array} = \begin{array}{c} \\ 1 \end{array} \begin{array}{c} 1 \\ \left[\, \infty \, \right] \end{array}$$

over the max-plus algebra, or more simply

$$\infty + v = \infty$$

The solution space is given by

$$(\text{-}\infty, \infty]$$

Example 14.7

Consider the system

$$\begin{array}{c} \\ 1 \\ 2 \end{array}\begin{array}{c} \begin{array}{cc} 1 & 2 \end{array} \\ \begin{bmatrix} \infty & \infty \\ \infty & \infty \end{bmatrix} \end{array} \begin{array}{c} \\ 1 \\ 2 \end{array}\begin{array}{c} 1 \\ \begin{bmatrix} v_1 \\ v_2 \end{bmatrix} \end{array} = \begin{array}{c} \\ 1 \\ 2 \end{array}\begin{array}{c} 1 \\ \begin{bmatrix} \infty \\ \infty \end{bmatrix} \end{array}$$

over the max-plus algebra, or more simply

$$\max(\infty + v_1, \infty + v_2) = \infty$$

The solution space is given by

$$((-\infty, \infty] \times [-\infty, \infty]) \cup ([-\infty, \infty] \times (-\infty, \infty])$$

14.3 Max-Blank Structure Theorem

One of the main properties that the max-min and max-plus algebras have that the more general supremum-blank algebras do not is that the underlying sets are totally ordered.

Definition 14.4

Max-Blank Algebra

A *max-blank algebra* is a supremum-blank algebra which is totally ordered.

Totally ordered supremum-blank algebras are so named because the binary supremum of a totally ordered set is the maximum operation.

The solution spaces in Example 14.6 and Example 14.7 cannot be written as finite unions of intervals in the product order. The key hypothesis that allows for the solution space to be written as a finite union of closed intervals is that the sets

$$f^{-1}_{\mathbf{A}(i,j)}(\mathbf{w}(i))$$

are closed intervals.

Example 14.8

In the max-min algebra

$$f^{-1}_{\mathbf{A}(i,j)}(\mathbf{w}(i)) = \begin{cases} \{\mathbf{w}(i)\} & \text{if } \mathbf{w}(i) \geq \mathbf{A}(i,j) \\ \emptyset & \text{otherwise} \end{cases}$$

Example 14.9

In the max-plus algebra $f^{-1}_{\mathbf{A}(i,j)}(\mathbf{w}(i))$ is determined by the following table.

		$\mathbf{w}(i)$		
		$-\infty$	$\mathbb{R}$	∞
	$-\infty$	$[-\infty,\infty]$	$\emptyset$	$\emptyset$
$\mathbf{A}(i,j)$	$\mathbb{R}$	$\{-\infty\}$	$\{\mathbf{w}(i)-\mathbf{A}(i,j)\}$	$\{\infty\}$
	∞	$\{-\infty\}$	$\emptyset$	$(-\infty,\infty]$

Theorem 14.12

Structure Theorem for Max-Blank Algebras

Suppose $\mathbf{A}$ is an $n\times m$ matrix and $\mathbf{w}$ is an element of V^n. Further suppose that for each i, j, the set

$$f^{-1}_{\mathbf{A}(i,j)}(\mathbf{w}(i)) = \{v \in V \mid \mathbf{A}(i,j)\otimes v = \mathbf{w}(i)\}$$

is a non-empty closed interval whenever it is non-empty.
Let

$$U_i = \{j \in \{1,\dots,m\} \mid f^{-1}_{\mathbf{A}(i,j)}(\mathbf{w}(i)) \neq \emptyset\}$$

For $j \in U_i$, write

$$f^{-1}_{\mathbf{A}(i,j)}(\mathbf{w}(i)) = [p^i_j, q^i_j]$$

For $j \notin U_i$, let q^i_j be the largest element such that $\mathbf{A}(i,j)\otimes q^i_j \leq \mathbf{w}(i)$. Write

$$\mathbf{p}_{i,j'}(j) = \begin{cases} p^i_{j'} & \text{if } j = j' \\ -\infty & \text{otherwise} \end{cases} \quad \text{and} \quad \mathbf{q}_i(j) = q^i_j$$

Then

$$X(\mathbf{A},\mathbf{w}) = \bigcup_{j'_1\in U_1,\dots,j'_n\in U_n} \left[\bigvee_{i=1}^{n} \mathbf{p}_{i,j'_i}, \bigwedge_{i=1}^{n} \mathbf{q}_i\right]$$

In particular, $X(\mathbf{A},\mathbf{w})$ is a finite union of closed intervals with common endpoint

$$x(\mathbf{A},\mathbf{w}) = \bigwedge_{i=1}^{n} \mathbf{q}_i$$

Proof. The proof will consist of finding the solution space of the equation

$$\max_{j\in\{1,\dots,m\}} (\mathbf{A}(i,j)\otimes \mathbf{v}(j)) = \mathbf{w}(i)$$

for each $i \in \{1,\ldots,n\}$ and showing that it is the union of closed intervals with a common terminal point. The intersection of these solution spaces is $X(\mathbf{A},\mathbf{w})$. Then Lemma 14.9 and Lemma 14.10 allow for writing the intersection in the desired form.

Let

$$U_i = \{j \in \{1,\ldots,m\} \mid f^{-1}_{\mathbf{A}(i,j)}(\mathbf{w}(i)) \neq \emptyset\}$$

For $j \in U_i$, write

$$f^{-1}_{\mathbf{A}(i,j)}(\mathbf{w}(i)) = [p^i_j, q^i_j]$$

For $j \notin U_i$ let q^i_j be the largest element such that $\mathbf{A}(i,j) \otimes q^i_j \leq \mathbf{w}(i)$. Such an element exists since $-\infty \otimes v = -\infty$ for any $v \in V$. Define

$$\mathbf{p}_{i,j'}(j) = \begin{cases} p^i_{j'} & \text{if } j = j' \\ -\infty & \text{otherwise} \end{cases}$$

and

$$\mathbf{q}_i(j) = q^i_j$$

Now the solution space to

$$\max_{j \in \{1,\ldots,m\}} (\mathbf{A}(i,j) \otimes \mathbf{v}(j)) = \mathbf{w}(i) \tag{14.1}$$

can be found. A given $\mathbf{v}$ satisfies Equation 14.1 if and only if there exists a $j' \in \{1,\ldots,m\}$ such that

$$\mathbf{A}(i,j') \otimes \mathbf{v}(j') = \mathbf{w}(i)$$

and for all $j \in \{1,\ldots,m\}$

$$\mathbf{A}(i,j) \otimes \mathbf{v}(j) \leq \mathbf{w}(i)$$

The first condition is equivalent to

$$\mathbf{v}(j') \in [p^i_{j'}, q^i_{j'}]$$

and the second condition is equivalent to

$$\mathbf{v}(j) \in [-\infty, q^i_j]$$

because multiplication by a fixed element is a monotonic function. Then the solution space of Equation 14.1 can be written as

$$\bigcup_{j' \in U_i} \left([p^i_{j'}, q^i_{j'}] \times \prod_{j \in \{1,\ldots,m\}, j \neq j'} [-\infty, q^i_j] \right) = \bigcup_{j' \in U_i} [\mathbf{p}_{i,j'}, \mathbf{q}_i]$$

Then using Lemma 14.9 and Lemma 14.10 gives

$$\begin{aligned} X(\mathbf{A},\mathbf{w}) &= \bigcap_{i=1}^{n} \bigcup_{j' \in U_i} [\mathbf{p}_{i,j'}, \mathbf{q}_i] \\ &= \bigcup_{j'_1 \in U_1, \ldots, j'_n \in U_n} \bigcap_{i=1}^{n} [\mathbf{p}_{i,j'_i}, \mathbf{q}_i] \\ &= \bigcup_{j'_1 \in U_1, \ldots, j'_n \in U_n} \left[\bigvee_{i=1}^{n} \mathbf{p}_{i,j'_1}, \bigwedge_{i=1}^{n} \mathbf{q}_i \right] \end{aligned}$$

which is precisely a finite union of closed intervals with the common terminal point

$$\bigwedge_{i=1}^{n} \mathbf{q}_i = x(\mathbf{A},\mathbf{w})$$

□

The above proof both verifies the existence of a maximum solution, if any exists at all, and provides an algorithm by which the solution set may be computed, so long as the sets $f_r^{-1}(s)$ can be computed. It also provides the following corollary that tells precisely when a solution exists.

Corollary 14.13

Existence of Solution in Max-Blank Algebra

Suppose $\mathbf{A}$ is an $n \times m$ matrix and $\mathbf{w}$ an element of V^m satisfying the conditions of Theorem 14.12. Then there exists a solution to

$$\mathbf{A}\mathbf{v} = \mathbf{w}$$

if and only if

(i) for each i, j, there exists q such that

$$\mathbf{A}(i,j) \otimes q \leq \mathbf{w}(i) \tag{14.2}$$

(ii) if q^i_j is the largest such q satisfying Equation 14.2, at least one of the inequalities

$$\mathbf{A}(i,j) \otimes q^i_j \leq \mathbf{w}(i)$$

for each $i \in \{1,\ldots,n\}$, and

(iii) $\bigwedge_{i=1}^{n} \mathbf{q}_i$ is a solution to $\mathbf{A}\mathbf{v} = \mathbf{w}$ where $\mathbf{q}_i(j) = q^i_j$.

14.4 Examples of Supremum-Blank Algebras

In this section, the fact that the semirings of interest are supremum-blank algebras is verified, and hence are amenable to the methods and results of the last section.

$\mathbb{R} \cup \{-\infty, +\infty\}$ is a complete linearly ordered set with binary supremum given by max. Thus, to show that the max-plus algebra is a supremum-blank algebra, + must satisfy the suprema-preservation condition.

Lemma 14.14

+ Satisfies the Suprema-Preservation Condition

Suppose U is a subset of $\mathbb{R} \cup \{-\infty, \infty\}$ and $v \in \mathbb{R} \cup \{-\infty, \infty\}$. Then

$$v + \sup_{u \in U} u = \sup_{u \in U} (v + u) \tag{14.3}$$

Proof. First note that if $v = -\infty$, then Equation 14.3 holds true since $-\infty$ is an annihilator for +.

Now suppose $v = \infty$. If U contains an element of $\mathbb{R} \cup \{\infty\}$, then Equation 14.3 holds true since ∞ is an annihilator for + on $\mathbb{R} \cup \{\infty\}$. Otherwise, U is either empty or only contains $-\infty$, in which case both sides of Equation 14.3 are equal to $-\infty$.

Thus, assume $v \in \mathbb{R}$. If U is unbounded above in $\mathbb{R}$ or contains ∞, then the set

$$\{u + v \mid u \in U \text{ and } v \in V\}$$

is also unbounded above or contains ∞, so Equation 14.3 becomes $v + \infty = \infty$, which is true since $v \neq -\infty$ by hypothesis.

If U is bounded above in $\mathbb{R}$, let $u^* = \sup U$. Then

$$v + u \leq v + u^*$$

for all $u \in U$, so $v + u^*$ is an upper bound of $\{v + u \mid u \in U\}$.

To see that it is the upper bound, suppose for the sake of contradiction that there is w such that

$$v + u \leq w < v + u^*$$

for all $u \in U$. There exists $\epsilon > 0$ such that

$$w + \epsilon < v + u^*$$

Because u^* is the least upper bound of U, there exists $u' \in U$ such that

$$u^* - \epsilon < u' \leq u^*$$

and hence

$$v + u^* - \epsilon < v + u' \leq v + u^*$$

But then

$$w < v + u'$$

contradicting the hypothesis that w was an upper bound of $\{v + u \mid u \in U\}$. Thus $v + u^*$ is the least upper bound of $\{v + u \mid u \in U\}$ and Equation 14.3 holds. □

That the max-min algebra and Boolean algebras are supremum-blank algebras follows from a more general fact concerning when the bounded complete lattice

$$(V, \vee, \wedge, \text{-}\infty, \infty)$$

is a supremum-blank algebra, or namely when $\wedge$ satisfies the suprema-preservation condition

$$v \wedge \bigvee_{u \in U} u = \bigvee_{u \in U} (v \wedge u)$$

Definition 14.5

Heyting Algebra

[12, p. 5] A *Heyting algebra* is a quintuple

$$(V, \vee, \wedge, \Rightarrow, 0, 1)$$

such that

$$(V, \vee, \wedge, 0, 1)$$

is a bounded distributive lattice and for each u, v in V

$$w = u \Rightarrow v$$

is the greatest element satisfying

$$u \wedge w \leq v$$

Proposition 14.15

Heyting Algebra and the Supremum-Preservation Condition

Suppose V is a bounded complete lattice. Then

$$v \wedge \bigvee_{u \in U} u = \bigvee_{u \in U} (v \wedge u)$$

for every $v \in V$ and $U \subset V$ if and only if V is a Heyting algebra.

Proof. First suppose that V is a Heyting algebra. Then there is the following sequence of equivalences

$$
\begin{aligned}
\left(\bigvee_{u \in U} u\right) \wedge v \leq w \quad & \text{if and only if} \quad \bigvee_{u \in U} u \leq v \Rightarrow w \\
& \text{if and only if} \quad \forall u \in U\,(u \leq v \Rightarrow w) \\
& \text{if and only if} \quad \forall u \in U\,(u \wedge v \leq w) \\
& \text{if and only if} \quad \bigvee_{u \in U} (u \wedge v) \leq w
\end{aligned}
$$

Now, letting

$$w = \left(\bigvee_{u \in U} u\right) \wedge v$$

shows that

$$v \wedge \bigvee_{u \in U} u \geq \bigvee_{u \in U} (v \wedge u)$$

and letting

$$w = \bigvee_{u \in U} (u \wedge v)$$

shows that

$$v \wedge \bigvee_{u \in U} u \leq \bigvee_{u \in U} (v \wedge u)$$

from which the result follows. Now suppose that the suprema-preservation condition holds. Suppose $v, u \in V$. Let U be the set of all elements w of V such that

$$v \wedge w \leq u$$

Then define

$$v \Rightarrow u = \bigvee_{w \in U} w$$

To see that this fulfills the necessary condition, if

$$v \wedge w' \leq u$$

then

$$w' \leq \bigvee_{w \in U} w = v \Rightarrow u$$

and if

$$w' \leq v \Rightarrow u$$

then

$$
\begin{aligned}
v \wedge w' &\leq v \wedge \bigvee_{w \in U} w \\
&= \bigvee_{w \in U} (v \wedge w) \\
&\leq \bigvee_{w \in U} u \\
&= u
\end{aligned}
$$

□

Corollary 14.16

Heyting Algebra and Supremum-Blank Algebra

A bounded complete distributive lattice

$$(V, \vee, \wedge, -\infty, \infty)$$

is a supremum-blank algebra with multiplication as the binary infimum if and only if it is a complete Heyting algebra.

In order to apply the above corollary to the max-min algebra, it is first verified that it is a Heyting algebra.

Proposition 14.17

Max-Min Algebras are Heyting Algebras

A max-min algebra

$$(V, \max, \min, -\infty, \infty)$$

forms a Heyting algebra when

$$\Rightarrow : V \times V \to V$$

is defined such that

$$v \Rightarrow u = \begin{cases} u & \text{if } v > u \\ \infty & \text{otherwise} \end{cases}$$

otherwise.

Proof. See Exercise 14.11.

can be explicitly computed in certain cases for supremum-blank algebras. For Heyting algebras, including Boolean algebras, power set algebras, and max-min algebras, these solutions can be largely subsumed in the following theorem.

Theorem 14.20

Maximum Solution of Linear System in Heyting Algebras

[13, p. 845] Let

$$(V, \vee, \wedge, \Rightarrow, 0, 1)$$

be a Heyting algebra, and let $\mathbf{A}$ be an $n \times m$ matrix over V and $\mathbf{w} \in V^n$. If $X(\mathbf{A}, \mathbf{w})$ is non-empty, then the maximum element

$$\mathbf{x} = x(\mathbf{A}, \mathbf{w})$$

of $X(\mathbf{A}, \mathbf{w})$ is defined by

$$\mathbf{x}(j) = \bigwedge_{i=1}^{n} (\mathbf{A}(i, j) \Rightarrow \mathbf{w}(i))$$

Corollary 14.21

Maximum Solution of Linear System in Tropical Max-Min Algebras

Let

$$(V, \max, \min, -\infty, \infty)$$

be a max-min algebra, $\mathbf{A}$ an $n \times m$ matrix over V, and $\mathbf{w}$ an element of V^n. If $X(\mathbf{A}, \mathbf{w})$ is non-empty, then the maximum element

$$\mathbf{x} = x(\mathbf{A}, \mathbf{w})$$

of $X(\mathbf{A}, \mathbf{w})$ is defined by

$$\mathbf{x}(j) = \min_{1 \leq i \leq n} q_{i,j}$$

where

$$q_{i,j} = \begin{cases} \infty & \text{if} \quad \mathbf{A}(\mathrm{i,j}) \leq \mathbf{w}(\mathrm{i}) \\ \mathbf{w}(i) & \text{if} \quad \mathbf{A}(\mathrm{i,j}) > \mathbf{w}(\mathrm{i}) \end{cases}$$

Proof. The proof follows immediately from the above corollary since max-min algebras are Heyting algebras, but it can also be verified that this is the solution by combining Theorem 14.12 and Example 14.8.

Example 14.10

Consider the following subset of tracks in the Track column of the array in Figure 4.1

{Kill The Light, Christina, Junk, Sugar, G#, Japanese Eyes, Cut It Out}

This set can be made into a semiring by ordering the set as above from left to right (so Cut It Out is the greatest element) and by equipping it with the tropical max-min algebra. Then consider the system

$$\begin{array}{c} \\ 1 \\ 2 \end{array}\begin{array}{c} \begin{array}{cc} 1 & 2 \end{array} \\ \begin{bmatrix} \text{G\#} & \text{Kill The Light} \\ \text{Sugar} & \text{Sugar} \end{bmatrix} \end{array} \mathbf{v} = \begin{array}{c} \\ 1 \\ 2 \end{array}\begin{array}{c} 1 \\ \begin{bmatrix} \text{Christina} \\ \text{Junk} \end{bmatrix} \end{array}$$

and apply Corollary 14.21 to show that if a solution exists, then the maximum solution to the system is given by

$$x(\mathbf{A}, \mathbf{w}) = \begin{array}{c} \\ 1 \\ 2 \end{array}\begin{array}{c} 1 \\ \begin{bmatrix} \min(\text{Christina}, \text{Junk}) \\ \min(\text{Cut It Out}, \text{Junk}) \end{bmatrix} \end{array} = \begin{array}{c} \\ 1 \\ 2 \end{array}\begin{array}{c} 1 \\ \begin{bmatrix} \text{Christina} \\ \text{Junk} \end{bmatrix} \end{array}$$

This can be explicity checked to be an actual solution, so it is the maximum such solution.

Corollary 14.22

Maximum Solution of Linear System in Power Set Algebras

Let

$$(\mathcal{P}(V), \vee, \wedge, \emptyset, V)$$

be a power set algebra, $\mathbf{A}$ an $n \times m$ matrix over V, and $\mathbf{w}$ an element of

$$\mathcal{P}(V)^n$$

If $X(\mathbf{A}, \mathbf{w})$ is non-empty, then the maximum element

$$\mathbf{x} = x(\mathbf{A}, \mathbf{w})$$

of $X(\mathbf{A}, \mathbf{w})$ is defined by

$$\mathbf{x}(j) = \bigcap_{i=1}^{n} (\mathbf{A}(i, j)^{\mathsf{c}} \cup \mathbf{w}(i))$$

The following result deals with the remaining semiring of interest, the max-plus algebra.

Proposition 14.23

Maximum Solution of Linear System in Max-Plus Algebras

Suppose

$$(\mathbb{R} \cup \{-\infty, \infty\}, \max, +, -\infty, 0)$$

is the max-plus algebra , $\mathbf{A}$ is an $n \times m$ matrix, and $\mathbf{w}$ an element of $(\mathbb{R} \cup \{-\infty, \infty\})^n$. Then

$$\mathbf{x} = x(\mathbf{A}, \mathbf{w})$$

defined by

$$\mathbf{x}(j) = \min_{1 \leq i \leq n} (\mathbf{w}(i) - \mathbf{A}(i, j))$$

is the maximum solution of the linear system $\mathbf{A}\mathbf{v} = \mathbf{w}$ (if any solutions exist) when writing

$$\begin{aligned} -\infty - v &= -\infty && \text{for } v \in (-\infty, \infty] \\ -\infty - -\infty &= \infty && \\ \infty - v &= \infty && \text{for } v \in [-\infty, \infty] \end{aligned}$$

Proof. By Example 14.9, the maximum element q^i_j of $f^{-1}_{\mathbf{A}(i,j)}(\mathbf{w}(i))$ is given by

		$\mathbf{w}(i)$		
		$-\infty$	$\mathbb{R}$	∞
	$-\infty$	∞	none	none
$\mathbf{A}(i, j)$	$\mathbb{R}$	$-\infty$	$\mathbf{w}(i) - \mathbf{A}(i, j)$	∞
	∞	$-\infty$	$\emptyset$	∞

where Corollary 14.13 implies that the above covers all of the cases when a solution exists. By defining

$$\begin{aligned} -\infty - v &= -\infty && \text{for } v \in (-\infty, \infty] \\ -\infty - -\infty &= \infty && \\ \infty - v &= \infty && \text{for } v \in [-\infty, \infty] \end{aligned}$$

q^i_j can be written as

$$q^i_j = \mathbf{w}(i) - \mathbf{A}(i, j)$$

Then Theorem 14.12 implies that

$$\mathbf{x}(j) = \min_{1 \leq i \leq n} (\mathbf{w}(i) - \mathbf{A}(i,j))$$

gives the maximum solution.

(Note that the case where $\mathbf{A}(i,j) = \mathbf{w}(i) = \infty$ occurs is not strictly covered by Theorem 14.12, but its proof does extend to showing that $\bigwedge \mathbf{q}_i$ is the maximum solution, if any exists.) □

Example 14.11

The system

$$\begin{array}{c} \\ 1 \\ 2 \end{array} \begin{array}{c} \begin{array}{cc} 1 & 2 \end{array} \\ \begin{bmatrix} 0 & 0 \\ 0 & 1 \end{bmatrix} \end{array} \begin{array}{c} \\ 1 \\ 2 \end{array} \begin{array}{c} 1 \\ \begin{bmatrix} x \\ y \end{bmatrix} \end{array} = \begin{array}{c} \\ 1 \\ 2 \end{array} \begin{array}{c} 1 \\ \begin{bmatrix} 0 \\ 0 \end{bmatrix} \end{array}$$

solved in Example 14.5 was given by

$$X(\mathbf{A},\mathbf{w}) = \left[\begin{array}{c} \\ 1 \\ 2 \end{array} \begin{array}{c} 1 \\ \begin{bmatrix} 0 \\ -\infty \end{bmatrix} \end{array}, \begin{array}{c} \\ 1 \\ 2 \end{array} \begin{array}{c} 1 \\ \begin{bmatrix} 0 \\ -1 \end{bmatrix} \end{array} \right]$$

has a maximum solution (0,-1). This solution can be verified with Proposition 14.23, which states that $\mathbf{x} = x(\mathbf{A},\mathbf{w})$ has components given by

$$\mathbf{x}(j) = \min_{1 \leq i \leq n} (\mathbf{w}(i) - \mathbf{A}(i,j))$$

so that computing each of the quantities $\mathbf{w}(i) - \mathbf{A}(i,j)$ results in the quantities

$$\begin{aligned} \mathbf{w}(1) - \mathbf{A}(1,1) &= 0 \\ \mathbf{w}(1) - \mathbf{A}(1,2) &= 0 \\ \mathbf{w}(2) - \mathbf{A}(2,1) &= 0 \\ \mathbf{w}(2) - \mathbf{A}(2,2) &= -1 \end{aligned}$$

Thus

$$\mathbf{x} = \begin{array}{c} \\ 1 \\ 2 \end{array} \begin{array}{c} 1 \\ \begin{bmatrix} 0 \\ -1 \end{bmatrix} \end{array}$$

For power set algebras, uniqueness of solutions can also be determined by applying the following proposition.

Proposition 14.24

Uniqueness of Solution to Linear System in Power Set Algebras

Let $\mathbf{A}$ be an $n \times m$ matrix over some power set algebra $\mathcal{P}(V)$ and

$$\mathbf{w} \in \mathcal{P}(V)^n$$

For each

$$u \in x(\mathbf{A}, \mathbf{w})(i) \subset V$$

define $\mathbf{v}^{i,u}$ to have components

$$\mathbf{v}^{i,u}(j) = \begin{cases} x(\mathbf{A}, \mathbf{w})(j) \setminus \{u\} & \text{if } i = j \\ x(\mathbf{A}, \mathbf{w})(j) & \text{otherwise} \end{cases}$$

In short, $\mathbf{v}^{i,u}$ is the matrix that differs from $x(\mathbf{A}, \mathbf{w})$ by removing u from the i-th entry. Then $X(\mathbf{A}, \mathbf{w})$ has more than one element if and only if $\mathbf{v}^{i,u}$ is a solution for some $i \in \{1, \ldots, m\}$ and

$$u \in x(\mathbf{A}, \mathbf{w})(i)$$

Proof. If $\mathbf{v}^{i,u}$ is a solution for some $i \in \{1, \ldots, m\}$ and

$$u \in x(\mathbf{A}, \mathbf{w})(i)$$

then

$$\mathbf{A}\mathbf{v} = \mathbf{b}$$

has more than one solution since by construction

$$\mathbf{v}^{i,u} \neq x(\mathbf{A}, \mathbf{w})$$

for every $i \in \{1, \ldots, m\}$ and

$$u \in x(\mathbf{A}, \mathbf{w})(i)$$

Conversely, if there exists a solution

$$\mathbf{v} \neq x(\mathbf{A}, \mathbf{w})$$

to the system, then

$$\mathbf{v} < x(\mathbf{A}, \mathbf{w})$$

Thus $\mathbf{v}(k)$ is a proper subset of $x(\mathbf{A},\mathbf{w})(k)$ for some k. If an element

$$u \in x(\mathbf{A},\mathbf{w})(k) \setminus \mathbf{v}(k)$$

is picked, then

$$\mathbf{v} < \mathbf{v}^{k,u} < x(\mathbf{A},\mathbf{w})$$

By proposition 14.7 it follows that $\mathbf{v}^{k,u}$ is a solution. □

For any finite set V, it is easy to compute the (finite) set of vectors $\{\mathbf{v}^{i,u}\}$. Thus, Proposition 14.24 immediately yields a means of checking uniqueness of a solution.

14.6 Conclusions, Exercises, and References

The null space is the starting point of many results in standard matrix mathematics. However, tn the case of semirings, alternative approaches are needed. The concept of a supremum-blank algebra provides an alternative foundation for associative array algebra. The conditions for solutions to equations in supremum-blank algebras can then be described so as to cover a large number of the semirings of interest for associative arrays. These solutions lay the foundation for the discussion of eigenvalues and eigenvectors that are invaluable to the analysis of real data.

Exercises

Exercise 14.1 — Find the null space for the linear map $f : \mathcal{P}(\{0,1\})^{2\times 1} \to \mathcal{P}(\{0,1\})^{2\times 1}$ defined by

$$f\left(\begin{array}{cc} & 1 \\ 1 & \left[\begin{array}{c} a \\ b \end{array}\right] \\ 2 & \end{array}\right) = \begin{array}{c} \\ 1 \\ 2 \end{array}\begin{array}{c} \begin{array}{cc} 1 & 2 \end{array} \\ \left[\begin{array}{cc} \{1\} & \{0\} \\ \{0\} & \{1\} \end{array}\right] \end{array} \begin{array}{c} \\ 1 \\ 2 \end{array}\begin{array}{c} 1 \\ \left[\begin{array}{c} a \\ b \end{array}\right] \end{array}$$

Exercise 14.2 — Suppose M and N are semimodules over the semiring V, and that $f : M \to N$ is a linear map. Show that the null space of f is a subspace of M and the image of f is a subspace of N.

Exercise 14.3 — In the case where V is a ring, which allows subtraction, and M and N are modules over V, it makes sense to quotient M by the null space of a linear map $f : M \to N$ by first defining an equivalence relation $\sim$ on M such that $\mathbf{v} \sim \mathbf{w}$ if and only if $\mathbf{v} - \mathbf{w}$ is in the null space of f. Then the set of equivalence classes of M, denoted $M/\sim$, forms a module over V, and f is constant on each equivalence class so that if $\mathbf{v} \sim \mathbf{w}$, then $f(\mathbf{v}) = f(\mathbf{w})$, so f induces a linear map $f' : M/\sim \to N$. Show that this is an isomorphism of $M/\sim$ onto its image.

Is there a way to extend this idea to the case where V is only a semiring?

Exercise 14.4 — Prove Lemma 14.4.

Exercise 14.5 — Prove Proposition 14.5.

Exercise 14.6 — Find $X(\mathbf{A}, \mathbf{w})$ for

$$\mathbf{A} = \begin{array}{c} \\ 1 \\ 2 \end{array} \begin{array}{c} \begin{array}{cc} 1 & 2 \end{array} \\ \begin{bmatrix} \{0\} & \{1\} \\ \{1\} & \{0\} \end{bmatrix} \end{array} \quad \text{and} \quad \mathbf{w} = \begin{array}{c} \\ 1 \\ 2 \end{array} \begin{array}{c} 1 \\ \begin{bmatrix} \emptyset \\ \{0,1\} \end{bmatrix} \end{array}$$

over the semiring $\mathcal{P}(\{0,1\})$ and write it in the form given in Theorem 14.8.

Exercise 14.7 — Prove Lemma 14.9.

Exercise 14.8 — Prove Lemma 14.10.

Exercise 14.9 — Prove Theorem 14.11.

Exercise 14.10 — Use Theorem 14.12 to find $X(\mathbf{A}, \mathbf{w})$ for

$$\mathbf{A} = \begin{array}{c} \\ 1 \\ 2 \end{array} \begin{array}{c} \begin{array}{cc} 1 & 2 \end{array} \\ \begin{bmatrix} 4 & -\infty \\ -1 & 0 \end{bmatrix} \end{array} \quad \text{and} \quad \mathbf{w} = \begin{array}{c} \\ 1 \\ 2 \end{array} \begin{array}{c} 1 \\ \begin{bmatrix} 0 \\ 1 \end{bmatrix} \end{array}$$

over the (complete) max-plus algebra.

Exercise 14.11 — Prove Proposition 14.17.

Exercise 14.12 — Prove Proposition 14.18.

Exercise 14.13 — Prove Proposition 14.19.

Exercise 14.14 — Use Corollary 14.22 and Proposition 14.24 to determine if the system $\mathbf{A}\mathbf{v} = \mathbf{w}$ where

$$\mathbf{A} = \begin{array}{c} \\ 1 \\ 2 \end{array} \begin{array}{c} \begin{array}{cc} 1 & 2 \end{array} \\ \begin{bmatrix} \{0,1\} & \{1\} \\ \{0,1\} & \emptyset \end{bmatrix} \end{array} \quad \text{and} \quad \mathbf{w} = \begin{array}{c} \\ 1 \\ 2 \end{array} \begin{array}{c} 1 \\ \begin{bmatrix} \{1\} \\ \emptyset \end{bmatrix} \end{array}$$

has a unique solution in the power set algebra $\mathcal{P}(\{0,1\})$.

References

[1] E. Sanchez, *Solutions in Composite Fuzzy Relation Equations: Application to Medical Diagnosis in Brouwerian Logic*. Faculté de Médecine de Marseille, 1977.

[2] P. Butkovič, "Necessary solvability conditions of systems of linear extremal equations," *Discrete Applied Mathematics*, vol. 10, no. 1, pp. 19–26, 1985.

[3] K. Cechlárová, "Unique solvability of max-min fuzzy equations and strong regularity of matrices over fuzzy algebra," *Fuzzy Sets and Systems*, vol. 75, no. 2, pp. 165–177, 1995.

[4] P. Butkovič and F. Hevery, "A condition for the strong regularity of matrices in the minimax algebra," *Discrete Applied Mathematics*, vol. 11, no. 3, pp. 209–222, 1985.

[5] P. Butkovič, "Strong regularity of matrices–A survey of results," *Discrete Applied Mathematics*, vol. 48, no. 1, pp. 45–68, 1994.

[6] M. Gavalec and J. Plávka, "Strong regularity of matrices in general max–min algebra," *Linear Algebra and Its Applications*, vol. 371, pp. 241–254, 2003.

[7] K. Cechlárová, "Strong regularity of matrices in a discrete bottleneck algebra," *Linear Algebra and Its Applications*, vol. 128, pp. 35–50, 1990.

[8] K. Cechlárová and K. Kolesár, "Strong regularity of matrices in a discrete bounded bottleneck algebra," *Linear Algebra and Its Applications*, vol. 256, pp. 141–152, 1997.

[9] M. Gavalec, "Solvability and unique solvability of max–min fuzzy equations," *Fuzzy Sets and Systems*, vol. 124, no. 3, pp. 385–393, 2001.

[10] M. Gavalec, "The general trapezoidal algorithm for strongly regular max–min matrices," *Linear Algebra and Its Applications*, vol. 369, pp. 319–338, 2003.

[11] J. S. Golan, *Semirings and Their Applications*. New York: Springer Science & Business Media, 2013.

[12] F. Borceux, *Handbook of Categorical Algebra: Volume 3, Sheaf Theory*. No. volume 53 in Encyclopedia of Mathematics and Its Applications, Cambridge University Press, 1994.

[13] I. Perfilieva, "Fixed points and solvability of systems of fuzzy relation equations," in *Theoretical Advances and Applications of Fuzzy Logic and Soft Computing*, pp. 841–849, Berlin-Heidelberg: Springer-Verlag, 2007.

15 Eigenvalues and Eigenvectors

Summary

This chapter describes the eigenvalue problem and explores its properties from the perspective of associative arrays. Eigenvalues and eigenvectors are among the most impactful results from matrix mathematics. Extending eigenvalues and eigenvectors to associative arrays allows these results to be applied to the broader range of data representable by associative arrays. The process for applying these results begins with defining quasi-inverses over join-blank algebras and leads to eigen-semimodules. From this perspective, it is possible to make some remarks on another important result from matrix mathematics: the singular value decomposition.

15.1 Introduction

Eigenvectors are a useful tool for capturing the structure of an array and the transformations it can perform. The PageRank algorithm's computation of the first eigenvector of the adjacency matrix of a graph is one example (see Section 7.10). Likewise the implications of array multiplication can also be understood in terms of array eigenvectors (see Figure 2.7). Understanding the meaning of eigenvectors in the context of diverse semirings requires revisiting the basic definitions that describe linear transformations.

Explorations of the eigenvalue problem over max.min, min.max, and min.+ algebras have been conducted by a number of researchers [1–3]. Like the linear systems problem, the eigenvalue problem is seen as particularly important to matrix theory. This section will primarily be a survey of eigenvalues and eigenvectors in the context of tropical and power set algebras.

Definition 15.1

Eigenvalues and Eigenvectors

Let $\mathbf{A}$ be an array. A vector $\mathbf{v}$ is an *eigenvector* if $\mathbf{v} \neq \mathbb{0}$ and there exists a $\lambda \in V$ such that

$$\mathbf{A}\mathbf{v} = \lambda\mathbf{v}$$

The scalar λ is called the *eigenvalue* associated with the eigenvector $\mathbf{v}$.

Geometrically, multiplication by $\mathbf{A}$ scales and rotates a column vector $\mathbf{v}$. If $\mathbf{v}$ is an eigenvector of $\mathbf{A}$, then it is *only* scaled and the scale factor is the eigenvalue associated with $\mathbf{v}$.

Eigenvalues and eigenvectors over semirings provide more general information about a matrix when compared to fields such as $\mathbb{R}$ and $\mathbb{C}$. This fact will become apparent with the presentation of the relevant theorems of eigenanalysis.

Example 15.1

Suppose $\mathbf{A}$ is given by an array with entries in the field $\mathbb{R}$

$$\mathbf{A} = \begin{array}{c} \\ 1 \\ 2 \end{array} \begin{array}{c} \begin{array}{cc} 1 & 2 \end{array} \\ \begin{bmatrix} 1 & 0 \\ 0 & -1 \end{bmatrix} \end{array}$$

Then by direct computation, it can be checked that

$$\mathbf{v} = \begin{array}{c} \\ 1 \\ 2 \end{array} \begin{array}{c} 1 \\ \begin{bmatrix} 1 \\ 0 \end{bmatrix} \end{array} \quad \text{and} \quad \mathbf{v} = \begin{array}{c} \\ 1 \\ 2 \end{array} \begin{array}{c} 1 \\ \begin{bmatrix} 0 \\ 1 \end{bmatrix} \end{array}$$

are eigenvectors with respective associated eigenvalues

$$\lambda = 1 \quad \text{and} \quad \lambda = -1$$

Example 15.2

Suppose $\mathbf{A}$ is given by an array with entries in the power set algebra $\mathscr{P}(\{0,1\})$

$$\mathbf{A} = \begin{array}{c} \\ 1 \\ 2 \end{array} \begin{array}{c} \begin{array}{cc} 1 & 2 \end{array} \\ \begin{bmatrix} \{0\} & \emptyset \\ \{0,1\} & \{0\} \end{bmatrix} \end{array}$$

Then by direct computation, it can be checked that

$$\mathbf{v} = \begin{array}{c} \\ 1 \\ 2 \end{array} \begin{array}{c} 1 \\ \begin{bmatrix} \{0\} \\ \{0,1\} \end{bmatrix} \end{array}$$

is an eigenvector with two eigenvalues

$$\lambda = \{0\} \quad \text{and} \quad \lambda = \{0,1\}$$

15.2 Quasi-Inverses

The eigenproblem for the semirings of interest, all of which have their addition operation being binary supremum with respect to a partial order and whose multiplication operations are compatible with those orderings, is subsumed into an approach from [4], which uses objects called *quasi-inverses*. Suppose that

$$(V, \oplus, \otimes, 0, 1)$$

is a semiring in which

$$\oplus \equiv \sup$$

is the (binary) supremum for a complete partial order $\leq$, which encompasses the semirings of interest, except for the incomplete max-plus algebra.

Now suppose that $\mathbf{A}$ is an $n \times n$ matrix with entries in V. Recall that the collection of $n \times n$ matrices over V is ordered by the product order where

$$\mathbf{A} \leq \mathbf{B} \quad \text{if and only if} \quad \mathbf{A}(i,j) \leq \mathbf{B}(i,j)$$

for every i, j. In particular, the collection of $n \times n$ matrices over V is a complete lattice, so collections of matrices can be added.

Definition 15.2

Quasi-Inverse

[4, p. 121] Suppose $\mathbf{A}$ is an $n \times n$ array. Define

$$\mathbf{A}^{(k)} = \mathbb{I} \oplus \mathbf{A} \oplus \mathbf{A}^2 \oplus \cdots \oplus \mathbf{A}^k = \bigoplus_{i=0}^{k} \mathbf{A}^i$$

and then define the **quasi-inverse** of $\mathbf{A}$, denoted $\mathbf{A}^*$, by

$$\mathbf{A}^* = \bigvee_{k=1}^{\infty} \mathbf{A}^{(k)}$$

Ensuring that the definition of a quasi-inverse is well-defined and exists for every matrix is the main motivating factor behind requiring that all semirings have addition as the supremum for a complete partial order. However, in Section 15.5 the incomplete max-plus algebra $\mathbb{R} \cup \{-\infty\}$ is considered, where the existence of the quasi-inverse is not automatic.

One property of the quasi-inverse is that

$$\mathbf{A}^* = \mathbb{I}_n \oplus \mathbf{A}\mathbf{A}^*$$

Closely related to the quasi-inverse $\mathbf{A}^*$ of $\mathbf{A}$ is the array

$$\mathbf{A}^+ = \mathbf{A} \oplus.\otimes \mathbf{A}^* = \mathbf{A} \oplus \mathbf{A}^2 \oplus \cdots$$

Given $\mathbf{A}^+$, the quasi-inverse $\mathbf{A}^*$ is determined by

$$\mathbf{A}^* = \mathbb{I} \oplus \mathbf{A}^+$$

Example 15.3

Consider

$$\mathbf{A} = \begin{array}{c} \\ 1 \\ 2 \end{array} \begin{array}{c} \begin{array}{cc} 1 & 2 \end{array} \\ \begin{bmatrix} \{1\} & \emptyset \\ \{0\} & \{0,1\} \end{bmatrix} \end{array}$$

over the power set algebra $\mathcal{P}(\{0,1\})$. Then

$$\mathbf{A}^2 = \mathbf{A}$$

so using idempotence

$$\mathbf{A}^* = \mathbb{I}_2 \oplus \mathbf{A} \oplus \mathbf{A} \oplus \cdots = \mathbb{I}_2 \oplus \mathbf{A} = \begin{array}{c} \\ 1 \\ 2 \end{array} \begin{array}{c} \begin{array}{cc} 1 & 2 \end{array} \\ \begin{bmatrix} \{0,1\} & \{1\} \\ \{1\} & \{0,1\} \end{bmatrix} \end{array}$$

Example 15.4

Consider

$$\mathbf{B} = \begin{array}{c} \\ 1 \\ 2 \end{array} \begin{array}{c} \begin{array}{cc} 1 & 2 \end{array} \\ \begin{bmatrix} \{0\} & \{1\} \\ \{1\} & \{0\} \end{bmatrix} \end{array}$$

over the power set algebra $\mathcal{P}(\{0,1\})$. Then

$$\mathbf{B}^2 = \mathbb{I}_2$$

so by using idempotence

$$\mathbf{B}^* = \mathbb{I}_2 \oplus \mathbf{B} \oplus \mathbb{I}_2 \oplus \mathbf{B} \oplus \cdots = \mathbb{I}_2 \oplus \mathbf{B} = \begin{array}{c} \\ 1 \\ 2 \end{array} \begin{array}{c} \begin{array}{cc} 1 & 2 \end{array} \\ \begin{bmatrix} \{0,1\} & \{1\} \\ \{1\} & \{0,1\} \end{bmatrix} \end{array}$$

Example 15.5

$$\mathbf{C} = \begin{array}{c} 1 \\ 2 \end{array} \overset{\begin{array}{cc} 1 & 2 \end{array}}{\begin{bmatrix} 1 & 0 \\ 1 & 1 \end{bmatrix}}$$

over the (complete) max-plus algebra has quasi-inverse

$$\mathbf{C}^* = \begin{bmatrix} 0 & -\infty \\ -\infty & 0 \end{bmatrix} \oplus \begin{bmatrix} 1 & 0 \\ 1 & 1 \end{bmatrix} \oplus \begin{bmatrix} 2 & 1 \\ 2 & 2 \end{bmatrix} \oplus \cdots \oplus \begin{bmatrix} n & n-1 \\ n & n \end{bmatrix} \oplus \cdots$$

$$= \begin{bmatrix} \infty & \infty \\ \infty & \infty \end{bmatrix}$$

Example 15.6

$$\mathbf{D} = \begin{array}{c} 1 \\ 2 \end{array} \overset{\begin{array}{cc} 1 & 2 \end{array}}{\begin{bmatrix} -1 & 0 \\ 0 & -1 \end{bmatrix}}$$

over the (complete) max-plus algebra has quasi-inverse

$$\mathbf{D}^* = \begin{bmatrix} 0 & -\infty \\ -\infty & 0 \end{bmatrix} \oplus \begin{bmatrix} -1 & 0 \\ 0 & -1 \end{bmatrix} \oplus \begin{bmatrix} 0 & -1 \\ -1 & 0 \end{bmatrix} \oplus \begin{bmatrix} -1 & 0 \\ 0 & -1 \end{bmatrix} \oplus \cdots$$

$$= \begin{bmatrix} 0 & -\infty \\ -\infty & 0 \end{bmatrix} \oplus \begin{bmatrix} -1 & 0 \\ 0 & -1 \end{bmatrix} \oplus \begin{bmatrix} 0 & -1 \\ -1 & 0 \end{bmatrix}$$

$$= \begin{bmatrix} 0 & 0 \\ 0 & 0 \end{bmatrix}$$

(All matrices have rows and columns labelled 1, 2.)

The quasi-inverse of an array **A** is closely related to its graph. Recall that the graph of **A** is the weighted directed graph

$$G_{\mathbf{A}} = (V, E, \mathrm{W})$$

where $V = \{1,\dots,n\}$ and

$$E = \{(i, j) \in \{1,\dots,n\}^2 \mid \mathbf{A}(i, j) \neq 0\}$$

and where

$$\mathrm{W}(i, j) = \mathbf{A}(i, j)$$

for each $(i, j) \in E$. Also recall that a *walk* of *length* $k-1$ was a k-tuple $(v_1,\dots,v_k)$ such that

$$(v_j, v_{j+1}) \in E$$

for each j. A *cycle* is a walk $(v_1,\dots,v_k)$ in which $v_1 = v_k$.

Let $G = (V, E)$ be any directed graph, and let $v, w \in V$ and k any non-negative integer. Then let

$$P^k_{v,w}$$

be the set of all walks in G starting at v and ending at w of length k. Likewise, let

$$P^{(k)}_{v,w}$$

be the set of all walks in G starting at v and ending at w of length at most k. Finally, let

$$P_{v,w}$$

be the set of all walks in G starting at v and ending at w of length at least 1.

The weight function w can be extended to apply to any walk, and in fact any k-tuple $(v_1,\dots,v_k)$.

Definition 15.3

Weight of a Walk

Suppose that $G = (V, E, w)$ is a weighted directed graph and $c = (v_1,\dots,v_k)$ is a walk in G. Then the *weight of* c is defined by

$$\begin{aligned}\mathrm{W}(c) &= \mathrm{W}(v_1,\dots,v_k)\\ &= \bigotimes_{j=1}^{k-1} \mathrm{W}(v_j, v_{j+1})\\ &= \mathrm{W}(v_1, v_2) \otimes \cdots \otimes \mathrm{W}(v_{k-1}, v_k)\end{aligned}$$

Note the above formula agrees with the original w for walks consisting of a single edge. Finally, given *any* k-tuple $(v_1,\dots,v_k)$ (which may or may not correspond to an actual walk), define its weight to be $\mathrm{W}(v_1,\dots,v_k)$ as above if $(v_1,\dots,v_k)$ is a walk, and 0 otherwise.

When G is the graph of an array $\mathbf{A}$, the weight of a walk $(v_1,\ldots,v_k)$ can be rewritten as

$$\mathrm{W}(v_1,\ldots,v_k) = \mathbf{A}(v_1,v_2)\otimes\cdots\otimes\mathbf{A}(v_{k-1},v_k)$$

Example 15.7

Consider

$$\mathbf{A} = \begin{array}{c} \\ 1 \\ 2 \\ 3 \end{array}\begin{array}{c} \begin{array}{ccc} 1 & 2 & 3 \end{array} \\ \begin{bmatrix} 0 & 1 & 0 \\ -\infty & 0 & \infty \\ 0 & -1 & 2 \end{bmatrix} \end{array}$$

over the max-plus algebra.

Then $G_{\mathbf{A}} = (V,E,\mathrm{W})$ where $V = \{1,2,3\}$, E consists of every pair (i,j) except $(2,1)$, and W is the restriction of $\mathbf{A}$ to E.

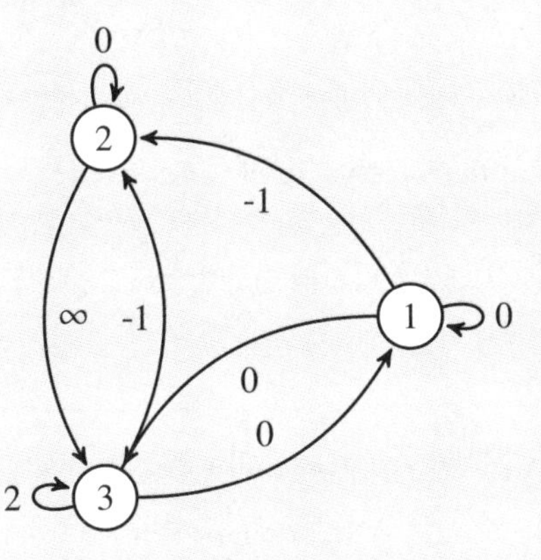

An example walk in $G_{\mathbf{A}}$ is $(1,2,2,3,2)$. The weight of this walk is given by

$$\mathbf{A}(1,2)\otimes\mathbf{A}(2,2)\otimes\mathbf{A}(2,3)\otimes\mathbf{A}(3,2) = 1+0+\infty+\text{-}1 = \infty$$

Another example walk in $G_{\mathbf{A}}$ is $(1,1,3,2,2)$, which has weight

$$\mathbf{A}(1,1)\otimes\mathbf{A}(1,3)\otimes\mathbf{A}(3,2)\otimes\mathbf{A}(2,2) = 0+0+\text{-}1+0 = \text{-}1$$

The concept of the weight of a walk gives an equivalent definition of the quasi-inverse of an array $\mathbf{A}$. Starting with $\mathbf{A}^{(k)}$, regard $\mathbf{A}$ as the adjacency array for a weighted directed graph. By induction on k, the (i,j)-th entry of $\mathbf{A}^k$ gives the sum of weights of all walks from i to j of length exactly k. Since

$$\mathbf{A}^{(k)} = \mathbb{I}\oplus\mathbf{A}\oplus\mathbf{A}^2\oplus\cdots\mathbf{A}^k$$

it follows that the (i, j)-th entry of $\mathbf{A}^{(k)}$ gives the sum of weights of all walks from i to j of length at most k. The (i, j)-th entry of $\mathbf{A}^*$ is the sum of weights of all walks from i to j.

Proposition 15.1

Graph-Theoretic Definitions of $\mathbf{A}^k$, $\mathbf{A}^{(k)}$, $\mathbf{A}^*$

[4, p. 122] If $\mathbf{A}$ is an array and $G_\mathbf{A}$ is its associated weighted directed graph, then

(i) $\mathbf{A}^k(i, j)$ is the sum of the weights of all walks from i to j of length exactly k

$$\mathbf{A}^k(i,j) = \bigoplus_{c \in P^k_{i,j}} \mathrm{W}(c) = \bigoplus_{(v_0,\ldots,v_k) \in P^k_{i,j}} \bigotimes_{\ell=0}^{k-1} \mathbf{A}(v_\ell, v_{\ell+1})$$

(ii) $\mathbf{A}^{(k)}(i, j)$ is the sum of the weights of all walks from i to j of length at most k

$$\mathbf{A}^{(k)}(i,j) = \bigoplus_{c \in P^{(k)}_{i,j}} \mathrm{W}(c) = \bigoplus_{(v_0,\ldots,v_{k_c}) \in P^{(k)}_{i,j}} \bigotimes_{\ell=0}^{k_c-1} \mathbf{A}(v_\ell, v_{\ell+1})$$

(iii) $\mathbf{A}^*(i, j)$ is the sum of the weights of all walks from i to j

$$\mathbf{A}^*(i,j) = \bigoplus_{c \in P_{i,j}} \mathrm{W}(c) = \bigoplus_{(v_0,\ldots,v_k) \in P_{i,j}} \bigotimes_{\ell=0}^{k-1} \mathbf{A}(v_\ell, v_{\ell+1})$$

Proof. $\mathbb{I}(i, j)$ gives the sum of the weights of all walks from i to j of length 0. Now suppose

$$\mathbf{A}^k(i,j) = \bigoplus_{c \in P^k_{i,j}} \mathrm{W}(c)$$

Then

$$\mathbf{A}^{k+1}(i,j) = \bigoplus_{\ell=1}^{n} \mathbf{A}^k(i,\ell) \otimes \mathbf{A}(\ell, j) = \bigoplus_{\ell=1}^{n} \bigoplus_{c \in P^k_{i,\ell}} \mathrm{W}(c) \otimes \mathrm{W}(\ell, j) = \bigoplus_{c \in P^{k+1}_{i,j}} \mathrm{W}(c)$$

so by induction

$$\mathbf{A}^k(i,j) = \bigoplus_{c \in P^k_{i,j}} \mathrm{W}(c)$$

for all k.

The graph theoretic interpretations for $\mathbf{A}^{(k)}(i, j)$ and $\mathbf{A}^*(i, j)$ then follow from the definitions

$$\mathbf{A}^{(k)} = \mathbf{A}^0 \oplus \mathbf{A} \oplus \mathbf{A}^2 \oplus \cdots$$

and

$$\mathbf{A}^* = \bigvee_{k=0}^{\infty} \mathbf{A}^{(k)} = \mathbf{A}^0 \oplus \mathbf{A} \oplus \mathbf{A}^2 \oplus \cdots$$

□

15.3 Existence of Eigenvalues for Idempotent Multiplication

Now that the necessary machinery has been built up, the topic of the existence of eigenvectors/values can be approached. Let $\mathbf{A}(:,i)$ denote the i-th column vector of $\mathbf{A}$. The following theorem gives the major relationship between eigenvectors/values and quasi-inverses.

Theorem 15.2

Equivalent Condition to be Eigenvector when $\oplus = \sup$

[4, p. 210] Let $\mathbf{A}$ be an array with entries in the semiring V with $\oplus = \sup$. Then the following are equivalent:

(i) For some i and some λ

$$\lambda \bigoplus_{c \in P_{i,i}} \mathrm{W}(c) \oplus \lambda = \lambda \bigoplus_{c \in P_{i,i}} \mathrm{W}(c)$$

(ii) For some i and some λ, the column vectors $\lambda\mathbf{A}^+(:,i)$ and $\lambda\mathbf{A}^*(:,i)$ are eigenvectors with eigenvalue 1.

Proof. Suppose (i) holds for some i and λ. Then

$$\begin{aligned}\lambda\mathbf{A}^*(:,i) &= \lambda\mathbb{I}(:,i) \oplus \lambda\mathbf{A}^+(:,i)\\ &= \lambda\mathbf{A}^+(:,i)\end{aligned}$$

where the first equality follows from the definition of $\mathbf{A}^+$ and $\mathbf{A}^*$, and the second equality follows because for the j-th entry for $j \neq i$ of the column this is immediate, and for the i-th entry of the column this is exactly the statement (i). Finally, this means

$$\begin{aligned}\mathbf{A}(\lambda\mathbf{A}^+(:,i)) &= \lambda\mathbf{A}^*(:,i)\\ &= \lambda\mathbf{A}^+(:,i)\end{aligned}$$

Conversely, suppose (ii) holds for some λ and some i. In other words

$$\mathbf{A}(\lambda\mathbf{A}^*(:,i)) = \lambda\mathbf{A}^*(:,i)$$

then

$$\lambda\mathbf{A}^+(:,i) = \lambda\mathbf{A}^*(:,i)$$

so that

$$\lambda(\mathbb{I}(:,i) \oplus \mathbf{A}^+(:,i)) = \lambda \mathbf{A}^+(:,i)$$

Examining this equality at the i-th entry of these column vectors gives the statement of (i). □

The above theorem is effective at establishing the existence of eigenvectors with any given λ as an eigenvalue in the case where $\otimes$ is idempotent.

Lemma 15.3

Conversion of Eigenvectors to Other Eigenvalues with Idempotent $\otimes$

[4, p. 209] Suppose $\otimes$ is idempotent. Then if $\mathbf{v}$ is an eigenvector of $\mathbf{A}$ with eigenvalue 1, then $\lambda\mathbf{v}$ is an eigenvector of $\mathbf{A}$ with eigenvalue λ.

Proof. See Exercise 15.5. □

The above result is in contrast to the case of eigenvalues when V is a field, as in that case there are only a finite number of eigenvalues. In the case of idempotent operations, being an eigenvalue is straightforward and is described in the following corollary.

Corollary 15.4

Eigenvectors and Quasi-Inverse for Idempotent $\oplus$ and $\otimes$

[4, p. 212] Suppose $\oplus$ and $\otimes$ are idempotent. Let $\mathbf{A}$ be an array and i arbitrary. Let

$$\mu = \mathbf{A}^+(i,i) = \bigoplus_{c \in P_{i,i}} \mathbf{W}(c)$$

Then for any λ and i, the vector

$$\lambda\mu\mathbf{A}^*(:,i)$$

is an eigenvector of $\mathbf{A}$ with eigenvalue λ.

Proof. See Exercise 15.6. □

15.4 Strong Dependence and Characteristic Bipolynomial

Using the determinant is a common approach for finding eigenvalues and their eigenvectors in the case of a square array $\mathbf{A}$ with entries in a ring or even field, like $\mathbb{R}$ or $\mathbb{C}$. Or, more precisely, the eigenvalues can be computed from the determinant of

$$\det(\mathbf{A} - \lambda\mathbb{I}) = 0$$

in order to solve for λ. This equation gives a polynomial in λ whose solutions are the eigenvalues of $\mathbf{A}$. Then, given an eigenvalue λ, the system of equations

$$(\mathbf{A} - \lambda\mathbb{I})\mathbf{v} = \mathbf{0}$$

is solved to find the associated eigenvectors. In a semiring, this method cannot be used because $\mathbf{A} - \lambda\mathbb{I}$ does not necessarily make sense. However, another array can take the place of $\mathbf{A} - \lambda\mathbb{I}$ to achieve similar results.

Given the $n \times n$ array $\mathbf{A}$, define the $2n \times 2n$ block array by

$$\tilde{A}(\lambda) = \begin{bmatrix} \mathbf{A} & \lambda\mathbb{I} \\ \mathbb{I} & \mathbb{I} \end{bmatrix}$$

Recall that a set of A vectors in a semimodule M are said to be (linearly) *dependent* if there exists a finite subset

$$A' = \{\mathbf{v}_1, \ldots, \mathbf{v}_n\}$$

of A and nonzero scalars

$$\alpha_1, \ldots, \alpha_n, \beta_1, \ldots, \beta_m$$

such that

$$\bigoplus_{i=1}^{n} \alpha_i \mathbf{v}_i = \bigoplus_{i=1}^{n} \beta_i \mathbf{v}_i$$

but $\alpha_i \neq \beta_i$ for some i. A closely related notion is given in the following definition.

Definition 15.4

Strong Dependence

[4, p. 177] A set A of vectors in a semimodule M is *strongly dependent*, or *dependent* as in [4], if there exist disjoint finite subsets

$$A' = \{\mathbf{v}_1, \ldots, \mathbf{v}_n\} \quad \text{and} \quad A'' = \{\mathbf{w}_1, \ldots, \mathbf{w}_m\}$$

of A and nonzero scalars

$$\alpha_1, \ldots, \alpha_n, \beta_1, \ldots, \beta_m$$

such that

$$\bigoplus_{i=1}^{n} \alpha_i \mathbf{v}_i = \bigoplus_{j=1}^{m} \beta_j \mathbf{w}_j$$

Theorem 15.5

Eigenvalues and Strong Dependence

[4, p. 231] Let $\mathbf{A}$ be an array. Then λ is an eigenvalue of $\mathbf{A}$ if and only if the columns of $\tilde{\mathbf{A}}(\lambda)$ are strongly dependent.

Proof. First suppose that λ is an eigenvalue and $\mathbf{v}$ an associated eigenvector. Let

$$J_1 = \{1, \ldots, n\}$$

and

$$J_2 = \{n+1, \ldots, 2n\}$$

and let

$$\mu_j = \mathbf{v}(j)$$

for $j \in J_1$ and

$$\mu_j = \mathbf{v}(j-n)$$

for $j \in J_2$. Then because

$$\mathbf{A}\mathbf{v} = \lambda\mathbf{v}$$

it follows that

$$\bigoplus_{j \in J_1} \mu_j \otimes \tilde{\mathbf{A}}(\lambda)(:, j) = \bigoplus_{j \in J_2} \mu_j \otimes \tilde{\mathbf{A}}(\lambda)(:, j)$$

or, in other words, that the columns of $\tilde{\mathbf{A}}(\lambda)$ are strongly dependent. Conversely, suppose that the columns of $\tilde{\mathbf{A}}(\lambda)$ are strongly dependent so that there are disjoint subsets J_1, J_2 of $\{1, \ldots, n\}$ and (nonzero) scalars μ_j for

$$j \in J_1 \cup J_2$$

such that

$$\bigoplus_{j \in J_1} \mu_j \otimes \tilde{\mathbf{A}}(\lambda)(:, j) = \bigoplus_{j \in J_2} \mu_j \otimes \tilde{\mathbf{A}}(\lambda)(:, j)$$

Extend the definition of μ_j to all $j \in \{1, \ldots, n\}$ by defining $\mu_j = 0$ for

$$j \notin J_1 \cup J_2$$

By using the above (strong) dependence and looking at the bottom n entries of the j-th and $(j+n)$-th columns, if both j and $j+n$ are in J_i for some fixed i, then

$$\mu_j \oplus \mu_{j+n} = 0$$

Since $\oplus$ is the (binary) supremum for a partial ordering $\leq$ with 0 the minimum element, it follows that $\mu_j = \mu_{j+n}$. Thus, if $j \in J_1$ or J_2, then $j+n$ is in the remaining set of indices.

Moreover, the same (strong) dependence shows that $\mu_j = \mu_{j+n}$. Then define $\mathbf{v}$ by setting

$$\mathbf{v}(j) = \mu_j$$

for $j \in J_1$ and

$$\mathbf{v}(j) = 0$$

for $j \in \{1,\ldots,n\} \setminus J_1$. The above (strong) dependence then shows that $\mathbf{v}$ is an eigenvector with eigenvalue λ. □

When the semiring is also a ring, then

$$\det(\tilde{\mathbf{A}}(\lambda)) = \det(\mathbf{A} - \lambda\mathbb{I})$$

motivating

$$\det(\tilde{\mathbf{A}}(\lambda))$$

to take the role of

$$\det(\mathbf{A} - \lambda\mathbb{I})$$

to define the characteristic polynomial. But since the determinant makes explicit use of negatives, "positive" determinants

$$\det{}^+(\mathbf{A})$$

and "negative" determinants

$$\det{}^-(\mathbf{A})$$

are introduced so that

$$\det{}^+(\mathbf{A}) - \det{}^-(\mathbf{A}) = \det(\mathbf{A})$$

when in a ring.

To motivate the definition of the bideterminant, recall the following formula for the determinant when in a ring

$$\det(\mathbf{A}) = \sum_{\sigma \text{ a permutation of } \{1,\ldots,n\}} \operatorname{sgn}(\sigma) \prod_{i=1}^{n} \mathbf{A}(i,\sigma(i))$$

Recall that a permutation of $\{1,\ldots,n\}$ is a bijection

$$\sigma : \{1,\ldots,n\} \to \{1,\ldots,n\}$$

A permutation σ gives rise to an $n \times n$ array P_σ whose (i,j)-th entry is 1 if $\sigma(i) = j$ and 0 otherwise. A permutation σ is said to be *even* if

$$\det(P_\sigma) = 1$$

and *odd* if

$$\det(P_\sigma) = -1$$

It can be shown that these are the only cases. The map sgn sends a permutation σ to $\det(P_\sigma)$, and $\mathrm{sgn}(\sigma)$ is called the *sign of* σ. Denote by $\mathrm{Per}(n)$ the set of all permutations of $\{1,\ldots,n\}$, $\mathrm{Per}^+(n)$ the set of all even permutations, and $\mathrm{Per}^-(n)$ the set of all odd permutations.

Example 15.8

Permutations of $\{1,2,3,4\}$

This example illustrates the even and odd permutations of $\{1,2,3,4\}$. The permutation

$$\sigma : \{1,2,3,4\} \rightarrow \{1,2,3,4\}$$

which fixes 1 and sends 2 to 3, 3 to 4, and 4 to 2, is written as

$$(1)(2\ 3\ 4)$$

In other words, a permutation is written as a product of *cycles*

$$(i\ \ \sigma(i)\ \ \sigma(\sigma(i))\ \ \cdots)$$

by continuing to apply σ to the previous element of the cycle until i is reached again. The identity map is typically written as 1, and cycles like (1) are dropped, so the *cycle decomposition* of σ above is

$$(2\ 3\ 4)$$

Then $\mathrm{Per}^+(4)$ is given by

$$\begin{array}{cccccc} 1, & (1\ 2\ 3), & (1\ 3\ 2), & (1\ 2\ 4), & (1\ 4\ 2), & (1\ 3\ 4), \\ (1\ 4\ 3), & (2\ 3\ 4), & (2\ 4\ 3), & (1\ 2)(3\ 4), & (1\ 3)(2\ 4), & (1\ 4)(2\ 3) \end{array}$$

$\mathrm{Per}^-(4)$ consists of the remaining permutations

$$\begin{array}{cccccc} (1\ 2\ 3\ 4), & (1\ 2\ 4\ 3), & (1\ 3\ 2\ 4), & (1\ 3\ 4\ 2), & (1\ 4\ 2\ 3), & (1\ 4\ 3\ 2), \\ (1\ 2), & (1\ 3), & (1\ 4), & (2\ 3), & (2\ 4), & (3\ 4) \end{array}$$

The above notation gives a suggestive form for $\det(\mathbf{A})$ when in a ring

$$\det(\mathbf{A}) = \sum_{\sigma \in \mathrm{Per}^+(n)} \prod_{i=1}^{n} \mathbf{A}(i,\sigma(i)) - \sum_{\sigma \in \mathrm{Per}^-(n)} \prod_{i=1}^{n} \mathbf{A}(i,\sigma(i))$$

and leads to the following definition.

Definition 15.5

Bideterminant

[4, p. 59] Given an $n \times n$ array $\mathbf{A}$, the *bideterminant* is the pair $(\det^+(\mathbf{A}), \det^-(\mathbf{A}))$ where

$$\det^+(\mathbf{A}) = \bigoplus_{\sigma \in \mathrm{Per}^+(n)} \bigotimes_{i=1}^{n} \mathbf{A}(i, \sigma(i)) \quad \text{and} \quad \det^-(\mathbf{A}) = \bigoplus_{\sigma \in \mathrm{Per}^-(n)} \bigotimes_{i=1}^{n} \mathbf{A}(i, \sigma(i))$$

called the *positive determinant* and *negative determinant*, respectively.

Definition 15.6

Characteristic Bipolynomial

[4, p. 233] If $\mathbf{A}$ is an array, then the *characteristic bipolynomial* is the pair

$$(P^+(\lambda), P^-(\lambda)) = \left(\det^+(\tilde{\mathbf{A}}(\lambda)), \det^-(\tilde{\mathbf{A}}(\lambda))\right)$$

Example 15.9

Consider the following array over the max-plus algebra

$$\mathbf{A} = \begin{array}{c} \\ 1 \\ 2 \end{array} \begin{array}{c} \begin{array}{cc} 1 & 2 \end{array} \\ \begin{bmatrix} 1 & 2 \\ 0 & -1 \end{bmatrix} \end{array}$$

then $\tilde{\mathbf{A}}(\lambda)$ is given by

$$\begin{array}{c} \\ 1 \\ 2 \\ 3 \\ 4 \end{array} \begin{array}{c} \begin{array}{cccc} 1 & 2 & 3 & 4 \end{array} \\ \begin{bmatrix} 1 & 2 & \lambda & -\infty \\ 0 & -1 & -\infty & \lambda \\ 0 & -\infty & 0 & -\infty \\ -\infty & 0 & -\infty & 0 \end{bmatrix} \end{array}$$

Computing the positive and negative determinants of $\tilde{\mathbf{A}}(\lambda)$ gives

$$\det^+(\tilde{\mathbf{A}}(\lambda)) = \max(0, 2\lambda)$$
$$\det^-(\tilde{\mathbf{A}}(\lambda)) = \max(\lambda + 1, 2)$$

When $\oplus$ is not just a binary supremum but also maximum so that the ordering is total, and $(V, \otimes, 1)$ is a group, then the eigenvalues λ for an array $\mathbf{A}$ are exactly the solutions to

$$P^+(\lambda) = P^-(\lambda)$$

The crux of the above equation is the relationship between the bideterminant of an array and the strong dependence of its columns in semirings of the above form.

Theorem 15.6

Bideterminant and Strong Dependence

[4, p. 191,194] Suppose

$$(V, \oplus, \otimes, 0, 1)$$

is such that $\oplus$ is the (binary) max function for a total order $\leq$ on V and

$$(V \setminus \{0\}, \otimes, 1)$$

is a commutative group. If $\mathbf{A}$ is an array with entries in $V \setminus \{0\}$, then

$$\det{}^+(\mathbf{A}) = \det{}^-(\mathbf{A})$$

if and only if the columns of $\mathbf{A}$ are strongly dependent.

Corollary 15.7

Characteristic Bipolynomial for $\oplus = \max$ and Commutative Group $\otimes$

[4, p. 233] Suppose

$$(V, \oplus, \otimes, 0, 1)$$

is such that $\oplus$ is the (binary) max function for a total order $\leq$ on V and

$$(V \setminus \{0\}, \otimes, 1)$$

is a commutative group. If $\mathbf{A}$ is an array with entries in V, then λ is an eigenvalue for $\mathbf{A}$ if and only if λ satisfies

$$P^+(\lambda) = P^-(\lambda)$$

where $(P^+(\lambda), P^-(\lambda))$ is the characteristic bipolynomial of $\mathbf{A}$.

Proof. From Theorem 15.6, it follows that

$$P^+(\lambda) = \det{}^+(\tilde{\mathbf{A}}(\lambda)) = \det{}^-(\tilde{\mathbf{A}}(\lambda)) = P^-(\lambda)$$

if and only if the columns of $\tilde{\mathbf{A}}(\lambda)$ are strongly dependent. By Theorem 15.5, the latter occurs if and only if λ is an eigenvalue of $\mathbf{A}$. □

Corollary 15.7 is analogous to the standard way of finding eigenvalues for arrays over fields. This is because $P^+(\lambda) = P^-(\lambda)$ occurs for arrays over a field (where the determinant is now well-defined) exactly when $P(\lambda) = \det(\mathbf{A}) = 0$.

Since the max-plus algebra fulfills the hypotheses of Corollary 15.7, that result applies to the max-plus algebra.

Example 15.10

Using the array

$$\mathbf{A} = \begin{array}{c} \\ 1 \\ 2 \end{array} \begin{array}{c} \begin{matrix} 1 & 2 \end{matrix} \\ \begin{bmatrix} 1 & 2 \\ 0 & -1 \end{bmatrix} \end{array}$$

from Example 15.9, the characteristic bipolynomial was found to be

$$(\max(0, 2\lambda), \max(\lambda + 1, 2))$$

For

$$\max(0, 2\lambda) = \max(\lambda + 1, 2)$$

to hold, either

$$0 = \lambda + 1$$

for $\lambda \leq 0$, or

$$2\lambda = \lambda + 1$$

for $1 \leq \lambda$, or

$$2\lambda = 2$$

for $0 \leq \lambda \leq 1$. These equations result in $\lambda = -1$, $\lambda = 1$, or $\lambda = 1/2$, so these three values are the eigenvalues of $\mathbf{A}$ by Corollary 15.7.

15.5 Eigenanalysis for Irreducible Matrices for Invertible Multiplication

While the results of Section 15.3 showed that when ⊗ is idempotent, every element of the underlying semiring is an eigenvalue. This case is in stark contrast with Example 15.10, in which $\mathbf{A}$ has only three eigenvalues. Moreover, while Theorem 15.6 gives a condition for λ to be an eigenvalue, it does not prove that an eigenvalue always exists. By restricting to a smaller class of matrices, the guaranteed existence of a *unique* eigenvalue can be shown. The algebraic property of the max-plus algebra that distinguishes itself from the other semirings of interest is that the multiplication is not binary infimum. Also, by removing $+\infty$, the multiplicative operation $+$ on $\mathbb{R}$ is invertible.

Definition 15.7

Incomplete Max-Plus and Min-Plus Algebras

$(\mathbb{R} \cup \{-\infty\}, \max, +, -\infty, 0)$ is the *incomplete max-plus algebra.*
$(\mathbb{R} \cup \{\infty\}, \min, +, \infty, 0)$ is the *incomplete min-plus algebra.*
$(\mathbb{R}_{\geq 0}, \max, \times, 0, 1)$ is the *incomplete max-times algebra.*
$(\mathbb{R}_{>0} \cup \{\infty\}, \min, \times, \infty, 1)$ is the *incomplete min-times algebra.*

Definition 15.8

Irreducible Matrix

A square array $\mathbf{A}$ with entries in the semiring $(V, \oplus, \otimes, 0, 1)$ is *reducible* if there exists a permutation matrix $\mathbf{P}$ such that

$$\mathbf{PAP}^\mathsf{T} = \begin{bmatrix} \mathbf{B} & \mathbf{0} \\ \mathbf{C} & \mathbf{D} \end{bmatrix}$$

with $\mathbf{B}$ and $\mathbf{D}$ square and $\mathbf{0}$ the zero array of appropriate dimensions. In other words, $\mathbf{A}$ is reducible if it is similar to a block lower-triangular array. If $\mathbf{A}$ is not reducible, it is *irreducible.*

There is also a graph-theoretic equivalent to the above definition of irreducible.

Definition 15.9

Strongly Connected Graph

Suppose G is a directed graph. G is said to be *strongly connected* if, given any two vertices v, w, there exists a directed walk from v to w, in other words a walk

$$c = (v_0, \ldots, v_n)$$

such that $v_0 = v$, $w = v_n$, and for each i, (v_i, v_{i+1}) is a (directed) edge in G. The *strong components* of a directed graph G are the maximal strongly connected full subgraphs of G. The sets of vertices of the strong components partition the set of vertices of G.

Lemma 15.8

Irreducibility and Strong-Connectedness Equivalence

If $\mathbf{A}$ is a square array and G the directed graph with $\mathbf{A}$ its adjacency array, then the following are equivalent

(i) $\mathbf{A}$ is irreducible.
(ii) G is strongly connected.

Proof. Without loss of generality, it can be assumed that the nonzero entries of $\mathbf{A}$ are 1, as this neither affects whether $\mathbf{A}$ is reducible or not, nor whether G is strongly connected or not. The effect of $\mathbf{P}$ on the directed graph G is to simply permute the vertices of G, with

$$\mathbf{PAP}^\intercal$$

the new adjacency array. Let $\mathbf{P}G$ be this new directed graph. $\mathbf{P}G$ is strongly connected if and only if G is. The main observation to make is that

$$\begin{bmatrix}\mathbf{B} & \mathbf{0}\\ \mathbf{C} & \mathbf{D}\end{bmatrix}\begin{bmatrix}\mathbf{E} & \mathbf{0}\\ \mathbf{F} & \mathbf{G}\end{bmatrix} = \begin{bmatrix}\mathbf{EB} & \mathbf{0}\\ \mathbf{CE}+\mathbf{DF} & \mathbf{DG}\end{bmatrix}$$

As such, it follows that if $\mathbf{A}$ is of the form

$$\begin{bmatrix}\mathbf{B} & \mathbf{0}\\ \mathbf{C} & \mathbf{D}\end{bmatrix}$$

then $\mathbf{A}^k$ is of the same form for every k. Recall that the (i, j)-th entry of $\mathbf{A}^k$ is the sum of the weights of the (directed) walks from i to j. From the assumption that the nonzero entries of $\mathbf{A}$ are 1, the weight of any directed walk is 1, so the sum of the weights of the directed walks is

$$1 \oplus \cdots \oplus 1$$

where 1 occurs the same number of times as there are directed walks from i to j. With the assumption that $\oplus$ is the join for some partial order $\leq$, this sum is 0 only when the sum is empty. Thus, $\mathbf{A}^k(i, j) = 0$ if and only if there is no path of length k from i to j.

Therefore, if $\mathbf{A}$ is reducible, then

$$\mathbf{PAP}^\intercal$$

is block lower-triangular and $\mathbf{P}G$ is not strongly connected and likewise G is not strongly connected. Since if i, j are chosen such that the (i, j)-th entry is in the $\mathbf{0}$ block, then there can be no directed walk from i to j. Conversely, suppose G is not strongly connected, so there is i, j such that there is no directed walk from i to j. Let

$$E_1, \ldots, E_m$$

be the strong components of G. Writing the elements of each

$$E_i \subset \{1, \dots, n\}$$

in order from least to greatest defines a permutation σ of $\{1, \dots, n\}$ such that if $i < j$ and $p \in E_i, q \in E_j$, then $\sigma(p) < \sigma(q)$. The permutation σ defines a permutation array $\mathbf{P}$. Then $\mathbf{PAP}^{\mathsf{T}}$ is of the form

$$\begin{bmatrix} \mathbf{B} & \mathbf{0} \\ \mathbf{C} & \mathbf{D} \end{bmatrix}$$

where $\mathbf{B}$ is the sub-array of $\mathbf{A}$ with entries drawn from $E_1 \times E_1$ and $\mathbf{D}$ is the sub-array of $\mathbf{A}$ with entries drawn from

$$(E_2 \cup \cdots \cup E_n) \times (E_2 \cup \cdots \cup E_n)$$

Thus, $\mathbf{A}$ is reducible. □

Example 15.11

Consider the array

$$\mathbf{A} = \begin{array}{c} \\ 1 \\ 2 \\ 3 \end{array} \begin{array}{c} \begin{array}{ccc} 1 & 2 & 3 \end{array} \\ \begin{bmatrix} 0 & 3 & -1 \\ 1 & 0 & 1 \\ 0 & -1 & 0 \end{bmatrix} \end{array}$$

This has weighted directed graph

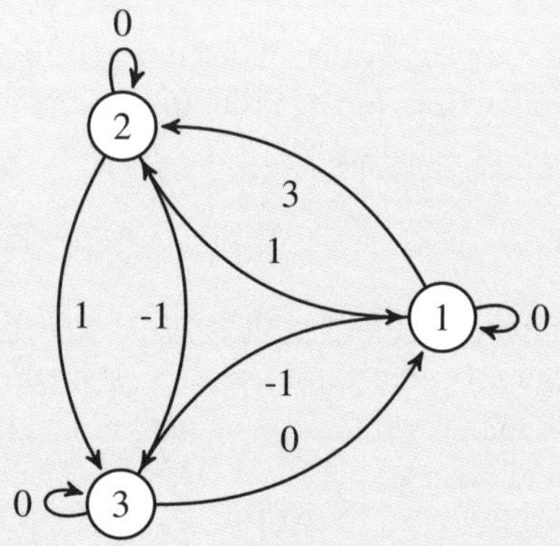

which is strongly connected, so $\mathbf{A}$ is irreducible by Lemma 15.8.

From this point onward it is assumed that the semiring $(V, \oplus, \otimes, 0, 1)$ under consideration satisfies

(i) $\oplus$ is the (binary) max for a total order $\leq$,
(ii) $(V \setminus \{0\}, \otimes, 1)$ is a group, and
(iii) for every positive integer n and $v \in V$, there exists a unique u such that $u^n = v$, denoted by $u = v^{1/n}$.

Example 15.12

The incomplete max-plus algebra, incomplete min-plus algebra, incomplete max-times algebra, and incomplete min-times algebra all satisfy the above properties.

Definition 15.10

Spectral Radius

Suppose $(V, \oplus, \otimes, 0, 1)$ satisfies the specified properties above. The *spectral radius* of

$$\mathbf{A} \in M_n(V)$$

is defined as

$$\rho(\mathbf{A}) = \bigoplus_{k=1}^{n} \left(\bigoplus_{i=1}^{n} \mathbf{A}^k(i,i) \right)^{1/k}$$

Note that $a^{1/k}$ is an element for which

$$(a^{1/k})^k = \bigotimes_{i=1}^{k} a^{1/k} = a$$

The spectral radius can be cast in graph-theoretic terms.

Lemma 15.9

Graph-Theoretic Formula for Spectral Radius

[4, p. 227] Suppose $(V, \oplus, \otimes, 0, 1)$ satisfies the specified properties above. Then

$$\rho(\mathbf{A}) = \bigoplus_{c \in \Gamma} \mathrm{W}(c)^{1/|c|}$$

where $\rho(\mathbf{A})$ is the spectral radius of $\mathbf{A}$, Γ is the set of all closed paths or walks $(v_0, \ldots, v_n)$ where all vertices are distinct except for the first and last, and $|c|$ is the length of walk c.

Theorem 15.10

Spectral Radius of Irreducible Matrix is Unique Eigenvalue

[4, p. 229] Suppose $(V,\oplus,\otimes,0,1)$ satisfies the specified properties above. If $\mathbf{A}$ is an irreducible array with spectral radius $\rho(\mathbf{A})$, then $\rho(\mathbf{A})$ is the unique eigenvalue of $\mathbf{A}$.

Example 15.13

Working in the incomplete max-plus algebra, the spectral radius $\rho(\mathbf{A})$ is the maximum of $\mathrm{W}(c)/|c|$ for c, a closed path in G. $\mathrm{W}(c)/|c|$ is called the *average weight* of the closed path c.
Consider the array

$$\mathbf{A} = \begin{array}{c} \\ 1 \\ 2 \\ 3 \end{array} \begin{array}{c} \begin{array}{ccc} 1 & 2 & 3 \end{array} \\ \begin{bmatrix} 0 & 3 & -1 \\ 1 & 0 & 1 \\ 0 & -1 & 0 \end{bmatrix} \end{array}$$

whose graph is given in Example 15.11 and which has closed paths

$$(1,1),(2,2),(3,3),$$
$$(1,2,1),(1,3,1),(2,1,2),(2,3,2),(3,1,3),(3,2,3),$$
$$(1,2,3,1),(1,3,2,1)$$

(up to a cyclic shift of vertices, which does not change the weight).
Taking the maximum of $\mathrm{W}(c)/|c|$ gives the spectral radius $\rho(\mathbf{A})$ as

$$2 = \frac{4}{2} = \frac{3+1+0}{2}$$

from the path $(1,2,3,1)$, and hence the unique eigenvalue for $\mathbf{A}$ is

$$\rho(\mathbf{A}) = 2$$

Indeed, some eigenvectors associated with the unique eigenvalue 2 are

$$\begin{array}{c} \\ 1 \\ 2 \\ 3 \end{array} \begin{array}{c} 1 \\ \begin{bmatrix} 0 \\ -1 \\ -2 \end{bmatrix} \end{array}, \quad \begin{array}{c} \\ 1 \\ 2 \\ 3 \end{array} \begin{array}{c} 1 \\ \begin{bmatrix} 1 \\ 0 \\ -1 \end{bmatrix} \end{array}$$

15.6 Eigen-Semimodules

Given an array **A** with entries in the semiring V and an eigenvalue λ of **A**, the space of all eigenvectors associated with λ has additional structure on it, motivating the following definition.

Definition 15.11

Eigen-semimodule

Given **A** an $n \times n$ array over V and λ an eigenvalue, $\mathcal{V}(\lambda)$ is the sub-semimodule of V^n consisting of all eigenvectors of **A** with corresponding eigenvalue λ.
The sub-semimodules $\mathcal{V}(\lambda)$ are called *eigen-semimodules* of **A**.

As before, $\oplus$ continues to be the binary supremum for a partial order $\leq$ on V. It is also assumed, as in the case of supremum-blank algebras, that infinite distributivity of $\otimes$ over $\oplus$ holds. This assumption ensures that array multiplication with the quasi-inverse works well.

Lemma 15.11

Eigenvector is Eigenvector of Quasi-Inverse

[4, p. 212] Suppose **A** has eigenvalue 1. Then

$$\mathbf{v} \in \mathcal{V}(1)$$

implies

$$\mathbf{v} = \mathbf{A}^* \mathbf{v}$$

Proof. See Exercise 15.9. □

Corollary 15.12

Quasi-Inverse and $\mathcal{V}(1)$

Suppose **A** has eigenvalue 1. Then

$$\mathcal{V}(1) \subset \text{span}\{\mathbf{A}^*(:,i) \mid 1 \leq i \leq n\}$$

Proof. See Exercise 15.10. □

The above corollary puts an upper bound on $\mathcal{V}(1)$. Likewise, a lower bound can be put on $\mathcal{V}(1)$ when $\otimes$ is idempotent, such as in the case of power set algebras, by making use of Corollary 15.4.

Lemma 15.13

Quasi-Inverse and $\mathcal{V}(1)$ for Idempotent $\otimes$

Suppose $\otimes$ is idempotent and $\mathbf{A}$ has eigenvalue 1. If

$$\mu_i = \mathbf{A}^*(i,i)$$

then

$$\text{span}\{\mu_i \mathbf{A}^*(:,i) \mid 1 \leq i \leq n\} \subset \mathcal{V}(1)$$

When $\oplus$ is not just binary supremum, but also a binary maximum, so $\leq$ is total as in the max-plus and max-min tropical algebras, the specific form of $\mathcal{V}(1)$ can be given.

Theorem 15.14

Quasi-Inverse and Structure of $\mathcal{V}(1)$ When $\oplus = \max$

[4, p. 213] Suppose $\oplus$ is max and $\mathbf{A}$ has eigenvalue 1. Then there exists a subset $\{i_1,\ldots,i_K\}$ of $\{1,\ldots,n\}$ and scalars μ_k such that

$$\mathcal{V}(1) = \text{span}\{\mu_k \mathbf{A}^*(:,i_k) \mid 1 \leq k \leq K\}$$

When 1 is the greatest element of V under $\leq$, which is the case for the max-min tropical algebra, Theorem 15.14 can be strengthened to remove the necessity of the subset $\{i_1,\ldots,i_K\}$ to simply be $\{1,\ldots,n\}$ and to give concrete values to the scalars μ_i:

Corollary 15.15

Quasi-Inverse and Structure of $\mathcal{V}(1)$ When $\oplus = \max$, $\otimes$ Idempotent, $1 = \sup(V)$

[4, p. 215] Suppose $\oplus$ is max, $\otimes$ is idempotent, 1 is the maximum of V under $\leq$, and $\mathbf{A}$ has eigenvalue 1. Then

$$\mathcal{V}(1) = \text{span}\{\mu_i \mathbf{A}^*(:,i) \mid 1 \leq i \leq n\}$$

where $\mu_i = \mathbf{A}^+(i,i)$.

Example 15.14

Consider the array

$$\mathbf{A} = \begin{array}{c} \\ 1 \\ 2 \\ 3 \end{array} \begin{array}{c} \begin{array}{ccc} 1 & 2 & 3 \end{array} \\ \begin{bmatrix} 0 & 3 & -1 \\ \infty & 1 & 2 \\ -\infty & 1 & 1 \end{bmatrix} \end{array}$$

with entries in the max-min algebra. Then $\mathbf{A}^*$ and $\mathbf{A}^+$ are given by

$$\mathbf{A}^* = \begin{array}{c} \\ 1 \\ 2 \\ 3 \end{array} \begin{array}{c} \begin{array}{ccc} 1 & 2 & 3 \end{array} \\ \begin{bmatrix} \infty & 3 & 2 \\ \infty & \infty & 2 \\ 1 & 1 & \infty \end{bmatrix} \end{array} \quad \text{and} \quad \mathbf{A}^+ = \mathbf{A}\mathbf{A}^* = \begin{array}{c} \\ 1 \\ 2 \\ 3 \end{array} \begin{array}{c} \begin{array}{ccc} 1 & 2 & 3 \end{array} \\ \begin{bmatrix} 3 & 3 & 2 \\ \infty & 3 & 2 \\ 1 & 1 & 1 \end{bmatrix} \end{array}$$

Then by Corollary 15.15, it follows that $\mathcal{V}(1)$ is spanned by the vectors

$$\begin{array}{c} \\ 1 \\ 2 \\ 3 \end{array} \begin{array}{c} 1 \\ \begin{bmatrix} 3 \\ 3 \\ 1 \end{bmatrix} \end{array}, \begin{array}{c} \\ 1 \\ 2 \\ 3 \end{array} \begin{array}{c} 1 \\ \begin{bmatrix} 2 \\ 2 \\ 1 \end{bmatrix} \end{array}$$

So far only the case of $\mathcal{V}(1)$ has been dealt with. When $\otimes$ is idempotent and 1 is the greatest element, as in power set algebras or the max-min tropical algebra, the results of Lemma 15.11 continue to hold:

Lemma 15.16

Eigenvector is Eigenvector of Quasi-Inverse when $1 = \sup(V)$

[4, p. 218] Suppose $\mathbf{A}$ has eigenvalue λ and 1 is the greatest element. Then

$$\mathbf{v} \in \mathcal{V}(\lambda)$$

implies

$$\mathbf{v} = \mathbf{A}^*\mathbf{v}$$

and hence

$$\text{span}\{\lambda\mu_i\mathbf{A}^*(:,i) \mid 1 \leq i \leq n\} \subset \mathcal{V}(\lambda) \subset \text{span}\{\mathbf{A}^*(:,i) \mid 1 \leq i \leq n\}$$

Proof. See Exercise 15.11. □

Likewise, Theorem 15.14 can be extended when $\oplus$ is assumed to be max.

Corollary 15.17

Quasi-Inverse and Structure of $\mathcal{V}(\lambda)$ When $\oplus$ = max, $\otimes$ Idempotent, 1 = sup(V)

[4, p. 219] Suppose $\oplus$ is max, $\otimes$ is idempotent, 1 is the greatest element, and **A** has eigenvalue λ. Then there exists a subset $\{i_1,\ldots,i_K\}$ of $\{1,\ldots,n\}$ such that

$$\mathcal{V}(\lambda) = \text{span}\{\lambda\mu_k\mathbf{A}^*(:,i_k) \mid 1 \leq k \leq K\}$$

where

$$\mu_k = \mathbf{A}^+(i_k,i_k)$$

To end the section, the case where $\otimes$ is invertible on $V \setminus \{0\}$ and **A** is irreducible is discussed.

Lemma 15.18

Existence of $(\lambda^{-1}\mathbf{A})^*$ for Irreducible A, Idempotent $\oplus$, and $\otimes$ is a Group

[4, p. 222] Suppose

$$(V \setminus \{0\},\otimes,1)$$

is a group, $\oplus$ is idempotent, and **A** is an irreducible array with eigenvalue λ. Then $(\lambda^{-1}\mathbf{A})^*$ exists.

The reason for stating this lemma is that the motivating example of the incomplete max-plus algebra is, as the name suggests, incomplete, and so the existence of $(\lambda^{-1}\mathbf{A})^*$ cannot be guaranteed by a completeness argument.

Theorem 15.19

Eigensemimodules of Irreducible Matrix When $\oplus$ = max and $\otimes$ is a Group

[4, p. 225] Suppose $\oplus$ is max, $(V \setminus \{0\},\otimes,1)$ is a group, and **A** is irreducible. Let

$$\tilde{\mathbf{A}} = \lambda^{-1}\mathbf{A}$$

where λ is the unique eigenvalue of **A**. Then there are $j_1,\ldots,j_p$ such that

$$\mathcal{V}(\lambda) = \text{span}\{\tilde{\mathbf{A}}^*(:,j_1),\ldots,\tilde{\mathbf{A}}^*(:,j_p)\}$$

The above theorem means that once $\tilde{\mathbf{A}}^*$ has been calculated, it just remains to check which of its columns are eigenvectors of $\mathbf{A}$ to get a generating set for $\mathcal{V}(\lambda)$.

Example 15.15

Recall Example 15.13 where

$$\mathbf{A} = \begin{array}{c} \\ 1 \\ 2 \\ 3 \end{array}\begin{array}{c} \begin{array}{ccc} 1 & 2 & 3 \end{array} \\ \begin{bmatrix} 0 & 3 & -1 \\ 1 & 0 & 1 \\ 0 & -1 & 0 \end{bmatrix} \end{array}$$

was shown to have the unique eigenvalue 2. The associated graph G of $\mathbf{A}$

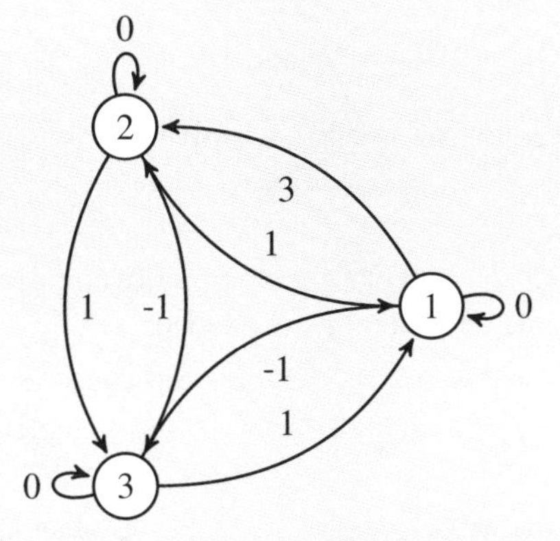

is strongly connected, so $\mathbf{A}$ is irreducible. Thus, Theorem 15.19 applies and

$$(\lambda^{-1}\mathbf{A})^* = (-2+\mathbf{A})^* = \begin{array}{c} \\ 1 \\ 2 \\ 3 \end{array}\begin{array}{c} \begin{array}{ccc} 1 & 2 & 3 \end{array} \\ \begin{bmatrix} 0 & 1 & 0 \\ -1 & 0 & -1 \\ -2 & -1 & 0 \end{bmatrix} \end{array}$$

Thus, from Theorem 15.19, it follows that $\mathcal{V}(2)$ is the span of the column vectors of $(\lambda^{-1}\mathbf{A})^*$, which are actually eigenvectors. As in Example 15.13

$$\mathcal{V}(2) = \mathrm{span}\left(\begin{array}{c} \\ 1 \\ 2 \\ 3 \end{array}\begin{array}{c} 1 \\ \begin{bmatrix} 0 \\ -1 \\ -2 \end{bmatrix} \end{array}, \begin{array}{c} \\ 1 \\ 2 \\ 3 \end{array}\begin{array}{c} 1 \\ \begin{bmatrix} 1 \\ 0 \\ -1 \end{bmatrix} \end{array} \right)$$

15.7 Singular Value Decomposition

A common factorization in traditional linear algebra is that of the singular value decomposition (SVD).

Definition 15.12

Singular Value Decomposition

Let $\mathbf{A}$ be a complex valued $m \times n$ array. Then the *singular value decomposition*, or *SVD*, is a factorization of the form

$$\mathbf{U\Sigma V}^{\dagger}$$

where $\mathbf{U}$ and $\mathbf{V}$ are unitary matrices such that

$$\mathbf{U}^{-1} = \mathbf{U}^{\dagger} \quad \text{and} \quad \mathbf{V}^{-1} = \mathbf{V}^{\dagger}$$

and $\mathbf{\Sigma}$ is diagonal whose entries are non-negative.

To prove the existence of the SVD and provide motivation for its algebraic necessities, some notation and additional terminology are needed.

Definition 15.13

Conjugate Transpose

The *conjugate transpose* of a complex $n \times n$ array $\mathbf{A}$, written $\mathbf{A}^{\dagger}$, is defined by

$$(\mathbf{A}^{\dagger})(i,j) = \overline{\mathbf{A}(j,i)}$$

The array $\mathbf{A}$ is said to be *Hermitian* if

$$\mathbf{A} = \mathbf{A}^{\dagger}$$

and is *unitary* if

$$\mathbf{A}^{\dagger} = \mathbf{A}^{-1}$$

A set S of vectors is *orthonormal* if

$$\sum_{i=1}^{n} \mathbf{u}(i)\overline{\mathbf{v}(i)} = 0$$

for all distinct $\mathbf{u}, \mathbf{v} \in S$ and

$$|\mathbf{v}| = 1$$

for all $\mathbf{v} \in S$.

One of the main ways to find a SVD of $\mathbf{A}$ is by finding eigenvectors and eigenvalues, not of $\mathbf{A}$ but of $\mathbf{A}\mathbf{A}^\dagger$ and $\mathbf{A}^\dagger\mathbf{A}$. The main result that makes this possible is the spectral theorem.

Theorem 15.20

Spectral Theorem

[5, p. 330] Suppose $\mathbf{A}$ is a Hermitian complex array. Then there exists a unitary array $\mathbf{U}$ such that

$$\mathbf{U}^\dagger\mathbf{A}\mathbf{U}$$

is diagonal. Equivalently, there exists an orthonormal eigenbasis for $\mathbf{A}$.

The proof of Theorem 15.20 makes necessary use of the algebraic properties of $\mathbb{C}$, particularly the fact that it is an algebraically closed field and that there is a notion of positivity which allows things like an inner product to make sense.

Proposition 15.21

Existence of Singular Value Decomposition

[5, p. 368] Suppose $\mathbf{A}$ is a complex $m \times n$ matrix. Then there exist a unitary $m \times m$ matrix $\mathbf{U}$, a diagonal $m \times n$ array $\mathbf{\Sigma}$, and a unitary $n \times n$ array $\mathbf{V}$ such that

$$\mathbf{A} = \mathbf{U}\mathbf{\Sigma}\mathbf{V}^\dagger$$

Proof. $\mathbf{A}^\dagger\mathbf{A}$ is positive semi-definite, as

$$\mathbf{v}^\dagger\mathbf{A}^\dagger\mathbf{A}\mathbf{v} = ||\mathbf{A}\mathbf{v}|| \geq 0$$

Thus, its eigenvalues are non-negative real numbers. Let $\sigma_1, \ldots, \sigma_r$ be the square roots of positive eigenvalues and take $\sigma_i = 0$ for $i > r$. Since $\mathbf{A}^\dagger\mathbf{A}$ is Hermitian so that it is equal to its conjugate transpose, it follows from Theorem 15.20 that there exists an orthonormal eigenbasis of $\mathbb{C}^n$, say $\mathbf{v}_1, \ldots, \mathbf{v}_n$. It can be assumed that the eigenvalues associated with $\mathbf{v}_1, \ldots, \mathbf{v}_n$ are $\sigma_1^2, \ldots, \sigma_n^2$ by permuting as necessary. r is the rank of $\mathbf{A}^\dagger\mathbf{A}$, and it can be shown that this is also the rank of $\mathbf{A}$. To see this, it suffices to show that their null spaces agree by the rank-nullity theorem. If $\mathbf{x}$ is in the null space of $\mathbf{A}$, then

$$\mathbf{A}^\dagger\mathbf{A}\mathbf{x} = \mathbf{A}^\dagger\mathbf{0} = \mathbf{0}$$

and if $\mathbf{x}$ is in the null space of $\mathbf{A}^\dagger\mathbf{A}$, then

$$\mathbf{x}^\dagger\mathbf{A}^\dagger\mathbf{A}\mathbf{x} = 0 = ||\mathbf{A}\mathbf{x}||$$

so $\mathbf{Ax} = \mathbf{0}$. By definition

$$\mathbf{A}^\dagger \mathbf{A}\mathbf{v}_i = \sigma_i^2 \mathbf{v}_i$$

for each i, so by multiplying $\mathbf{A}$ on both sides it follows that

$$\mathbf{A}\mathbf{A}^\dagger (\mathbf{A}\mathbf{v}_i) = \sigma_i^2 (\mathbf{A}\mathbf{v}_i)$$

Let

$$\mathbf{u}_i = \sigma_i^{-1} \mathbf{A}\mathbf{v}_i$$

and note that for $1 \le i \le r$ the vector $\mathbf{u}_i$ is nonzero, as otherwise

$$\sigma_i \mathbf{v}_i = \mathbf{0}$$

contradicts the fact that both σ_i and $\mathbf{v}_i$ are nonzero. Thus, $\mathbf{u}_i$ is an eigenvector of $\mathbf{A}\mathbf{A}^\dagger$ with eigenvalue σ_i^2. The set of vectors $\mathbf{u}_1, \ldots, \mathbf{u}_r$ is orthonormal, as

$$\mathbf{u}_i^\dagger \mathbf{u}_j = \sigma_i^{-1} \sigma_j^{-1} \mathbf{v}_i^\dagger \mathbf{A}^\dagger \mathbf{A}\mathbf{v}_j = \sigma_j \sigma_i^{-1} \mathbf{v}_i^\dagger \mathbf{v}_j$$

which is 1 if $i = j$ and 0 otherwise, since the set of vectors $\mathbf{v}_1, \ldots, \mathbf{v}_n$ are orthonormal. Since the set $\mathbf{u}_1, \ldots, \mathbf{u}_r$ is orthonormal, it can be extended to an orthonormal eigenbasis $\mathbf{u}_1, \ldots, \mathbf{u}_m$ of $\mathbf{A}\mathbf{A}^\dagger$. The above shows that the nonzero eigenvalues of $\mathbf{A}^\dagger \mathbf{A}$ are eigenvalues of $\mathbf{A}\mathbf{A}^\dagger$, and the same argument shows the converse. Hence, the remaining vectors $\mathbf{u}_{r+1}, \ldots, \mathbf{u}_m$ are associated with the eigenvalue 0. Now define $\mathbf{U}$ and $\mathbf{V}$ to be the block matrices

$$\mathbf{U} = \begin{bmatrix} \mathbf{u}_1 & \cdots & \mathbf{u}_m \end{bmatrix}$$
$$\mathbf{V} = \begin{bmatrix} \mathbf{v}_1 & \cdots & \mathbf{v}_n \end{bmatrix}$$

Because their columns form orthonormal bases, they are unitary matrices. Define $\mathbf{\Sigma}$ to be the $m \times n$ array whose first r diagonal entries are $\sigma_1, \ldots, \sigma_r$ and the remaining entries 0. Finally, it is shown that

$$\mathbf{A} = \mathbf{U}\mathbf{\Sigma}\mathbf{V}^\dagger$$

or equivalently that

$$\mathbf{A}\mathbf{V} = \mathbf{U}\mathbf{\Sigma}$$

By the linear independence of the columns of $\mathbf{V}$ and the fact that

$$\mathbf{A}\mathbf{v}_i \neq \mathbf{0}$$

for $i \le r$, it follows that

$$\mathbf{A}\mathbf{v}_i = \mathbf{0}$$

for $i > r$. Thus, the last $n - r$ columns of $\mathbf{AV}$ are $\mathbf{0}$. Likewise, the last $n - r$ columns of $\mathbf{U\Sigma}$ are $\mathbf{0}$ since the same holds true for $\mathbf{\Sigma}$. This means that it suffices to check that

$$\mathbf{A}\mathbf{v}_i = \sigma_i \mathbf{u}_i$$

for $i \leq r$, or that

$$\mathbf{u}_i = \sigma_i^{-1} \mathbf{A} \mathbf{v}_i$$

But this is exactly the definition of $\mathbf{u}_i$ for $i \leq r$, completing the proof. □

A certain converse can be given to the construction of the SVD given in the proof of Proposition 15.21, by showing that the squares of the diagonal entries of $\mathbf{\Sigma}$ are eigenvalues for $\mathbf{A}\mathbf{A}^\dagger$ and $\mathbf{A}^\dagger\mathbf{A}$, that the columns of $\mathbf{U}$ are eigenvectors for $\mathbf{A}\mathbf{A}^\dagger$, and that the columns of $\mathbf{V}$ are eigenvectors for $\mathbf{A}^\dagger\mathbf{A}$. For this reason, the existence of the SVD is closely related to the existence of eigenvalues and eigenvectors.

The SVD of an array is theoretically very nice, as the first r columns of $\mathbf{U}$ give a basis for the column space of $\mathbf{A}$ (equivalently, the row space of $\mathbf{A}^\dagger$), the last $m-r$ columns of $\mathbf{U}$ give a basis for the null space of $\mathbf{A}$, the first r columns of $\mathbf{U}$ give a basis for the row space of $\mathbf{A}$ (equivalently, the column space of $\mathbf{A}$), and the last $n-r$ columns of $\mathbf{U}$ give a basis for the null space of $\mathbf{A}^\dagger$.

The proof of Proposition 15.21 makes extensive use of the algebraic properties of $\mathbb{C}$ and suggests the following reasons why a generalization of the SVD to semirings has been elusive

(i) It is unclear what the appropriate generalization of the conjugate transpose should be, and with it the notions of an inner product and of Hermitian and unitary matrices.

(ii) Even if an inner product (or some generalization thereof) is created, division is not generally possible, and it is difficult to normalize a vector normalized by its magnitude.

(iii) In general, the eigenvalues and eigenvectors for $\mathbf{A}^\dagger\mathbf{A}$ need not exist.

(iv) Even if eigenvalues and eigenvectors for $\mathbf{A}^\dagger\mathbf{A}$ do exist, there seems no reason to suggest that they appear in the right numbers. There may not be enough to form the columns of an appropriately sized array and if they can be chosen to form the columns of a unitary array.

(v) Even if the eigenvalues and eigenvectors for $\mathbf{A}^\dagger\mathbf{A}$ exist in such a way that they can be chosen to form the columns of a unitary array, these eigenvalues need not have square roots, and so while $\mathbf{U}$ and $\mathbf{V}$ might exist, $\mathbf{\Sigma}$ might not.

Research on possible generalizations to the semirings of interest is ongoing, with the focus on the case of the max-plus algebra, as in [6]. In this paper, the existence of an SVD is given for an extension of the incomplete max-plus algebra, which seeks to address the issue of a lack of additive inverses. To gain motivation for this extension, consider

$$\mathbb{N} = \{0, 1, 2, \ldots\}$$

A standard approach to constructing $\mathbb{Z}$ from $\mathbb{N}$ is to define the set of equivalence classes of $\mathbb{N} \times \mathbb{N}$ under the equivalence relation

$$(n,m) \sim (p,q) \quad \text{if and only} \quad n+q = p+m$$

The intuition is that a pair (n,m) in $\mathbb{N}\times\mathbb{N}$ represents the difference $n-m$, and the equivalence relation identifies equal differences, and indeed

$$n-m=p-q \quad \text{if and only} \quad n+q=p+m$$

This intuition also suggests what the definitions of addition and multiplication should be between (n,m) and (p,q). Writing (n,m) as $n-m$ and (p,q) as $p-q$, then

$$(n-m)+(p-q)=(n+p)-(m+q)$$

so addition is defined by

$$(n,m)+(p,q)=(n+p,m+q)$$

Likewise

$$(n-m)(p-q)=(np+mq)-(mp+nq)$$

so multiplication is defined by

$$(n,m)(p,q)=(np+mq,mp+nq)$$

The equivalence relation $\sim$ is *compatible* with the operations addition and multiplication of pairs, which is to say that if

$$(n,m)\sim(n',m')$$

and

$$(p,q)\sim(p',q')$$

then

$$(n,m)+(p,q)\sim(n',m')+(p',q')$$

and

$$(n,m)(p,q)\sim(n',m')(p',q')$$

The above relation allows addition and multiplication to be defined between equivalence classes, giving the addition and multiplication operations on $\mathbb{Z}$. Showing that a semiring ($\mathbb{N}$) can be turned into a ring ($\mathbb{Z}$), motivates a similar approach for the incomplete max-plus algebra by considering

$$(\mathbb{R}\cup\{-\infty\})^2$$

and the relation

$$(t,u)\,\triangledown\,(v,w) \quad \text{if and only} \quad \max(v,u)=\max(t,w)$$

However, the relation $\triangledown$ is *not* an equivalence relation. While it is reflexive and symmetric, it fails to be transitive, as the following example indicates.

Example 15.16

[6, p. 424]

$$(3,3) \triangledown (1,3) \quad \text{and} \quad (3,3) \triangledown (1,3)$$

since

$$\max(3,3) = 3 = \max(3,0) \quad \text{and} \quad \max(3,3) = 3 = \max(3,1)$$

However

$$(3,0) \not\triangledown (1,3)$$

since

$$\max(3,3) = 3 \neq 1 = \max(0,1)$$

Definition 15.14

Symmetrized Max-Plus Algebra

De Shutter and De Moor correct this relation by defining a new relation B by

$$(t,u)\,\mathrm{B}\,(v,w) \quad \text{if and only if} \quad \begin{cases} (t,u) \triangledown (v,w) & \text{if } t \neq u \text{ and } v \neq w \\ (t,u) = (v,w) & \text{otherwise} \end{cases}$$

B is called the *balancing relation*. This relation is an equivalence relation. The equivalence relations of B are called *max-plus-positive* for

$$[(w,\text{-}\infty)] = \{(w,v) \mid v < w\}$$

and *max-plus-negative* for

$$[(\text{-}\infty,w)] = \{(v,w) \mid v < w\}$$

and *balanced* for

$$[(w,w)] = \{(w,w)\}$$

for any

$$w \in \mathbb{R} \cup \{\text{-}\infty\}$$

$\mathbb{S}$ is the collection of all equivalence classes, $\mathbb{S}^+$ is the collection of all max-plus-positive equivalence classes, $\mathbb{S}^-$ is the collection of all max-plus-negative equivalence classes, and $\mathbb{S}^\bullet$ is the collection of all balanced equivalence classes. Finally

$$\mathbb{S}^\vee = \mathbb{S}^+ \cup \mathbb{S}^-$$

The algebraic operations on $\mathbb{S}$ are defined analogously to how the algebraic operations were defined on $\mathbb{Z}$.

Definition 15.15

Operations and Relations on Symmetrized Max-Plus Algebra

For pairs

$$(x,y),(z,w) \in (\mathbb{R} \cup \{-\infty\})^2$$

define

$$\max((t,u),(v,w)) = (\max(t,v),\max(u,w))$$

and

$$(t,u)+(v,w) = (\max(t+v,u+w),\max(t+w,u+v))$$

B is compatible with both of these operations, and so they can be defined between equivalence classes. This definition gives rise to the semiring

$$(\mathbb{S},\max,+,[(-\infty,-\infty)],[(0,0)])$$

called the *symmetrized max-plus algebra*. Several other operations may be defined, like

$$\ominus(v,w) = (w,v)$$

which is the analog of negation,

$$(v,w)^{\bullet} = \max((v,w),\ominus(v,w)) = (\max(v,w),\max(v,w))$$

called the *balancing operation*, and

$$|(v,w)| = \max(v,w)$$

which is the analog of a norm, and B is compatible with them as well. Finally, B is compatible with ∇, in the sense that if

$$(t,u)\,\mathrm{B}\,(t',u')$$

and

$$(v,w)\,\mathrm{B}\,(v',w')$$

then

$$(t,u)\,\nabla\,(v,w)$$

implies

$$(t',u')\,\nabla\,(v',w')$$

The relation ∇ is finer than B, so even though two equivalence classes might be ∇-related, they need not be equal or B-related.

With this notation in mind, the main result concerning the SVD can be given.

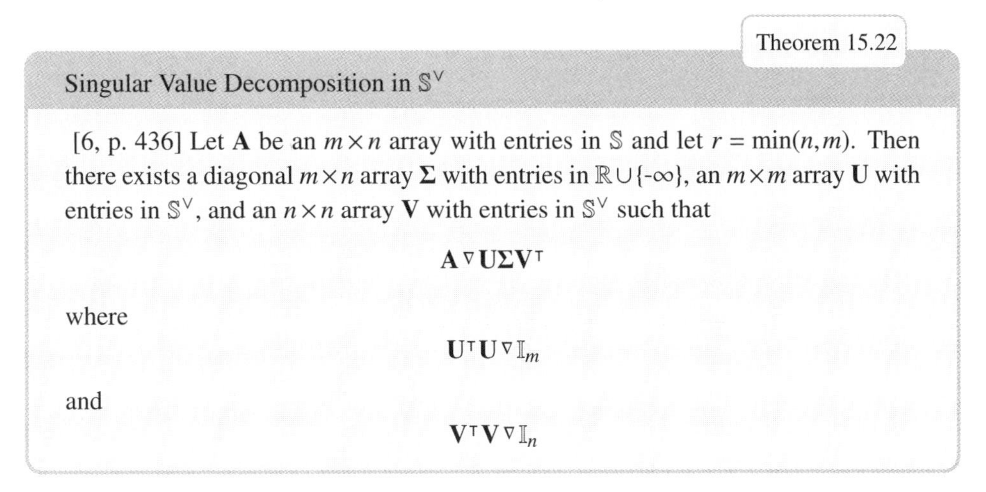

Theorem 15.22

Singular Value Decomposition in $\mathbb{S}^\vee$

[6, p. 436] Let $\mathbf{A}$ be an $m \times n$ array with entries in $\mathbb{S}$ and let $r = \min(n,m)$. Then there exists a diagonal $m \times n$ array $\mathbf{\Sigma}$ with entries in $\mathbb{R} \cup \{-\infty\}$, an $m \times m$ array $\mathbf{U}$ with entries in $\mathbb{S}^\vee$, and an $n \times n$ array $\mathbf{V}$ with entries in $\mathbb{S}^\vee$ such that

$$\mathbf{A} \nabla \mathbf{U}\mathbf{\Sigma}\mathbf{V}^\intercal$$

where

$$\mathbf{U}^\intercal \mathbf{U} \nabla \mathbb{I}_m$$

and

$$\mathbf{V}^\intercal \mathbf{V} \nabla \mathbb{I}_n$$

Here, many of the issues described above with capturing the SVD are bypassed by considering the SVD up to ∇ rather than requiring a strict equality.

There are also connections between singular values in the max-plus algebra and exponential functions [6, 7].

15.8 Conclusions, Exercises, and References

The eigenvalue problem has been extensively studied from the perspective of max.min and min.max algebras. Extending eigenvalues and eigenvectors to the broader class of associative arrays begins with defining quasi-inverses, which provides the necessary foundation for defining existence in the context of idempotent multiplication. The characteristic bipolynomial can be used to redefine the eigenvalue problem in terms that are well-defined for semirings. Additional structures can also be defined for eigen-semimodules.

Exercises

Exercise 15.1 — Find eigenvalues and eigenvectors of

$$\mathbf{A} = \begin{array}{c} \\ 1 \\ 2 \end{array} \begin{array}{c} \begin{array}{cc} 1 & 2 \end{array} \\ \left[\begin{array}{cc} 3 & -1 \\ 2 & 0 \end{array}\right] \end{array}$$

by direct computation in both the max-plus and tropical max-min algebras.

Exercise 15.2 — Assume $(V,\oplus,\otimes,0,1)$ is a join-blank algebra in which 1 is the largest element.
A *quasi-inverse* of a is the minimal solution a^* to

$$a^* \otimes a \oplus 1 = a \otimes a^* \oplus 1 = a^*$$

Show that the quasi-inverse is given by $a^* = 1 \oplus a \oplus a^2 \oplus \cdots = \bigvee_{k \geq 0} a^{(k)}$, where $a^{(k)} = 1 \oplus a \oplus \cdots \oplus a^k$.

Exercise 15.3 — Do quasi-inverses exist in general? How necessary were the order-theoretic completeness properties of the semirings of interest?

Exercise 15.4 — Calculate the quasi-inverse of

$$\mathbf{A} = \begin{array}{c} \\ 1 \\ 2 \end{array} \begin{array}{c} \begin{array}{cc} 1 & 2 \end{array} \\ \begin{bmatrix} -4 & 0 \\ 0 & -3 \end{bmatrix} \end{array}$$

(in the max-plus algebra).

Exercise 15.5 — Prove Lemma 15.3.

Exercise 15.6 — Prove Corollary 15.4.

Exercise 15.7 — Calculate the characteristic bipolynomial of

$$\mathbf{A} = \begin{array}{c} \\ 1 \\ 2 \end{array} \begin{array}{c} \begin{array}{cc} 1 & 2 \end{array} \\ \begin{bmatrix} 3 & -1 \\ 2 & 0 \end{bmatrix} \end{array}$$

by direct computation in the incomplete max-plus algebra and use it to find the eigenvalues of $\mathbf{A}$.

Exercise 15.8 — Show that the array

$$\mathbf{A} = \begin{array}{c} \\ 1 \\ 2 \end{array} \begin{array}{c} \begin{array}{cc} 1 & 2 \end{array} \\ \begin{bmatrix} -3 & 1 \\ 1 & -2 \end{bmatrix} \end{array}$$

over the incomplete max-plus algebra is irreducible and calculate its spectral radius. Confirm that it is an eigenvalue for $\mathbf{A}$.

Exercise 15.9 — Prove Lemma 15.11.

Exercise 15.10 — Prove Lemma 15.12.

Exercise 15.11 — Prove Lemma 15.16.

Exercise 15.12 — By calculating $\tilde{\mathbf{A}}^*$ (see Exercise 15.4) and using Exercise 15.8, find $\mathscr{V}(\lambda)$ (as a span of eigenvectors) of the incomplete max-plus algebraic array

$$\mathbf{A} = \begin{array}{c} \\ 1 \\ 2 \end{array} \begin{array}{c} \begin{array}{cc} 1 & 2 \end{array} \\ \begin{bmatrix} -3 & 1 \\ 1 & -2 \end{bmatrix} \end{array}$$

where λ is the unique eigenvalue for $\mathbf{A}$.

References

[1] E. Sanchez, "Resolution of eigen fuzzy sets equations," *Fuzzy Sets and Systems*, vol. 1, no. 1, pp. 69–74, 1978.

[2] K. Cechlárová, "Eigenvectors in bottleneck algebra," *Linear Algebra and Its Applications*, vol. 175, pp. 63–73, 1992.

[3] J. Plavka and P. Szabó, "On the λ-robustness of matrices over fuzzy algebra," *Discrete Applied Mathematics*, vol. 159, no. 5, pp. 381–388, 2011.

[4] M. Gondran and M. Minoux, *Graphs, Dioids and Semirings: New Models and Algorithms*, vol. 41. New York: Springer Science & Business Media, 2008.

[5] G. Strang, *Linear Algebra and Its Applications*, vol. 4. Wellesley, MA: Wellesley-Cambridge Press, 2006.

[6] B. De Schutter and B. De Moor, "The QR decomposition and the singular value decomposition in the symmetrized max-plus algebra revisited," *SIAM Review*, vol. 44, no. 3, pp. 417–454, 2002.

[7] J. Hook, "Max-plus singular values," *Linear Algebra and Its Applications*, vol. 486, pp. 419–442, 2015.

16 Higher Dimensions

Summary

Most of the previously described concepts for associative arrays are not dependent upon associative arrays being two-dimensional, but this is naturally where the intuition for associative array definitions resides because matrices are two-dimensional. d-dimensional associative arrays are a straightforward extension of their two-dimensional counterparts, and their ability to represent diverse data extends fully to higher dimensions. This chapter provides the necessary formal mathematics for extending associative arrays into arbitrary dimensions.

16.1 d-Dimensional Associative Arrays

The extension of matrix concepts to higher dimensional data has a long history [1–3]. These higher dimension matrices, referred to as tensors, are useful for a wide variety of applications, including psychology [4], chemistry [5], signal processing [6], data analysis [7, 8], and data compression [9]. A key challenge of tensors is the question of how to define tensor multiplication $\oplus.\otimes$. To the extent that associative arrays rely on matrix intuition, the challenge of extending array multiplication to higher dimensions is the same challenge presented by tensor multiplication and is left to that domain [10]. Fortunately, for many of the properties of associative arrays that broaden matrices, the extension of the definition of associative arrays to arbitrary dimensions greater than two is straightforward.

Definition 16.1

d-Dimensional Associative Array

A *d-dimensional associative array* is a map from a strict totally ordered set of keys to a value set with a semiring structure. Formally, an associative array is a map

$$\mathbf{A} : K_1 \times \cdots \times K_d \to V$$

that has a finite number of nonzero elements, or equivalently has finite support, and where V is the semiring

$$(V, \oplus, \otimes, 0, 1)$$

Additionally

$$\mathbb{A}(K_1,\ldots,K_d;V)$$

is used to denote the set of d-dimensional associative arrays with domain

$$K_1\times\cdots\times K_d$$

and codomain the commutative semiring

$$(V,\oplus,\otimes,0,1)$$

Unless otherwise mentioned, in this chapter $\mathbb{A}$ will refer to the general case

$$\mathbb{A}(K_1,\ldots,K_d;V)$$

Elements of

$$K_1\times\cdots\times K_d$$

are called *key tuples* and elements of V called *values*. One of the main differences between the d-dimensional and two-dimensional cases are that array multiplication is no longer as easily defined as it is in the one- and two-dimensional cases. However, element-wise addition, element-wise (Hadamard) multiplication, and scalar multiplication continue to be defined and have the same nice properties present in the one- and two-dimensional cases. The other main construction that remains is zero padding, which follows the same procedure as in the two-dimensional case. Given

$$\mathbf{A}: K_1\times\cdots\times K_d\to V$$

and

$$K_1\times\cdots\times K_d\subset K'_1\times\cdots\times K'_d$$

the unique zero padding of $\mathbf{A}$ to the new key tuples

$$K'_1\times\cdots\times K'_d$$

is denoted

$$\mathbf{A}' = \mathrm{pad}_{K'_1\times\cdots\times K'_d}\mathbf{A}$$

Similarly, in the d-dimensional case, the same procedure can be used to make an array $\mathbf{A}$ square.

A d-dimensional associative array

$$\mathbf{A}: K_1\times\cdots\times K_d\to V$$

is *square* if $K_j = K_i$ for every i, j, in which case

$$K_1\times\cdots\times K_d = K^d$$

where K is the common value of the sets K_j.

Example 16.1

Given any d-dimensional associative array

$$\mathbf{A} : K_1 \times \cdots \times K_d$$

a new d-dimensional associative array

$$\text{sq}\,\mathbf{A} : (K_1 \cup \cdots \cup K_d)^d \to V$$

can be defined where $\text{sq}\,\mathbf{A}$ is defined by

$$\text{sq}\,\mathbf{A}(\mathbf{k}) = \begin{cases} \mathbf{A}(\mathbf{k}) & \mathbf{k} \in K_1 \times \cdots \times K_d, \\ 0 & \text{otherwise.} \end{cases}$$

Then with $K = K_1 \cup \cdots \cup K_d$, it follows that

$$\text{sq}\,\mathbf{A} = \text{pad}_{K,\ldots,K}\mathbf{A}$$

Many of the properties defined for two-dimensional arrays and matrices continue to be defined here with straightforward extension to the d-dimensional case:

dim — dimension of a d-dimensional array $\mathbf{A} : K_1 \times \cdots \times K_d \to V$ is

$$\dim(\mathbf{A}) = (|K_1|, \ldots, |K_d|)$$

total — number of values

$$\text{total}(\mathbf{A}) = |K_1| \cdots |K_d|$$

support — tuples of keys corresponding to nonzero values

$$\text{support}(\mathbf{A}) = \{(k_1, \ldots, k_d) \in K_1 \times \ldots \times K_d : \mathbf{A}(k_1, \ldots, k_d) \neq 0\}$$

nnz — number of nonzero values

$$\text{nnz}(\mathbf{A}) = |\text{support}(\mathbf{A})|$$

density — fraction of values that are nonzero

$$\text{density}(\mathbf{A}) = \frac{\text{nnz}(\mathbf{A})}{\text{total}(\mathbf{A})}$$

sparsity — fraction of values that are zero

$$\text{sparsity}(\mathbf{A}) = 1 - \text{density}(\mathbf{A})$$

size — number of non-empty slices in the i-th dimension

$$\text{size}(\mathbf{A}, i) = |\pi_i(\text{support}(\mathbf{A}))|$$
$$\text{size}(\mathbf{A}) = (\text{size}(\mathbf{A}, 1), \ldots, \text{size}(\mathbf{A}, d))$$

total — number of values in the non-empty slices

$$\underline{\text{total}}(\mathbf{A}) = \text{size}(\mathbf{A}, 1) \cdots \text{size}(\mathbf{A}, d)$$

density — fraction of values in non-empty slices that are nonzero

$$\underline{\text{density}}(\mathbf{A}) = \frac{\text{nnz}(\mathbf{A})}{\underline{\text{total}}(\mathbf{A})}$$

sparsity — fraction of values in non-empty slices that are zero

$$\underline{\text{sparsity}}(\mathbf{A}) = 1 - \underline{\text{density}}(\mathbf{A})$$

Example 16.2

The *zero array* $\mathbb{0} \in \mathbb{A}$ has

$$\text{size}(\mathbb{0}) = (0, \ldots, 0)$$

Example 16.3

The *unit arrays* $\mathbf{e_k} \in \mathbb{A}$ have

$$\text{size}(\mathbf{e_k}) = (1, \ldots, 1)$$

16.2 Key Ordering and Two-Dimensional Projections

Many definitions of structures of an array can be reduced to the two-dimensional case.

Definition 16.2

Two-Dimensional Projections

Given an associative array $\mathbf{A} \in \mathbb{A}$, define a family

$$\mathbf{A}_{p,q} : K_p \times K_q \to V$$

of *two-dimensional projections* of $\mathbf{A}$ for $1 \leq p < q \leq d$ by

$$\mathbf{A}_{p,q}(i_1, i_2) = \begin{cases} 1 & \text{if there is } \mathbf{k} \in K_1 \times \cdots \times K_d \text{ such that } \mathbf{A}(\mathbf{k}) \neq 0 \\ & \quad \text{and } \mathbf{k}(p) = i_1 \text{ and } \mathbf{k}(q) = i_2 \\ 0 & \text{otherwise} \end{cases}$$

The associative arrays $\mathbf{A}_{p,q}$ can be thought of as "projections" of $\mathbf{A}$ onto each of the two dimensions p and q, but only recording if there was a nonzero entry in those dimensions so that

$$\mathbf{A}\Big(K_1 \times \cdots \times K_{p-1} \times \{i_1\} \times K_{p+1} \times \cdots \times K_{q-1} \times \{i_2\} \times K_{q+1} \times \cdots \times K_d\Big) \neq \{0\}$$

or that the above set contains a nonzero value. Then define the graph of an array

$$\mathbf{A} : K^d \to V$$

for $d > 1$ by the family of graphs of the two-dimensional projections $\mathbf{A}_{p,q}$.

Example 16.4

Let $\mathbf{A} : \{1,\ldots,3\}^3 \to \mathbb{N}$ be defined by each of the slices

$$\mathbf{A}(:,:,1) = \begin{array}{c} \\ 1 \\ 2 \\ 3 \end{array} \begin{array}{c} \begin{array}{ccc} 1 & 2 & 3 \end{array} \\ \left[\begin{array}{ccc} 1 & 2 & \\ & 7 & \\ & & 3 \end{array}\right] \end{array} \qquad \mathbf{A}(:,:,2) = \begin{array}{c} \\ 1 \\ 2 \\ 3 \end{array} \begin{array}{c} \begin{array}{ccc} 1 & 2 & 3 \end{array} \\ \left[\begin{array}{ccc} 8 & & 9 \\ & & \\ & 4 & \end{array}\right] \end{array}$$

$$\mathbf{A}(:,:,3) = \begin{array}{c} \\ 1 \\ 2 \\ 3 \end{array} \begin{array}{c} \begin{array}{ccc} 1 & 2 & 3 \end{array} \\ \left[\begin{array}{ccc} & & 1 \\ & 4 & \\ 1 & 1 & 1 \end{array}\right] \end{array}$$

Then the two-dimensional projection $\mathbf{A}_{1,2} : \{1,\ldots,3\}^2 \to \mathbb{N}$ is given by

$$\mathbf{A}_{1,2} = \begin{array}{c} \\ 1 \\ 2 \\ 3 \end{array} \begin{array}{c} \begin{array}{ccc} 1 & 2 & 3 \end{array} \\ \left[\begin{array}{ccc} 1 & 1 & 1 \\ & 1 & \\ 1 & 1 & 1 \end{array}\right] \end{array}$$

Example 16.5

The two-dimensional projection of a two-dimensional array

$$\mathbf{A} : K_1 \times K_2 \to V$$

is not $\mathbf{A}$ but is the result of replacing every nonzero entry of $\mathbf{A}$ with 1.

Example 16.6

The two-dimensional projections of the zero array $\mathbb{0} \in \mathbb{A}$ are the zero arrays

$$\mathbb{0} : K_p \times K_q \to V$$

Example 16.7

The two-dimensional projections of the unit array

$$\mathbf{e}_{(k_1,\ldots,k_d)} :\in \mathbb{A}$$

are the unit arrays

$$\mathbf{e}_{(k_p,k_q)} : K_p \times K_q \to V$$

An associative array $\mathbf{A} \in \mathbb{A}$ is *diagonal* if

$$\mathbf{A}(\mathbf{k}) \neq 0 \quad \text{if and only if} \quad \mathbf{k} = (k,\ldots,k)$$

where

$$k \in K_1 \cap \cdots \cap K_d$$

Likewise, $\mathbf{A}$ is *partially diagonal* if

$$\mathbf{A}(\mathbf{k}) \neq 0 \quad \text{implies} \quad \mathbf{K} = (k,\ldots,k)$$

for some

$$k \in K_1 \cap \cdots \cap K_d$$

(Partial) diagonality can be connected rather simply to the (partial) diagonality of the two-dimensional projections of $\mathbf{A}$.

Proposition 16.1

Diagonality and Two-Dimensional Projections

$\mathbf{A} \in \mathbb{A}$ is (partially) diagonal if and only if each of the two-dimensional projections

$$\mathbf{A}_{p,q} : K_p \times K_q \to V$$

is (partially) diagonal.

Proof. Suppose that $\mathbf{A} \in \mathbb{A}$ is (partially) diagonal, so that

$$\mathbf{A}(\mathbf{k}) \neq 0 \quad \text{implies} \quad \mathbf{k} = (k,\ldots,k)$$

$\mathbf{A}_{p,q}$ satisfies

$$\mathbf{A}_{p,q}(i_1, i_2) \neq 0$$

if and only if there exists $\mathbf{k}$ such that

$$\mathbf{A}(\mathbf{k}) \neq 0$$

and

$$\mathbf{k}(p) = i_1 \quad \text{and} \quad \mathbf{k}(q) = i_2$$

But

$$\mathbf{A}(\mathbf{k}) \neq 0 \quad \text{implies} \quad \mathbf{k} = (k, \ldots, k)$$

so

$$i_1 = i_2 = k$$

Moreover,

$$k \in K_p \cap K_q$$

since

$$K_1 \cap \cdots \cap K_d \subset K_p \cap K_q$$

and thus $\mathbf{A}_{p,q}$ is diagonal.

Conversely, suppose that each of the two-dimensional projections

$$\mathbf{A}_{p,q} : K_p \times K_q \to V$$

is diagonal. Suppose, furthermore, for the sake of a contradiction that there is $\mathbf{k}$ such that

$$\mathbf{A}(\mathbf{k}) \neq 0$$

but

$$\mathbf{k} \neq (k, \ldots, k)$$

for any k. Thus, there exists p and q in $\{1, \ldots, d\}$ with $p < q$ such that

$$\mathbf{k}(p) \neq \mathbf{k}(q)$$

Then

$$\mathbf{A}_{p,q}(\mathbf{k}(p), \mathbf{k}(q)) \neq 0$$

but

$$\mathbf{k}(p) \neq \mathbf{k}(q)$$

contradicting the assumption that each of the two-dimensional projections is diagonal. □

Weak diagonality also has a simple generalization as well: an associative array $\mathbf{A} \in \mathbb{A}$ is *weakly diagonal* if, given any element k of K_i for $1 \leq i \leq d$, the size of

$$\text{support}(\mathbf{A}) \cap (K_1 \times \cdots \times K_{i-1} \times \{k\} \times K_{i+1} \times \cdots \times K_d)$$

is at most 1.

Proposition 16.2

Weak Diagonality and Two-Dimensional Projections

$\mathbf{A} \in \mathbb{A}$ is weakly diagonal if and only if each of the two-dimensional projections

$$\mathbf{A}_{p,q} : K_p \times K_q \to V$$

is weakly diagonal.

Proof. Suppose that $\mathbf{A}$ is weakly diagonal, so that each set

$$\text{support}(\mathbf{A}) \cap (K_1 \times \cdots \times K_{i-1} \times \{k\} \times K_{i+1} \times \cdots \times K_d)$$

has size at most 1 for $k \in K_i$ and $i \in \{1, \ldots, d\}$. If a two-dimensional projection

$$\mathbf{A}_{p,q} : K_p \times K_q \to V$$

is not weakly diagonal, then there exists k_p in K_p such that

$$\text{support}(\mathbf{A}_{p,q}) \cap (\{k_p\} \times K_q)$$

has size at least 2. Let

$$a_p, b_p \in K_p \quad \text{and} \quad a_q, b_q \in K_q$$

be such that

$$\mathbf{A}_{p,q}(a_p, a_q) \neq 0 \quad \text{and} \quad \mathbf{A}_{p,q}(b_p, b_q) \neq 0$$

where either $a_p = b_p$ or $a_q = b_q$ (but not both). By definition

$$\mathbf{A}_{p,q}(a_p, a_q) \neq 0$$

if and only if there is

$$a \in K_1 \times \cdots \times K_d$$

such that $\mathbf{A}(a) \neq 0$ and $a(p) = a_p$ and $a(q) = a_q$. Similarly, there is

$$b \in K_1 \times \cdots \times K_d$$

such that $\mathbf{A}(b) \neq 0$ and $b(p) = b_p$ and $b(q) = b_q$. If $a_{j_n} = b_{j_n}$, then

$$\text{support}(\mathbf{A}) \cap \left(K_1 \times \cdots \times K_{j_n - 1} \times \{a_{j_n}\} \times K_{j_n + 1} \times \cdots \times K_d\right)$$

has cardinality at least 2, contradicting the fact that $\mathbf{A}$ was weakly diagonal. Thus, each of the two-dimensional projections is weakly diagonal. Now suppose that each of the two-dimensional projections is weakly diagonal. Further suppose that there exists i and $k \in K_i$ such that

$$\text{support}(\mathbf{A}) \cap (K_1 \times \cdots \times K_{i-1} \times \{k\} \times K_{i+1} \times \cdots \times K_d)$$

has cardinality at least 2, so that $\mathbf{A}$ is not weakly diagonal. Let $\mathbf{k}$ and $\mathbf{k}'$ be distinct elements of this set, and let i' be an element of $\{1,\ldots,d\}$ such that

$$\mathbf{k}(i') \neq \mathbf{k}'(i')$$

which must exist because κ and κ' are distinct. Note that $i' \neq i$ because it is necessarily the case that

$$\mathbf{k}(i) = \mathbf{k}'(i) = k$$

If $i < i'$, then let $p = i$ and $q = i'$. Otherwise, let $p = i'$ and $q = i$. Then the two-dimensional projection $\mathbf{A}_{p,q}$ maps either $(\mathbf{k}(i'),k)$ and $(\mathbf{k}'(i'),k)$ (if $i' < i$) or $(k,\mathbf{k}(i'))$ and $(k,\mathbf{k}'(i'))$ to 1, showing that $\mathbf{A}_{p,q}$ is not weakly diagonal. Thus, it must be the case that $\mathbf{A}$ is weakly diagonal. □

The issue of the key sets being unordered continues in higher dimensions, thus motivating the generalization of an ordered associative array.

Definition 16.3

Ordered Associative Array

An *ordered associative array* is defined as an associative array $\mathbf{A} \in \mathbb{A}$ where each K_j is equipped with a fixed total order $\leq_j$ and each K_j is finite.

Using the analogy with (weak) diagonality, say that an ordered associative array $\mathbf{A} \in \mathbb{A}$ is *(partially) upper triangular* if each two-dimensional projection is upper triangular. An analogous definition exists for (partial) lower triangularity. These notions can be extended to ordered associative arrays.

Suppose

$$K_1,\ldots,K_d$$

are finite totally ordered key sets with orderings

$$<_1,\ldots,<_d$$

and

$$\mathbf{A} : K_1 \times \cdots \times K_d \to V$$

is an associative array. K_j is order isomorphic to $\{1,\ldots,m_j\}$ for some m_j, and let

$$\mathbf{B} : \{1,\ldots,m_1\} \times \cdots \times \{1,\ldots,m_d\} \to V$$

be the associative array obtained by identifying

$$\{1,\ldots,m_1\} \times \cdots \times \{1,\ldots,m_d\}$$

with

$$K_1 \times \cdots \times K_d$$

$\mathbf{B}$ is the *matrix associated with* $\mathbf{A}$ (with respect to $<_1,\ldots,<_d$).

Definition 16.4

Ordered Diagonality, Triangularity

Suppose $\mathbf{A} : K_1 \times \cdots \times K_d \to V$ is an array with $K_1,\ldots,K_d$ finite and strict totally ordered by $<_1,\ldots,<_d$, respectively. Let $\mathbf{B}$ be the matrix associated with $\mathbf{A}$ with respect to $<_1,\ldots,<_d$, Then $\mathbf{A}$ is

- *ordered (partially) diagonal* if $\mathbf{B}$ is (partially) diagonal or equivalently that every two-dimensional projection of $\mathbf{B}$ is (partially) diagonal and
- *ordered (partially) upper/lower triangular* if $\mathbf{B}$ is (partially) upper/lower triangular or equivalently that every two-dimensional projection of $\mathbf{B}$ is (partially) upper/lower triangular.

Without ordering the key sets, the best that can be done is weak triangularity. Likewise with upper and lower triangularity, the associative array

$$\mathbf{A} : K_1 \times \cdots \times K_d \to V$$

is *weakly upper triangular* or *weakly lower triangular* if each of the two-dimensional projections

$$\mathbf{A}_{i,j} : K_i \times K_j \to V$$

is weakly upper triangular or weakly lower triangular, respectively.

16.3 Algebraic Properties

Every permutation of two elements is either the identity permutation or the transposition of the two elements. This fact motivates the following generalization of symmetry of an associative array. An associative array

$$\mathbf{A} : K^d \to V$$

is *symmetric* if for every

$$\mathbf{k} \in K^d$$

and any permutation σ of $\{1,\ldots,d\}$, the permuted element

$$\mathbf{k}' = (\mathbf{k}(\sigma(1)),\ldots,\mathbf{k}(\sigma(d)))$$

satisfies

$$\mathbf{A}(\mathbf{k}) = \mathbf{A}(\mathbf{k}')$$

Regardless of the dimension of the associative array, the notions of skew-symmetry and being Hermitian depend upon algebraic properties of V that are not included in the properties of being a semiring. When V is a ring, an array

$$\mathbf{A} : K^d \to V$$

is said to be *skew-symmetric* if for every

$$\mathbf{k} \in K^d$$

and any permutation σ of $\{1,\ldots,d\}$, the permuted element

$$\mathbf{k}' = (\mathbf{k}(\sigma(1)),\ldots,\mathbf{k}(\sigma(d)))$$

satisfies

$$\mathbf{A}(\mathbf{k}) = \mathrm{sgn}(\sigma)\mathbf{A}(\mathbf{k}')$$

where $\mathrm{sgn}(\sigma)$ is the additive inverse -1 of the multiplicative identity 1 of V when σ is an odd permutation, and the multiplicative identity 1 of V when σ is an even permutation.

Similarly, when V is a sub-semiring of $\mathbb{C}$ closed under complex conjugation, say that the associative array

$$\mathbf{A} : K^d \to V$$

is *Hermitian* if for every

$$\mathbf{k} \in K^d$$

and any permutation σ of $\{1,\ldots,d\}$, the permuted element

$$\mathbf{k}' = (\mathbf{k}(\sigma(1)),\ldots,\mathbf{k}(\sigma(d)))$$

satisfies

$$\mathbf{A}(\mathbf{k}) = \mathbf{A}(\mathbf{k}')$$

when σ is an even permutation and

$$\mathbf{A}(\mathbf{k}) = \overline{\mathbf{A}(\mathbf{k}')}$$

when σ is an odd permutation, where $\bar{z} = x - y\sqrt{i}$ is the complex conjugate of the complex number $z = x + y\sqrt{i}$.

16.4 Sub-Array Properties

Finally, the notion of sub-arrays can also be extended to arbitrary dimensions greater than 1.

Definition 16.5

Sub-Array

Given an associative array

$$\mathbf{A} : K_1 \times \cdots \times K_d \to V$$

the associative array

$$\mathbf{B} : K_1' \times \cdots \times K_d' \to V$$

is said to be a *sub-array* of $\mathbf{A}$ if

$$K_1' \times \cdots \times K_d' \subset K_1 \times \cdots \times K_d$$

and $\mathbf{B}$ agrees with $\mathbf{A}$ on

$$K_1' \times \cdots \times K_d'$$

As in the two-dimensional case, in considering block associative arrays the partitions of $K_1, \ldots, K_d$ are most meaningful when the partitions are into closed intervals with respect to strict total orders $<_1, \ldots, <_d$ on $K_1, \ldots, K_d$. Write

$$K_i = [k_1^i, k_2^i] \cup \cdots \cup [k_{2p_i-1}^i, k_{2p_i}^i]$$

where

$$k_1^i \leq_i k_2^i \leq_i \cdots \leq_i k_{2p_i-1} \leq_i k_{2p_i}$$

then the partitions

$$\mathcal{K}_i = \left\{[k_1^i, k_2^i], \ldots, [k_{2p_i-1}^i, k_{2p_i}^i]\right\}$$

can be ordered by

$$[k_1^i, k_2^i] \leq_i \cdots \leq_i [k_{2p_i-1}^i, k_{2p_i}^i]$$

for each $1 \leq i \leq d$.

Then the definition of a d-dimensional block associative array follows analogously to the two-dimensional case.

Definition 16.6

Block Associative Array and Associated Block Structure Map

Suppose $K_1,\ldots,K_d$ are finite totally ordered sets. A (d-dimensional) *block associative array*

$$\mathbf{A} : K_1 \times \cdots \times K_d \to V$$

has closed intervals tuple partitions $\mathcal{K}_1,\ldots,\mathcal{K}_d$ of $K_1,\ldots,K_d$. The *associated block structure map* is the associative array

$$\mathbf{A}' : \mathcal{K}_1 \times \cdots \times \mathcal{K}_d \to \mathbb{A}$$

defined by

$$\mathbf{A}'([k_1,k_1'],\ldots,[k_d,k_d']) = \text{pad}_{K_1\times\cdots\times K_d} \mathbf{A}|_{[k_1,k_1']\times\cdots\times[k_d,k_d']}$$

Because

$$\mathbb{A}(K_1,\ldots,K_d;V)$$

forms a semiring itself under element-wise (Hadamard) multiplication, the associated block structure map

$$\mathbf{A}' : \mathcal{K}_1 \times \cdots \times \mathcal{K}_d \to \mathbb{A}$$

is also a well-defined associative array. With this extra block structure in mind, it is possible to define additional partition concepts.

Definition 16.7

Block Diagonal, Triangular

A block associative array

$$\mathbf{A} : K_1 \times \cdots \times K_d \to V$$

with the partitions

$$(\mathcal{K}_1,\ldots,\mathcal{K}_d)$$

is

- *block (partially) diagonal* if the corresponding block structure map $\mathbf{A}' : \mathcal{K}_1 \times \cdots \times \mathcal{K}_d \to \mathbb{A}$ is (partially) diagonal and
- *block (partially) upper/lower triangular* if the corresponding block structure map $\mathbf{A}' : \mathcal{K}_1 \times \cdots \times \mathcal{K}_d \to \mathbb{A}$ is (partially) upper/lower triangular.

A further generalization can be made in which the elements of the partition need not be closed intervals, giving a well-defined associated block structure map in nearly the same way as above. However, it is not clear if there is a natural and general way to order these more general partitions.

16.5 Conclusions, Exercises, and References

Extending the concepts of matrix mathematics to tensors is a well-studied endeavor, and the challenges of defining tensor multiplication are similar for associative arrays. The diverse data representation concepts that give associative arrays their true power naturally extend to higher dimensions with little change.

Exercises

Exercise 16.1 — Why is the rank of a d-dimensional matrix no longer defined?

Exercise 16.2 — Why is only the existence of a nonzero term recorded in the two-dimensional projection of a d-dimensional associative array? Why not add up or multiply out those nonzero terms?

Exercise 16.3 — Find the two-dimensional projections of the three-dimensional array $\mathbf{A} : \{1,2,3\}^3 \to \mathbb{R} \cup \{\pm\infty\}$ with the following slices:

$$\mathbf{A}(:,:,1) = \begin{array}{c} \\ 1 \\ 2 \\ 3 \end{array} \begin{array}{c} \begin{array}{ccc} 1 & 2 & 3 \end{array} \\ \begin{bmatrix} -\infty & 1 & 0 \\ \infty & -\infty & -1 \\ 1 & 0 & 3 \end{bmatrix} \end{array} \qquad \mathbf{A}(:,:,2) = \begin{array}{c} \\ 1 \\ 2 \\ 3 \end{array} \begin{array}{c} \begin{array}{ccc} 1 & 2 & 3 \end{array} \\ \begin{bmatrix} -\infty & 0 & \infty \\ -\infty & 0 & 0 \\ 1 & 1 & 0 \end{bmatrix} \end{array}$$

$$\mathbf{A}(:,:,3) = \begin{array}{c} \\ 1 \\ 2 \\ 3 \end{array} \begin{array}{c} \begin{array}{ccc} 1 & 2 & 3 \end{array} \\ \begin{bmatrix} -\infty & -\infty & -\infty \\ 1 & 3 & -2 \\ -1 & \infty & 1 \end{bmatrix} \end{array}$$

where $\mathbb{R} \cup \{-\infty, \infty\}$ is the max-plus algebra. What about with the max-min tropical algebra? Explain the relationship between the two-dimensional projections in each of the two cases.

Exercise 16.4 — Write down an explicit example of a weakly diagonal three-dimensional array that is not diagonal.

Exercise 16.5 — Describe the associated block structure map for the associative array $\mathbf{A}$ defined in Exercise 16.3 with the partitions

$$K_1 = \{1,2,3\} = \{1\} \cup \{2,3\}$$
$$K_2 = \{1,2,3\} = \{1,2\} \cup \{3\}$$
$$K_3 = \{1,2,3\} = \{1,2\} \cup \{3\}$$

References

[1] F. L. Hitchcock, "Multiple invariants and generalized rank of a p-way matrix or tensor," *Studies in Applied Mathematics*, vol. 7, no. 1-4, pp. 39–79, 1928.

[2] R. B. Cattell, "The three basic factor-analytic research designs–their interrelations and derivatives," *Psychological Bulletin*, vol. 49, no. 5, p. 499, 1952.

[3] T. G. Kolda and B. W. Bader, "Tensor decompositions and applications," *SIAM Review*, vol. 51, no. 3, pp. 455–500, 2009.

[4] R. A. Harshman, "Foundations of the PARAFAC procedure: Models and conditions for an 'explanatory' multi-modal factor analysis," *UCLA Working Papers in Phonetics*, vol. 16, pp. 1–84, 1970.

[5] A. Smilde, R. Bro, and P. Geladi, *Multi-Way Analysis: Applications in the Chemical Sciences*. Chichester, U.K.: John Wiley & Sons, 2005.

[6] D. Muti and S. Bourennane, "Multidimensional filtering based on a tensor approach," *Signal Processing*, vol. 85, no. 12, pp. 2338–2353, 2005.

[7] T. G. Kolda, "Orthogonal tensor decompositions," *SIAM Journal on Matrix Analysis and Applications*, vol. 23, no. 1, pp. 243–255, 2001.

[8] P. M. Kroonenberg, *Applied Multiway Data Analysis*, vol. 702. Hoboken, N.J.: John Wiley & Sons, 2008.

[9] W. Austin, G. Ballard, and T. G. Kolda, "Parallel tensor compression for large-scale scientific data," in *International Parallel and Distributed Processing Symposium*, pp. 912–922, IEEE, 2016.

[10] B. W. Bader and T. G. Kolda, "Algorithm 862: Matlab tensor classes for fast algorithm prototyping," *ACM Transactions on Mathematical Software*, vol. 32, no. 4, pp. 635–653, 2006.

Appendix: Notation

MATLAB Operators

+	—	addition of real numbers
.*	—	multiplication of real numbers
max	—	the maximum of two real numbers
min	—	the minimum of two real numbers
union	—	the union of two sets
intersect	—	the intersection of two sets
|	—	logical OR
&	—	logical AND
.′	—	matrix transpose

Associative Arrays and Matrices

$\mathbf{A}, \mathbf{B}, \mathbf{C}, \ldots$	—	associative arrays, including matrices and column vectors
$\mathbf{u}, \mathbf{v}, \mathbf{w}, \ldots$	—	arbitrary elements of a semimodule, or more particularly column vectors in V^n
$a, b, c, u, v, w, \ldots$	—	arbitrary elements of the underlying semiring V
$\mathsf{A}, \mathsf{A1}, \ldots$	—	D4M arrays
K_i	—	an arbitrary key set
I, J, K	—	arbitrary index sets
i, j, k	—	arbitrary indices in I, J, K, respectively
$V^{n \times m}$	—	the set of all $n \times m$ matrices over the semiring V
$f(a,:)$, $f(:,b)$	—	given the function $f : A \times B \to C$, the functions $f(a,:) : B \to C$ and $f(:,b) : A \to C$ are defined by $f(a,:)(b) = f(a,b)$ and $f(:,b)(a) = f(a,b)$, respectively. Similar notation is extended to higher dimensions
$\mathbb{A}(K_1, \ldots, K_d; V)$	—	the set of associative arrays $\mathbf{A} : \prod_{j=1}^{d} K_j \to V$
$\mathcal{A}$	—	the set of sub-arrays of an associative array $\mathbf{A}$
$\mathcal{K}$	—	the set of rectangular subsets $\prod_{j=1}^{d} K'_j \subset \prod_{j=1}^{d} K_j$

$\mathbf{0}$	—	the zero matrix: $\mathbf{0}(i, j) = 0$ for all i, j
$\mathbb{I}, \mathbb{I}_n$	—	the $n \times n$ identity matrix: $\mathbb{I}_n(i, j) = \begin{cases} 1 & \text{if } i = j \\ 0 & \text{otherwise} \end{cases}$
$\mathbf{e}_i$	—	the column vector, that is zero everywhere except the i-th entry, where it is 1
$\mathbf{e}_{i,j}$	—	the matrix, that is zero everywhere except the i, j-th entry, where it is 1
$\mathbb{1}$	—	the matrix of all ones: $\mathbb{1}(i, j) = 1$ for all i, j
$-\infty$	—	a formal minimum of a linearly ordered set
∞	—	a formal maximum of a linearly ordered set
$\mathbf{A}^\intercal$	—	the transpose of the matrix $\mathbf{A}$, defined by $\mathbf{A}^\intercal(i, j) = \mathbf{A}(j, i)$

Graphs and Associative Arrays

G	—	an arbitrary (directed, weighted) graph
$u, v, w, \ldots$	—	arbitrary vertices in a graph
$K_{\text{out}}, K_{\text{in}}$	—	in a graph G, the set K_{out} of vertices that have outgoing edges and the set K_{in} of vertices that have incoming edges
c	—	an arbitrary walk in a graph
$P^k_{v,w}$	—	the set of all walks in a graph G starting at v and ending at w of length exactly k
$P^{(k)}_{v,w}$	—	the set of all walks in a graph G starting at v and ending at w of length at most k
$\mathrm{w}(c)$	—	the weight $\mathrm{w}(v_1, v_2) \otimes \cdots \otimes \mathrm{w}(v_{k-1}, v_k)$ of the walk $c = (v_1, \ldots, v_k)$ in a weighted (directed) graph
$G_{\mathbf{A}}$	—	the graph of the array $\mathbf{A}$
$\mathbf{E}_{\text{out}}$	—	an arbitrary out-vertex incidence array $\mathbf{E}_{\text{out}} : E \times K_{\text{out}} \to V$ with $\mathbf{E}_{\text{out}}(e, k_{\text{out}}) \neq 0$ if and only if edge e is directed outward from vertex k_{out}
$\mathbf{E}_{\text{in}}$	—	an arbitrary in-vertex incidence array $\mathbf{E}_{\text{in}} : E \times K_{\text{in}} \to V$ with $\mathbf{E}_{\text{in}}(e, k_{\text{in}}) \neq 0$ if and only if edge e is directed into vertex k_{in}
$\bar{G}$	—	the reverse of a graph G with arrows turned around

Commonly Used Operator Symbols

$\oplus$	—	the addition operation in a general semiring or array addition

$\bigoplus_{i=1}^{d} a_i$	—	the sum $a_1 \oplus a_2 \oplus \cdots \oplus a_d$
$\bigoplus_{i \in I} M_i$	—	the direct sum $\{(\mathbf{v}_i)_{i \in I} \in \prod_{i \in I} M_i \mid \mathbf{v}_i$ for all but a finite number of $i \in I\}$ of an indexed family of semimodules
$\otimes$	—	the multiplication operation in a general semiring or array Hadamard product
$\oplus.\otimes$	—	array product
$\wedge$	—	the meet operation in a meet-semilattice or lattice; the greatest lower bound
$\bigwedge_{s \in S} s$		the greatest lower bound of the set S
$\vee$	—	the join operation in a join-semilattice or lattice; the least upper bound
$\bigvee_{s \in S} s$	—	the least upper bound of the set S
$\max A$	—	the maximum element of the finite set A
$\min A$	—	the minimum element of the finite set A
$+$	—	the addition operation in a field or ring
$\sum_{i=1}^{n} a_i$	—	the sum $a_1 + a_2 + \cdots + a_n$
$\times$	—	either the Cartesian product or the multiplication operation in a field or ring
$\bar{z}$	—	the complex conjugate of $z = x + yi$ given by $\bar{z} = x - yi$

Commonly Used Symbols for Sets

V	—	an arbitrary semiring, unless otherwise stated
R	—	an arbitrary ring, unless otherwise stated
M, N	—	arbitrary semimodules or semialgebras, unless otherwise stated
$\mathscr{B}$	—	an arbitrary basis of a semimodule
$\mathbb{F}$	—	an arbitrary field, unless otherwise stated
$\mathbb{F}_p$	—	the field with underlying set $\{0, \ldots, p-1\}$ equipped with addition modulo p and multiplication modulo p, where p is a prime number
$\mathbb{C}$	—	the complex field
$\mathbb{R}$	—	the real field
$\mathbb{Q}$	—	the rational field
$\mathbb{Z}$	—	the ring of integers

$\mathbb{N}$	—	the semiring of natural numbers

Set-Theoretic Notation

$\{1,\ldots,n\}$	—	the set of all natural numbers m such that $1 \leq m \leq n$
$\{x \in X \mid \varphi(x)\}$	—	the set of elements of X satisfying the logical formula $\varphi(x)$
$U, V, W, \ldots$	—	arbitrary sets
$U \subset V$	—	the statement that U is a subset of V, so that every element of U is an element of V
$U \subsetneq V$	—	the statement that U is a proper subset of V, so that $U \subset V$ but $U \neq V$
$U \cap V$	—	the intersection of the sets U and V; the set of elements in both U and V
$\bigcap_{i=1}^{d} U_i$	—	the intersection $U_1 \cap U_2 \cap \cdots \cap U_d$.
$A \cup B$	—	the union of the sets A and B; the set of elements in either A, B, or both
$\bigcup_{i=1}^{d} A_i$	—	the union $A_1 \cup A_2 \cup \cdots \cup A_d$
$A \setminus B$	—	the set difference of A and B; the set of elements in A that are not elements of B
$\emptyset$	—	the unique set containing no elements
$\mathscr{P}(A)$	—	the power set of A, or the set of all subsets of A
$A \times B$	—	the Cartesian Product of A and B or the set of ordered pairs (a,b) where $a \in A$ and $b \in B$
$\prod_{j=1}^{d} A_j$	—	the Cartesian Product $A_1 \times A_2 \times \cdots \times A_d$
A^d	—	the Cartesian product $\prod_{j=1}^{d} A$
$\pi_i : \prod_{j=1}^{d} A_i$	—	projection onto the i-th coordinate; $\pi_i(a_1,\ldots,a_d) = a_i$
$\lvert A \rvert$	—	the size of, or number of elements in, A
$f : A \to B$	—	a function with domain A and codomain B
$f(a)$	—	the image of $a \in A$ under the function $f : A \to B$
$g(a,b)$	—	the image of $(a,b) \in A \times B$ under the function $g : A \times B \to C$
$f[A']$	—	the image of $A' \subset A$ under the function $f : A \to B$ or the set $\{f(a) \in B \mid a \in A'\}$
$f^{-1}[B']$	—	the pre-image of $B' \subset B$ under the function $f : A \to B$ or the set $\{a \in A \mid f(a) \in B'\}$
$\operatorname{dom} f$	—	the domain A of the function $f : A \to B$

More Array and Matrix-Related Notation

$\text{pad}_{K_1' \times K_2'}(\mathbf{A})$ — the zero padding of $\mathbf{A} : K_1 \times K_2 \to V$ to the new key tuples $K_1' \times K_2' \supset K_1 \times K_2$

$\dim(\mathbf{A})$ — the dimensions $m \times n = (m,n) = (|K_1|, |K_2|)$ of an associative array $\mathbf{A} : K_1 \times K_2 \to V$

$\text{total}(\mathbf{A})$ — the number of entries $mn = |K_1||K_2|$ of an associative array $\mathbf{A} : K_1 \times K_2 \to V$

$\text{support}(\mathbf{A})$ — the set of tuples of indices or keys corresponding to nonzero values $\{(k_1,k_2) \in K_1 \times K_2 \mid \mathbf{A}(k_1,k_2) \neq 0\}$

$\text{nnz}(\mathbf{A})$ — the number of nonzero values $|\text{support}(\mathbf{A})|$

$\text{density}(\mathbf{A})$ — the fraction of values that are nonzero $\text{nnz}(\mathbf{A})/\text{total}(\mathbf{A})$

$\text{sparsity}(\mathbf{A})$ — the fraction of values that are zero $1 - \text{density}(\mathbf{A})$

$\text{size}(\mathbf{A}, i)$ — the number of nonzero slices in the i-th dimension $|\pi_i(\text{support}(\mathbf{A}))|$

$\text{size}(\mathbf{A})$ — the tuple of sizes of $\mathbf{A}$ in each dimension $\underline{\text{m}} \times \underline{\text{n}} = (\underline{\text{m}}, \underline{\text{n}}) = (\text{size}(\mathbf{A},1), \text{size}(\mathbf{A},2))$

$\underline{\text{total}}(\mathbf{A})$ — the number of values in the non-empty slices $\underline{\text{mn}} = \text{size}(\mathbf{A},1)\text{size}(\mathbf{A},2)$

$\underline{\text{density}}(\mathbf{A})$ — fraction of values in non-empty slices that are nonzero $\text{nnz}(\mathbf{A})/\underline{\text{total}}(\mathbf{A})$

$\underline{\text{sparsity}}(\mathbf{A})$ — fraction of values in non-empty slices that are zero $1 - \underline{\text{density}}(\mathbf{A})$

$\text{image}(\mathbf{A})$ — set of vectors $\mathbf{Av}$ for all $\mathbf{v}$

$\text{rank}(\mathbf{A})$ — minimum number of linearly independent vectors needed to create $\text{image}(\mathbf{A})$

$\ker(\mathbf{A})$ — the set of vectors $\mathbf{v}$ for which $\mathbf{Av} = \mathbf{0}$

$X(\mathbf{A}, \mathbf{w})$ — the set of solutions $\mathbf{v}$ to the linear system $\mathbf{Av} = \mathbf{w}$

$x(\mathbf{A}, \mathbf{b})$ — the maximum solution to the linear system $\mathbf{Av} = \mathbf{b}$

$\text{span}(A)$ — the set of all linear combinations $\bigoplus_{\mathbf{a} \in B \subset A} c_{\mathbf{a}} \mathbf{a}$ where B is finite, $c_{\mathbf{a}} \in V$, and B is a subset of a semimodule over V

$\text{row}_{\mathbf{A}}(n)$ — the number of rows of $\mathbf{A}$ that have strictly more than n nonzero entries

$\text{col}_{\mathbf{A}}(n)$ — the number of columns of $\mathbf{A}$ that have strictly more than n nonzero entries

Eigenvalues and Eigenvectors

λ	—	an arbitrary eigenvalue
$\mathbf{A}^*$	—	the quasi-inverse of the matrix $\mathbf{A}$
$\mathbf{A}^+$	—	$\mathbf{A} \otimes \mathbf{A}^*$
$[\mathbf{A}]^i$	—	the i-th column $\mathbf{A}(:,i)$ of the array $\mathbf{A}$
$\mathrm{Per}^+(n)$	—	the set of even permutations of $\{1,\ldots,n\}$
$\mathrm{Per}^-(n)$	—	the set of odd permutations of $\{1,\ldots,n\}$
$\det^+(\mathbf{A})$	—	the positive determinant $\bigoplus_{\sigma \in \mathrm{Per}^+(n)} \bigotimes_{i=1}^n \mathbf{A}(i,\sigma(i))$
$\det^-(\mathbf{A})$	—	the negative determinant $\bigoplus_{\sigma \in \mathrm{Per}^-(n)} \bigotimes_{i=1}^n \mathbf{A}(i,\sigma(i))$
$\mathrm{trace}(\mathbf{A})$	—	the sum of the diagonal entries of $\mathbf{A}$
$\rho(\mathbf{A})$	—	the spectral radius $\bigoplus_{i=1}^n \mathrm{trace}(\mathbf{A}^k)^{1/k}$
$\mathscr{V}(\lambda)$	—	the eigen-semimodule of $\mathbf{A}$ corresponding to the eigenvalue λ

SVD and Symmetrized Max-Plus Algebra

$\mathbf{A}^\dagger$	—	the conjugate transpose of the complex matrix $\mathbf{A}$: $\mathbf{A}^\dagger(i,j) = \overline{\mathbf{A}(j,i)}$
∇	—	the binary relation on $(\mathbb{R} \cup \{\text{-}\infty\})^2$ defined by "$(x,y) \nabla (w,z)$ if $\max(w,y) = \max(x,z)$
B	—	the balancing relation on $(\mathbb{R} \cup \{\text{-}\infty\})^2$ defined by $(x,y)\, \mathrm{B}\, (w,z)$ if $\begin{cases} (x,y) \nabla (w,z) & \text{if } x \neq y \text{ and } w \neq z \\ (x,y) = (w,z) & \text{otherwise} \end{cases}$
$(\,)^\bullet$	—	the balancing operation on $(\mathbb{R} \cup \{\text{-}\infty\})^2$ defined by $(x,y)^\bullet = (\max(x,y), \max(x,y))$
$\ominus$	—	the operation on $(\mathbb{R} \cup \{\text{-}\infty\})^2$ defined by $\ominus(x,y) = (y,x)$
$\lvert\,\rvert$	—	the operation on $(\mathbb{R} \cup \{\text{-}\infty\})^2$ defined by $\lvert(x,y)\rvert = \max(x,y)$
$\mathbb{S}$	—	the set of equivalence classes of $(\mathbb{R} \cup \{\text{-}\infty\})^2$ under the equivalence relation B
$\mathbb{S}^\oplus$	—	the subset of $\mathbb{S}$ of max-plus-positive equivalence classes or equivalence classes of the form $\overline{(w,\text{-}\infty)} = \{(w,x) \mid x < w\}$
$\mathbb{S}^\ominus$	—	the subset of $\mathbb{S}$ of max-plus-negative equivalence classes or equivalence classes of the form $\overline{(\text{-}\infty,w)} = \{(x,w) \mid x < x\}$
$\mathbb{S}^\bullet$	—	the subset of $\mathbb{S}$ of balanced equivalence classes or equivalence classes of the form $\overline{(w,w)} = \{(w,w)\}$
$\mathbb{S}^\vee$	—	the subset of $\mathbb{S}$ defined by $\mathbb{S}^\oplus \cup \mathbb{S}^\ominus$

Higher-Dimensional Array Notation

$\text{pad}_{K_1'\times K_2'}(\mathbf{A})$ — the zero padding of $\mathbf{A} : K_1{\times}K_2 \to V$ to the new key tuples $K_1'{\times}K_2' \supset K_1{\times}K_2$

$\dim(\mathbf{A})$ — the dimensions $(|K_1|,\ldots,|K_d|)$ of an associative array $\mathbf{A} : K_1{\times}\cdots{\times}K_d \to V$

$\text{total}(\mathbf{A})$ — the number of entries $|K_1|\cdots|K_d|$ of an associative array $\mathbf{A} : K_1{\times}\cdots{\times}K_d \to V$

$\text{support}(\mathbf{A})$ — the set of tuples of indices or keys corresponding to nonzero values $\{(k_1,\ldots,k_d) \in K_1{\times}\cdots{\times}K_d \mid \mathbf{A}(k_1,\ldots,k_d) \neq 0\}$

$\text{nnz}(\mathbf{A})$ — the number of nonzero values $|\text{support}(\mathbf{A})|$

$\text{density}(\mathbf{A})$ — the fraction of values that are nonzero $\text{nnz}(\mathbf{A})/\text{total}(\mathbf{A})$

$\text{sparsity}(\mathbf{A})$ — the fraction of values that are zero $1 - \text{density}(\mathbf{A})$

$\text{size}(\mathbf{A}, i)$ — the number of nonzero slices in the i-th dimension $|\pi_i(\text{support}(\mathbf{A}))|$

$\text{size}(\mathbf{A})$ — the tuple of sizes of $\mathbf{A}$ in each dimension $(\text{size}(\mathbf{A},1),\ldots,\text{size}(\mathbf{A},d))$

$\underline{\text{total}}(\mathbf{A})$ — the number of values in the non-empty slices $\text{size}(\mathbf{A},1)\cdots\text{size}(\mathbf{A},d)$

$\underline{\text{density}}(\mathbf{A})$ — fraction of values in non-empty slices that are nonzero $\text{nnz}(\mathbf{A})/\underline{\text{total}}(\mathbf{A})$

$\underline{\text{sparsity}}(\mathbf{A})$ — fraction of values in non-empty slices that are zero $1 - \underline{\text{density}}(\mathbf{A})$

Index